Chemistry and Biological Activities of Ivermectin

Scrivener Publishing
100 Cummings Center, Suite 541J
Beverly, MA 01915-6106

Emerging Trends in Medicinal and Pharmaceutical Chemistry

Series Editor: Shahid-ul-Islam

Scope: The Emerging Trends in Medicinal Chemistry and Pharmacology Series is intended to provide recent trends, the state-of-the-art, and advancements particularly in the rapidly growing fields of drug design and synthesis, medicinal natural products, phytochemistry, pharmacology and applications. With a focus on generating means to combat different human diseases, the series addresses novel strategies and advanced methodology to circumvent the invasion from microbial infections and to ameliorate the effects caused by dreadful diseases.

Publishers at Scrivener
Martin Scrivener (martin@scrivenerpublishing.com)
Phillip Carmical (pcarmical@scrivenerpublishing.com)

Chemistry and Biological Activities of Ivermectin

Edited by
Rashid Ali
Department of Chemistry at Jamia Millia Islamia New Delhi, India
and
Shahid ul-Islam
Department of Textile & Fiber Engineering, Indian Institute of Technology Delhi, India

Scrivener Publishing

WILEY

This edition first published 2023 by John Wiley & Sons, Inc., 111 River Street, Hoboken, NJ 07030, USA and Scrivener Publishing LLC, 100 Cummings Center, Suite 541J, Beverly, MA 01915, USA

For more information about Scrivener publications please visit www.scrivenerpublishing.com.

Wiley Global Headquarters
111 River Street, Hoboken, NJ 07030, USA

For details of our global editorial offices, customer services, and more information about Wiley products visit us at www.wiley.com.

Library of Congress Cataloging-in-Publication Data

ISBN 978-1-394-16654-1

Cover image: Pixabay.Com
Cover design by Russell Richardson

Set in size of 11pt and Minion Pro by Manila Typesetting Company, Makati, Philippines

Printed in the USA

10 9 8 7 6 5 4 3 2 1

Contents

Preface

The Nobel Prize-winning Ivermectin (IVM), dubbed "the wonder drug," is a 16-membered macrolide that belongs to the family of avermectins (AVMs). Because of its widespread pharmaceutical applications, it has piqued the interest of scientists all over the world in recent years. Notably, initially, it was isolated by the fermentation of the soil microorganism *Streptomyces avermitilis* by S. Omura in Japan. Remarkably, it was approved by the FDA in 1997 for the treatment of Strongyliasis and crusted scabies, in patients with AIDS. After that, this wonder drug has played a significant role in diverse diseases, including its recent usage in COVID-19. Hence, taking into consideration the usefulness of the IVM, herewith we disclose the chemistry and biological activities of the IVM. With this particular book, our main intent is to spread a clear message to the readers about what has yet been known about IVM since its inception to date and where this emerging field might go in many years to come.

The first chapter of this book deals with the introduction, followed by a brief historical background and detailed chemistry about IVM in the second chapter. Though the treatment of the IVM for COVID-19 remains controversial and investigational, chapter 3 will be the heart of this book. The anti-parasitic and anti-malarial activities of the IVM drug will be discussed in detail in chapters 4 and 5. On the other hand, chapter 6 describes the anti-cancer activity, while the subsequent chapter 7 details the anti-inflammatory activity of IVM. The anthelminthic and insecticidal roles of IVM have been briefly described in Chapter 8. The chapter has exposed several cases of IVM in dermatology. Finally, chapters 10 and 11 close this volume of the book with anti-viral uses and safety, toxicology, and environmental aspects of IVM, respectively.

We are highly thankful to all the contributors for crafting their valuable knowledge and expertise and making it available in the form of their chapter(s)—helpful to those who really want to enjoy this subject and to those who are willing to enter into this research field.

Editors

1

Introduction to Ivermectin

Aeyaz Ahmad Bhat[1], Atif Khurshid Wani[2], Tahir ul Gani Mir[2]* and Nahid Akther[2]

[1]*School of Chemical Engineering and Physical Sciences, Lovely Professional University, Phagwara, Punjab, India*
[2]*School of Bioengineering and Biosciences, Lovely Professional University, Phagwara, Punjab, India*

Abstract
Ivermectin (IVM), an antiparasitic drug, has a wide range of biological applications. Taking its spectrum of actions into consideration along with the significant safety and efficacy it was approved by USFDA in 2015 and made available as a generic medicine. In addition to cancer, it is used to treat number of bacterial and viral infections in humans. IVM takes part in several biological operations, making it a potential treatment option for a variety of viruses, including SARS-CoV-2. The antiviral activity of IVM has been studied *in vitro* and *in vivo*. The target sites of IVM in its actions against viruses and cancerous cells include viral replication and cell cycle progression respectively. This chapter provides an overview of the sources and synthetic scheme involved in IVM building besides elucidating its therapeutic potential. There have been very few reports on the toxic effects of IVM that have been published so far. However, some of the concerns related to its toxicology have been delineated in this chapter.

Keywords: Ivermectin, pharmacology, antibacterial, anticancer, antiviral, COVID-19, toxicity

1.1 Introduction

Ivermectin (IVR) belongs to the group of broad-spectrum antiparasitic agents which have a unique mode of action and is currently authorized to be

**Corresponding author*: mirtahir4u@gmail.com

Rashid Ali and Shahid ul-Islam (eds.) Chemistry and Biological Activities of Ivermectin, (1–22)

used for the treatment of onchocerciasis, lymphatic filariasis, strongyloidiasis, scabies and head lice [1]. Recently, it has been shown that IVR can also exhibits a lot of new interesting activities such as antibacterial, antiviral and anticancer. IVR acts as a positive allosteric regulator of several channels including the glutamate-gated chloride channel (GluCl), γ-aminobutyric acid type-A receptor, glycine receptor, neuronal α7-nicotinic receptor, and purinergic P2X4 receptor. In most of the IVR-sensitive channels, the effects of IVR include the potentiation of agonist-induced currents at low concentrations and channel opening at higher concentrations [2]. In vertebrates, IVR also functions as a positive allosteric regulator of a number of ligand-gated ion channels. Glycine receptor (GlyR), γ-aminobutyric acid type-A receptor (GABAAR) and neuronal 7-nicotinic receptor are all activated or modulated by submicromolar concentrations of IVR (nAChR). The specific binding and high affinity of this compound to the GluCl channels found in the nerve and muscle cells of invertebrate animals is strictly related to its anti-parasitic activity. This causes the nerve or muscle cell to become hyperpolarized, increasing the permeability of the cell membrane to chloride ions, which causes the parasite to become paralyzed and die. Because some mammals lack GluCl channels, IVR has a very low affinity for mammalian GluCl channels and does not bind to them; these three facts serve as the foundation for its activity.

IVR appears to be safe for use in humans, but there have been reports linking the drug to parasympathetic disturbances (salivation, dilation of pupils) [3–5]. Along with its well-known anti-parasitic properties, IVR has also recently been shown to have strong anti-cancer properties, suggesting that it may be useful in the treatment of a number of cancers [6]. IVR exhibits anti-cancer properties due to its capacity to block the Wnt/TCF pathway, AKT/mTOR, and MDR (multidrug resistance) proteins (transcription factor of T-cells). One of the main oncogenic kinases, PAK-1 (p21-activated kinase), is degraded as a result of IVR [7]. By blocking Wnt-TCF, it is effective in treating lung, skin, and colon cancer as well as glioma multiforme, melanoma, and glioma [8]. IVR causes autophagy, a self-destructive effect in breast cancer, as demonstrated by Dou *et al.* studies using breast cancer cell lines, animal models, and 20 patient breast cancers have shown that reduced autophagy of breast cancer cells is associated with decreased expression of PAK-1 because of ubiquitin-mediated degradation. Inhibiting PAK-1 prevented Akt from becoming phosphorylated, which blocked the Akt/mTOR signaling pathway and slowed the growth of tumors [9].

1.2 Sources and Synthesis

Ivermectin (IVR) is a 22,23-dihydro derivative of avermectin B1 formed by the *Streptomyces avermitilis* bacterium from macrocyclic lactone (Figure 1.1). Besides IVR the other was approved for human use in the year 1987 by FDA [5, 10]. The general synthetic route is achievable by performing selective catalytic hydrogenation at C22–C23 as highlighted in Scheme 1.1.

IVR is a nonhygroscopic, crystalline powder that ranges in color from white to yellow-white and has a melting point of about 155°C. The respective empirical formulas are $C_{48}H_{74}O_{10}$ and $C_{47}H_{72}O_{14}$, with the molecular weights of 875.10 and 861.07.

Ivermectin B1a and B1b have very similar chemical structures; Ivermectin B1a has an ethyl group at the C-26 position, while B1b has a methyl group.

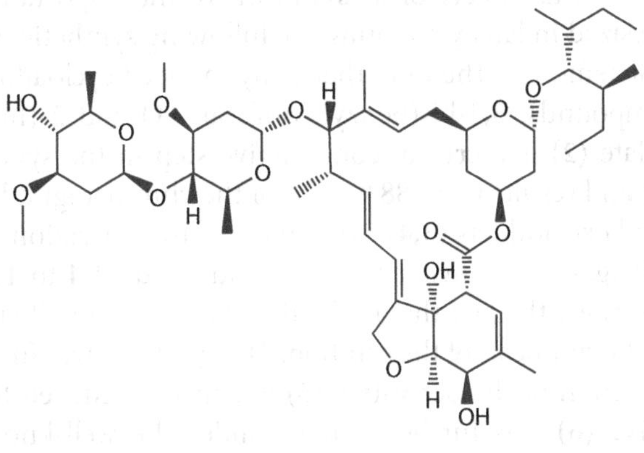

Figure 1.1 Chemical structure of Ivermectin.

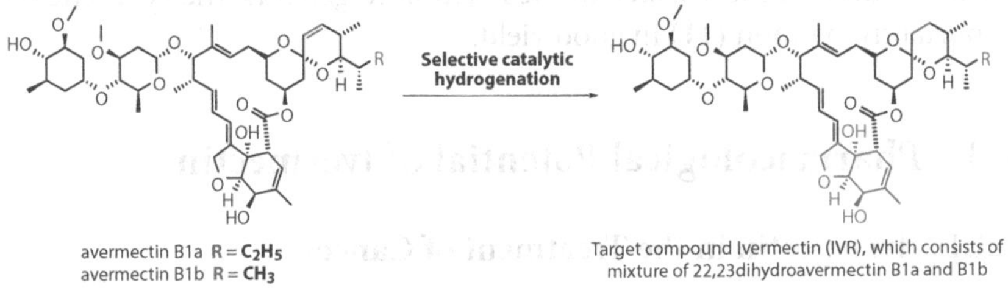

Scheme 1.1 Avermectin and IVR chemical structures.

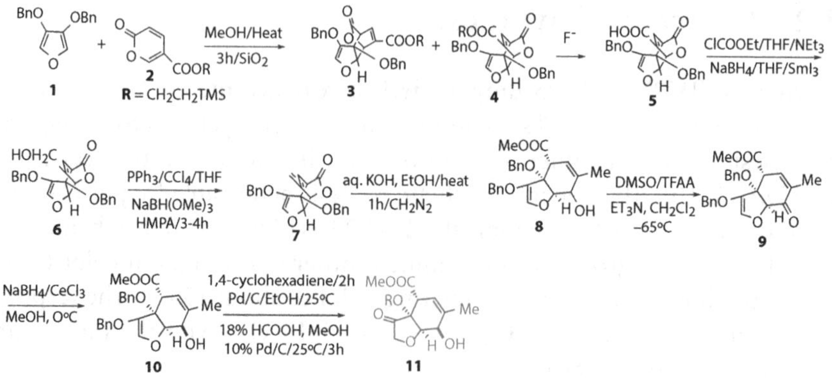

Scheme 1.2 Synthesis of synthetically important synthon of IVM.

IVR is made up of at least 80% B1a and no more than 20% B1b. A very important structural moiety or a "**synthon**" of the important drug IVM can be synthesized in lab by adopting the following synthetic procedure as depicted in Scheme 1.2. The mild thermally induced cycloaddition of two aromatic compounds, 3,4-bis(benzyloxy)furan **(1)** and 2-(himethylsilyl) ethyl coumalate **(2)** is a crucial constructive step in the synthesis of the synthon **(11)** and results in an 88% yield of the chromatographically separable endo and exo adducts **(3,4)**. Although the isomerization at C-24 may present challenges, the conversion of the endo adduct **4** to **11** is initially investigated despite the fact that both adducts are potentially useful intermediates for the synthesis of the synthon. The synthesis was further carried out by the reaction of the derivative **(5)** which was reduced to derivative **(6)**. **Derivative (6)** was further bought under the well-known concept of Witting reaction to yield the compound **(7)**. Reaction with base and further oxidation of the derivative **(7)** generated derivative **(10)** that was finally reduced in the last step of the reaction to generate the synthetically importantly synthon **(11)** in good yield.

1.3 Pharmacological Potential of Ivermectin

1.3.1 Ivermectin in the Treatment of Cancer

1.3.1.1 Ovarian Cancer

The fifth most common cause of cancer-related death in women is epithelial ovarian cancer. The therapeutic options for epithelial ovarian cancer have remained largely unchanged for more than 30 years, despite improvements in the treatment of many malignancies. Patients with ovarian cancer frequently receive platinum-based chemotherapy, such as cisplatin, but these treatments have a poor prognosis. This is probably because there is significant intra and inter tumor heterogeneity at the molecular and epigenetic levels [11, 12]. Sequencing has shown that TP53 mutations are frequently found in epithelial ovarian tumors, and that low-frequency mutations in the genes NF1, BRCA1, BRCA2 and RB1 are also present [13]. For better clinical management of ovarian cancer, new therapeutic approaches or agents are required. Let us start the pharmacological profile of the Ivermectin with the most important breakthrough which was provided by Zhang *et al.* The studies were carried out and it was found that by inhibiting Akt/mTOR signaling, Ivermectin improves the *in-vitro* and *in-vivo* effectiveness of cisplatin in treating epithelial ovarian cancer. In spite of specific cellular and molecular variations, Ivermectin inhibited growth in the G2/M phase and induced caspase-dependent apoptosis in ovarian cancer. The inhibitory effect of cisplatin on ovarian cancer cells was dramatically enhanced by Ivermectin in a dose-dependent manner. Ivermectin inhibited, in ovarian cancer cells, the phosphorylation of essential molecules in the Akt/mTOR signaling pathway. In addition, Ivermectin-induced inhibition of Akt/mTOR, growth arrest, and apoptosis were restored by overexpressing constitutively active Akt. Ivermectin alone significantly slowed tumor growth in a mouse model of ovarian cancer xenograft. Tumor growth was completely stopped when combined with cisplatin throughout the course of treatment without any side effects. Additionally, the Ivermectin concentrations used in our study are pharmacologically feasible.

1.3.1.2 Renal Cell Cancer

The proximal tubules of nephrons give rise to the epithelial tumor known as renal cell carcinoma (RCC), which is resistant to radiotherapy and chemotherapy [14, 15]. Patients with metastatic RCC still experience relapses as the disease worsens, despite the fact that targeted therapy significantly improved their clinical outcomes [16]. Therefore, patients who experience post-operative relapse or who have RCC that has spread to other organs require novel and potent therapeutic approaches. Recent research has shown that many cancers depend heavily on oxidative phosphorylation to meet their energy needs for survival and growth [17, 18]. Tumor stem cells are more reliant on mitochondrial metabolism than glycolysis when compared to differentiated tumor cells [19]. All of which imply that the special reliance on mitochondrial metabolism that cancer cells have can be utilized therapeutically. Zhu *et al.* reported that Ivermectin, an antibiotic, targets renal cancer with preference by causing mitochondrial dysfunction and oxidative damage. In numerous RCC cell lines that represent various histological subtypes and mutation statuses, Ivermectin significantly reduces proliferation and induces apoptosis. Importantly, Ivermectin is significantly less or completely ineffective in normal kidney cells than it is in RCC cells, proving that it is more toxic to RCC. Additionally significantly reducing RCC tumor growth *in vivo* is Ivermectin. Ivermectin causes mitochondrial dysfunction by reducing the potential of the mitochondrial membrane, the rate of respiration, and the production of ATP. Ivermectin-treated RCC cells and the xenograft mouse model exhibit oxidative stress and damage as a result of mitochondrial dysfunction. Ivermectin is confirmed to be targeting mitochondria in RCC cells by the rescue of Ivermectin's effect by acetyl-L-Carnitine (ALCAR, a mitochondrial fuel), or by the antioxidant N-acetyl-L-cysteine (NAC). Ivermectin may be more toxic to RCC than normal kidney cells because RCC cells have higher mitochondrial mass, respiration, and ATP production. This research suggests that Ivermectin is a promising treatment option for RCC and that targeting mitochondrial metabolism is an additional RCC treatment option.

1.3.2 Ivermectin Against Viruses

Ivermectin (IVM), an FDA approved drug, has shown great potential against viruses of different genetic nature (Figure 1.2). The mice infected with pseudorabies virus and piglets carrying circovirus infection have shown significant improvement in the survival rate upon treatment with IVM [20, 21]. Alphaviruses are dangerous viruses capable of causing neuroinvasive and arthralgia disease. IVM acts as a potential anti-alpha viral drug inhibiting chikungunya, sindbis, semliki forest, and flavivirus viruses. West Nile virus is also inhibited by IVM at low (μM) concentrations [22]. Zika virus, a positive sense RNA virus transmitted via mosquito, is inhibited by the IVM action during *in vitro* analysis [23]. Recently, IVM was found to be effective against SARS-CoV-2 when it was applied on engineered kidney cell of Vero/hSLAM [24]. Viral replication is one of the highly targeted molecular sites by drugs. Bovine herpesvirus 1 is inhibited by IVM treatment by targeting subunits of DNA polymerase [25]. Homosexuals infected with human immunodeficiency virus harboring strongyloides were given continuous IVM dosage. The patients showed significant parasitological and clinical improvement after receiving multiple IVM doses. Thus, results indicated effectiveness of IVM in strongyloidiasis in AIDS patients [26]. IVM is generally used for treating parasitosis, but off late it

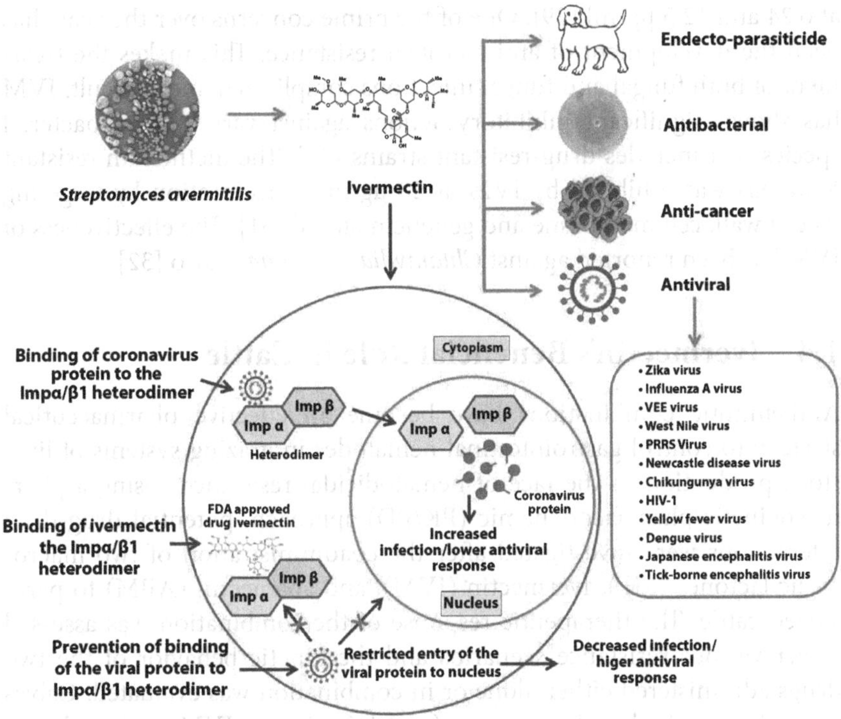

Figure 1.2 Antiviral properties of Ivermectin; figure adapted from [54].

has also shown efficiency in inhibiting importin protein which is essential for the replication of viruses with RNA genome like HIV and dengue virus [27]. Because of the approved status of IVM, it is imperative to explore its mechanistic details in target the dangerous pathogens to prevent pandemics like SARS-CoV-2.

1.3.3 Ivermectin in the Treatment of Bacterial Infections

Microorganisms dwell in everywhere environment owing to their adaptable nature. They adapt to extreme environments by regulating their different set of genes for survival [28]. The property of adapting to different environments is one of the major factors that allow pathogens to infect plants, humans, animals, and livestock. Several strategies have been developed to counteract bacterial and fungal infections. One of such drugs is the IVM which has been found to be effective against several pathogenic strains of both bacterial and fungal origin. IVM was tested for anti-staphylococcal activity that were isolated from bovine milk and it was found to be effective at 6.24 and 12.5 μg/ml [29]. One of the prime concerns over the years has been the development of antimicrobial resistance. This makes the treatment of both fungal and fungal infections complicated and difficult. IVM has shown significant inhibitory actions against various mycobacterial species that includes drug-resistant strains [30]. The methicillin resistant *S. aureus* was inhibited by IVM at 20 μg/ml concentration by targeting its cell wall, cell membrane and genetic material [31]. The effectiveness of IVM has been reported against *Chlamydia trachomatis* also [32].

1.4 Ivermectin's Beneficial Role in Cattle

Anthelmintic combinations have become an effective pharmaceutical strategy to control gastrointestinal nematodes in grazing systems of livestock production in the face of nematodicidal resistance. Using a pharmacokinetic/pharmacodynamic (PK/PD) approach, potential drug-drug interactions were investigated after the coadministration of two macrocyclic lactones (ML), Ivermectin (IVM), and abamectin (ABM) to parasitized cattle. The therapeutic response of the combination was assessed under various resistance scenarios and the kinetic behavior of the two drugs administered either alone or in combination was evaluated. Calves were given a single subcutaneous (s.c.) injection of IVM (100 μg/kg), a single s.c. injection of ABM (100 μg/kg), or a combination of IVM+ABM (50 μg/kg each) to achieve a final ML dose of 100 μg/kg (Farm 1). Up to 20 days after the treatment, plasma samples were taken from those animals. The plasma concentrations of IVM and ABM were measured using HPLC. In Farm 2, experimental animals were given IVM (200 μg/kg), ABM (200 μg/kg), or IVM+ABM (100 μg/kg each), while Farms 3 and 4 used IVM+ABM (200 μg/kg each). The fecal egg count reduction test (FECRT) was used to determine the anthelmintic effectiveness. Similar trends for IVM kinetic behavior following coadministration with ABM were revealed by PK analysis. In contrast, the presence of IVM increased the systemic exposure during the elimination phase and prolonged the half-life of ABM elimination. Despite the fact that IVM alone was ineffective at controlling *Cooperia* spp., IVM+ABM was the only treatment to have an efficacy of more than 95% against these organisms across all farms.

1.5 Ivermectin in the Treatment of COVID-19

Ivermectin, a widely used anti-parasitic drug, inhibits SARS-CoV-2 replication by preventing viral proteins from entering the host cell nucleus

[Figure 1.2] [24]. Doxycycline was recently identified as a potential inhibitor of SARS-CoV-2 papain-like protease in a virtual drug screening [43]. An observational study combining a single dose of Ivermectin with a multi-dose of doxycycline for the treatment of COVID-19 found that symptoms and viral response improved significantly [44]. A recent retrospective study discovered that hospitalized patients who received Ivermectin in combination with other treatments (e.g., azithromycin and hydroxychloroquine) had lower mortality than those who did not receive Ivermectin [44]. More research is needed to clarify these findings. The observations that SARS-CoV-2 rapidly multiplies in the respiratory tract, as well as evidence from animal models showing 3-fold higher levels of Ivermectin in pulmonary tissue than in plasma one week after oral dosing, highlight the importance of this need [1].

Methods: Patients were examined physically for COVID-19-related symptoms and vital signs (e.g., temperature, blood pressure, pulse rate, oxygen saturation, and respiratory rate) were taken. On the day of enrollment, as well as on days 3, 7, and 14, nasopharyngeal swabs were obtained to confirm the presence of SARS-CoV-2 using rRT-PCR. Patients were followed up on a weekly basis after day 14 until the test was found to be negative.

1.5.1 Mode of Action

IVM is quickly absorbed through the mouth, highly liposoluble, widely distributed throughout the body, metabolized in the liver (cytochrome P450 system) and almost entirely eliminated in feces [45]. It reaches peak plasma levels after a typical oral dose in healthy humans at 3.4–5 h and the plasma half-life has been reported to be 12-66 h [46]. There are not many studies on the pharmacokinetics of Ivermectin in humans, despite its widespread use [47]. In healthy subjects, Ivermectin binds to plasma proteins with a high affinity (93.2 %) [48]. When given in nations where malnutrition and hypoalbuminemia are prevalent, such a "avid binding" can be helpful because it increases the availability of "free fraction" of Ivermectin [45]. In patients with COVID-19, hypoalbuminemia is a common symptom that also seems to be related to the severity of lung injury [49]. Ivermectin may therefore have adequate bioavailability in this situation. Ivermectin's mode of action seems to be influenced by how it affects the ion channels in cell membranes. In particular, it results in an influx of negatively charged ions, which hyperpolarizes the affected cells and paralyses them. The neurotransmitter gamma-amino butyric acid (GABA) was initially believed to be its primary target because there is a wealth of evidence showing that Ivermectin opens

GABA-regulated Cl^- channels in a variety of *in vitro* preparations of invertebrate and, particularly, vertebrate membranes. Ivermectin, however, has recently been demonstrated to cause an influx of Cl^- through channels that are not controlled by GABA. In fact, compared to other types of channel, GABA-linked Cl channels in the cell membranes of the parasitic nematode *Ascaris suum* are less sensitive to Ivermectin [50]. In both mammalian brain tissue and the free-living nematode worm *Caenorhabditis elegans*, specific Ivermectin binding sites have been found. However, Ivermectin's affinity for the nematode site is roughly 100 times greater than it is for the mammalian brain site [51]. It appears that the GABA and Ivermectin receptors are very different because GABA was unable to compete with Ivermectin for binding to the *C. elegans* receptor protein [52]. Ivermectin is toxic to both nematodes and arthropods and photoaffinity labeling has revealed specific Ivermectin binding proteins in the cell membranes of the fruit fly *Drosophila melanogaster* and the worm *C. elegans* [53].

1.6 Toxicity of Ivermectin

The toxicity associated with use of Ivermectin is relatively low [55]. It has been subjected to *in vitro* mutagenesis experiments in both microorganisms and mammalian cells. Studies on the acute oral toxicity of Ivermectin in mice, dogs, and monkeys reveal species-specific variations in susceptibility. Human beings, like all other primates, are obviously less sensitive than rodents, especially mice [56]. Toxic effects of the medication are higher in newborn rats than in young adult rats. The treatment of animals with very high doses of Ivermectin resulted in embryotoxicity [57]. Off-label use has been linked to sporadic cases of toxicity in collies and other herding breeds. Ataxia, tremors, mydriasis, depression, and even coma and death might be among the side effects. Old English sheepdogs, Shetland sheepdogs, and Australian sheepdogs should not be given Ivermectin since it is toxic to them [58]. Animals should be clear of *Dirofilaria immitis* (heartworm) microfilaria before receiving large dosages of Ivermectin [59].

Oral administration of Ivermectin has been and continues to be the standard method of dosing [60]. Topical route is also one of the common routes of administration [61, 62]. Occasionally, the rectal route is used in humans, but the subcutaneous route is more common in cattle and the intravenous route in experimental veterinary medicine [63, 64]. Ivermectin is used to treat a wide range of parasite illnesses, and is usually given in a single dosage of 150–200 g/kg. Dosing may be repeated once or twice a few days or three to six months following the last oral dosage. Recommended dosing for individuals with crusted scabies, according to the Centers for Disease Control and Prevention [65] is 150 g/kg orally on days 1, 2, 8, 9, 15, 22, and 29. Several clinical studies and human studies have been completed or are now underway to assess Ivermectin's preventative or curative efficacy in COVID-19 . Most studies had been using a dosage of 0.2 mg/kg for 1 day and 0.6 mg/kg for 5 days [44, 66, 67]. Using a randomized, controlled design, researchers are testing the effects of giving healthy human volunteers Ivermectin every day for 28 days at doses of up to 100 micrograms per kilogram. As the study's end date drew closer, no safety concerns had been. However, there is a lack of clinical evidence on whether or not Ivermectin pharmacokinetics is affected by diet. Most of an oral dosage of Ivermectin is excreted intact in the feces (between 98 and 99 percent in most animal species) due to its low biotransformation. Biotransformation of Ivermectin is mostly done by CYP3A4, followed by CYP2E1 and CYP2D6 to a much smaller degree [68]. The significance of these newly reported findings in regards to Ivermectin's efficacy and safety must be validated, and the function, if any, of these metabolites must be explored further.

1.6.1 Acute Toxicity

Studies of acute (single dose) toxicity have been performed on mice, rats, rabbits, dogs, and monkeys. Tremors, bradypnea, ptosis, ataxia, and lack of the righting reflex were symptoms of Ivermectin's acute toxicity in rats. Ivermectin was shown to have caused these clinical symptoms due to its direct action on the central nervous system. It was observed that newborn rats were more vulnerable than young adults, and it was hypothesized that this was because the blood-brain barrier in rats only fully developed after birth. Mydriasis has been found most reliable sign of toxicity in Beagle dogs. Preceding deaths were characterized by prolonged comas. Vomiting was the most reliable sign of toxicity in Rhesus monkeys [69, 70]. When compared to rats, monkeys did not exhibit a high dose-response curve. Repeated doses of Ivermectin over three-month oral studies in mice, four-week dermal studies in Sprague Dawley rats, three- and nine-month oral studies in Beagle dogs, 2-week dermal studies in minipigs, and 2-week oral studies in rhesus monkeys have been conducted [69, 71]. Mortality was seen at a dosage 9 mg/kg/day when Ivermectin was orally administered to rats for 13 weeks. Death preceded by neurotoxic symptoms was detected exclusively in rats given the maximum daily dosage of 12 mg/kg after being fed 1, 3, or 12 mg/kg orally for 27 weeks. Mortality was highest among females and occurred mostly during the first two weeks of treatment. Dermally administering 20 mg/kg of Ivermectin to rats once daily for four weeks resulted in no observed toxicity. Excessive salivation and a loss in body weight were seen in Beagle dogs given Ivermectin by oral gavage at doses of 0.1, 0.25, 0.5, and 1.5 mg/kg/day for 14 weeks. In another research, 14 weeks of oral treatment with 0.5, 1, or 2 mg/kg/day resulted in neurotoxicity and poor health for four of the eight dogs in the high dosage group. Beagle dogs given Ivermectin orally at doses of 0.1, 0.5, or 1.5 mg/kg per day for 39 weeks showed no signs of death or serious side effects. After 2 weeks of once-daily Ivermectin treatments, rhesus monkeys showed no signs of toxicity. Finally, mice and minipigs administered daily through the dermal route with up to 13 and 20 mg/kg/day Ivermectin, respectively, showed no observable adverse effects (for 13 weeks in mice and up to 39 weeks in minipigs) [69, 70].

1.6.2 Developmental and Reproduction Toxicity

Repeated administration of Ivermectin at doses of 0.2, 8.1, and 4.5 times the maximum recommended human dosage was shown to be teratogenic in mice, rats, and rabbits, respectively [69]. There have been a number of studies spanning many generations. Teratogenicity was not seen in rats [72], but there was significant toxicity to both parents and offspring. It was suggested that the blood-brain barrier in young rats had a role. Dogs given 600 g/kg once a month for eight months and mated to untreated female dogs showed no reproductive toxicity [71].

1.6.3 General and Safety Pharmacology

In 1987, when Ivermectin was initially licensed as a human drug, general pharmacology was considered "optional" and not governed by formal regulations. There was no electrocardiogram (EKG) abnormality in dogs given Ivermectin orally at doses up to 1.5 mg/kg/day for 39 weeks. Marked hypotension during nonclinical toxicity testing was only found in moribund animals treated with Ivermectin. In 32 older Liberian males, treated with Ivermectin, EKG recording were done twice daily pretreatment and on five times post-treatment. Twenty of the subjects had aberrant heart rhythms before therapy. Neither substantial modifications nor new abnormalities were identified [73] Respiratory abnormalities were only seen in moribund animals which were given Ivermectin. No direct impact of Ivermectin on the respiratory system has yet been proven or proposed [74]. As was previously indicated, the most often reported adverse symptoms during preclinical tests with Ivermectin were neurotoxic consequences. A Serbian research team performed pharmacological tests to better describe the central and peripheral effects in rats dosed with Ivermectin. Although a single intravenous administration of 2.5, 5.0, or 7.5 mg/kg generated no obvious CNS depression however, drowsiness and stumbling were noticed between 10 to 40 min following a dosage of 10 mg/kg Ivermectin, and substantial CNS depression resulting to death in half of the animals treated with 15 mg/kg. Consistent with its GABAergic properties 14, Ivermectin was shown to prolong the time it took to fall asleep after being given thiopentone [75]. Furthermore, Ivermectin prevented convulsions generated by lidocaine and strychnine [76]. Anticonvulsive dosages in both cases were much less than the Ivermectin LD_{50} that was determined by calculations (18.2 mg/kg). Ivermectin generates diverse inhibitory effects in the CNS of animals through GABA-sensitive and GABA-insensitive pathways; moreover, flumazenil antagonized the effects of Ivermectin solely against lidocaine-induced convulsions. Safety and toxicity to the immune system Ivermectin was licensed for human use years before the first recommendations relating to immunotoxicity assessment were released by EM(E)A (2000), the US FDA (2002) and ICH (2005). No substantial investigation of Ivermectin immunological safety has so far been undertaken utilizing state-of-the-art technologies in line with current strategies. In male CD-1 mice, Ivermectin effects on T-dependent and T-independent antibody responses were evaluated after a single subcutaneous injection of either 0.2 or 20 mg/kg [77]. This study was an early example of the types of immunotoxicity studies commonly performed at the turn of the millennium. A statistically substantial augmentation of T-dependent antibody response was observed.

1.7 Conclusion

IVM has been proven effective in humans against a number of parasite diseases, but when compared to animals the pharmacokinetic characteristics of this drug are less acknowledged in people. IVM has potential food and drug interactions, which should be taken into account when using the medication for therapeutic purposes. IVM has shown promise in the treatment of a variety of bacterial, viral, and even fungal infections with activity frequently found to be satisfactory in these cases. Recently, the use of IVM has been expanded to encompass the treatment of COVID-19 and different cancers, together with the treatment of livestock grazing systems. Table 1.1 lists the medical conditions and infections for which this medication is used along with the critical concentration and mode of action. However, research is still needed to confirm and fully examine the drug's

Table 1.1 Pharmacological applications of IVM and its mode of action.

Pharmacological application	Concentration	Mechanisms of action	References
Wound healing	0.03-1 %	Decrease in exudation, edge edema, granulation, and hyperemia	[33]
Anti-inflammation	32mg/kg/24h	Blocks NF-kB pathway for lipopolysaccharide production	[34]
Antioxidant	1mg/kg	Increase in plasma nitric oxide and decrease in total antioxidant capacity	[35]
Cytotoxic	2.5–15 μM	Promotion of cellular apoptosis	[36]
Anti-fungal	210.0 mg/L	Reduction in the population growth by targeting replication	[37]
Anti-viral	2 mM	Inhibition of viral replication by targeting NS3 helicase enzyme	[38]
Anti-ovarian cancer	5 μM	Regulation of lncRNA-EIF_4A_3-mRNA	[39]
Ant-colorectal cancer	2.5 and 5 μM	Cell cycle arrest	[40]
Anti-malaria	300 μg/kg/day	Reduction in prevalence and incidence	[41]
Anti-dengue	400 μg/kg/day	Inhibition of viral replication	[42]

promising human applications, while also taking into account its pivotal toxicity and marking on well-established earlier reports.

References

1. Campbell, W.C., History of avermectin and ivermectin, with notes on the history of other macrocyclic lactone antiparasitic agents. *Curr. Pharm. Biotechnol.*, 13, 853–865, 2012.
2. Zemkova, H., Tvrdonova, V., Bhattacharya, A., Jindrichova, M., Allosteric modulation of ligand gated ion channels by ivermectin. *Physiol. Res.*, 63, 2014.
3. Crump, A., Ivermectin: Enigmatic multifaceted 'wonder'drug continues to surprise and exceed expectations. *J. Antibiot. (Tokyo)*, 70, 495–505, 2017.
4. Guzzo, C.A., Furtek, C.I., Porras, A.G., Chen, C., Tipping, R., Clineschmidt, C.M., Sciberras, D.G., Hsieh, J.Y., Lasseter, K.C., Safety, tolerability, and pharmacokinetics of escalating high doses of ivermectin in healthy adult subjects. *J. Clin. Pharmacol.*, 42, 1122–1133, 2002.
5. Ōmura, S. and Crump, A., Ivermectin and malaria control. *Malar. J.*, 16, 1–3, 2017.
6. Juarez, M., Schcolnik-Cabrera, A., Dueñas-Gonzalez, A., The multitargeted drug ivermectin: From an antiparasitic agent to a repositioned cancer drug. *Am. J. Cancer Res.*, 8, 317, 2018.
7. Melotti, A., Mas, C., Kuciak, M., Lorente-Trigos, A., Borges, I., Ruiz i Altaba, A., The river blindness drug I vermectin and related macrocyclic lactones inhibit WNT-TCF pathway responses in human cancer. *EMBO Mol. Med.*, 6, 1263–1278, 2014.
8. Dominguez-Gomez, G., Chavez-Blanco, A., Medina-Franco, J.L., Saldivar-Gonzalez, F., Flores-Torrontegui, Y., Juarez, M., Díaz-Chávez, J., Gonzalez-Fierro, A., Dueñas-González, A., Ivermectin as an inhibitor of cancer stem-like cells. *Mol. Med. Rep.*, 17, 3397–3403, 2018.
9. Dou, Q., Chen, H.-N., Wang, K., Yuan, K., Lei, Y., Li, K., Lan, J., Chen, Y., Huang, Z., Xie, N., Ivermectin induces cytostatic autophagy by blocking the PAK1/Akt axis in breast cancerivermectin induces cytostatic autophagy. *Cancer Res.*, 76, 4457–4469, 2016.
10. Laing, R., Gillan, V., Devaney, E., Ivermectin–old drug, new tricks? *Trends Parasitol.*, 33, 463–472, 2017.
11. Blagden, S.P., Harnessing pandemonium: The clinical implications of tumor heterogeneity in ovarian cancer. *Front. Oncol.*, 5, 149, 2015.
12. Siegel, R.L., Miller, K.D., Jemal, A., Cancer statistics, 2015. *CA. Cancer J. Clin.*, 65, 5–29, 2015.
13. Verhaak, R., Cancer genome atlas research network: Integrated genomic analysis identifies clinically relevant subtypes of glioblastoma characterized

by abnormalities in PDGFRA, IDH1, EGFR, and NF1. *Cancer Cell*, 17, 98–110, 2010.

14. Bukowski, R.M., Natural history and therapy of metastatic renal cell carcinoma: The role of interleukin-2. *Cancer Interdiscip. Int. J. Am. Cancer Soc.*, 80, 1198–1220, 1997.
15. Kanesvaran, R. and Tan, M.-H., Targeted therapy for renal cell carcinoma: The next lap. *J. Carcinog.*, 13, 1–8, 2014.
16. Posadas, E.M., Limvorasak, S., Figlin, R.A., Targeted therapies for renal cell carcinoma. *Nat. Rev. Nephrol.*, 13, 496–511, 2017.
17. Lagadinou, E.D., Sach, A., Callahan, K., Rossi, R.M., Neering, S.J., Minhajuddin, M., Ashton, J.M., Pei, S., Grose, V., O'Dwyer, K.M., BCL-2 inhibition targets oxidative phosphorylation and selectively eradicates quiescent human leukemia stem cells. *Cell Stem Cell*, 12, 329–341, 2013.
18. Miranda, S., Foncea, R., Guerrero, J., Leighton, F., Oxidative stress and upregulation of mitochondrial biogenesis genes in mitochondrial DNA-depleted HeLa cells. *Biochem. Biophys. Res. Commun.*, 258, 44–49, 1999.
19. De Luca, A., Fiorillo, M., Peiris-Pagès, M., Ozsvari, B., Smith, D.L., Sanchez-Alvarez, R., Martinez-Outschoorn, U.E., Cappello, A.R., Pezzi, V., Lisanti, M.P., Mitochondrial biogenesis is required for the anchorage-independent survival and propagation of stem-like cancer cells. *Oncotarget.*, 6, 14777, 2015.
20. Lv, C., Liu, W., Wang, B., Dang, R., Qiu, L., Ren, J., Yan, C., Yang, Z., Wang, X., Ivermectin inhibits DNA polymerase UL42 of pseudorabies virus entrance into the nucleus and proliferation of the virus *in vitro* and *vivo*. *Antivir. Res.*, 159, 55–62, 2018.
21. Wang, X., Lv, C., Ji, X., Wang, B., Qiu, L., Yang, Z., Ivermectin treatment inhibits the replication of Porcine circovirus 2 (PCV2) *in vitro* and mitigates the impact of viral infection in piglets. *Virus Res.*, 263, 80–86, 2019, https://doi.org/10.1016/j.virusres.2019.01.010.
22. Yang, S.N.Y., Atkinson, S.C., Wang, C., Lee, A., Bogoyevitch, M.A., Borg, N.A., Jans, D.A., The broad spectrum antiviral ivermectin targets the host nuclear transport importin α/β1 heterodimer. *Antivir. Res.*, 177, 104760, 2020, https://doi.org/10.1016/j.antiviral.2020.104760.
23. Ketkar, H., Yang, L., Wormser, G.P., Wang, P., Lack of efficacy of ivermectin for prevention of a lethal Zika virus infection in a murine system. *Diagn. Microbiol. Infect. Dis.*, 95, 38–40, 2019, https://doi.org/10.1016/j.diagmicrobio.2019.03.012.
24. Caly, L., Druce, J.D., Catton, M.G., Jans, D.A., Wagstaff, K.M., The FDA-approved drug ivermectin inhibits the replication of SARS-CoV-2 *in vitro*. *Antivir. Res.*, 178, 104787, 2020, https://doi.org/10.1016/j.antiviral.2020.104787.
25. Raza, S., Shahin, F., Zhai, W., Li, H., Alvisi, G., Yang, K., Chen, X., Chen, Y., Chen, J., Hu, C., Chen, H., Guo, A., Ivermectin inhibits bovine herpesvirus 1 DNA polymerase nuclear import and interferes with viral

replication. *Microorganisms*, 8, 409, 2020, https://doi.org/10.3390/microorganisms8030409.

26. Torres, J.R., Isturiz, R., Murillo, J., Guzman, M., Contreras, R., Efficacy of ivermectin in the treatment of strongyloidiasis complicating AIDS. *Clin. Infect. Dis. Off. Publ. Infect. Dis. Soc. Am.*, 17, 900–902, 1993, https://doi.org/10.1093/clinids/17.5.900.
27. Wagstaff, K.M., Sivakumaran, H., Heaton, S.M., Harrich, D., Jans, D.A., Ivermectin is a specific inhibitor of importin α/β-mediated nuclear import able to inhibit replication of HIV-1 and dengue virus. *Biochem. J.*, 443, 851–856, 2012, https://doi.org/10.1042/BJ20120150.
28. Wani, A.K., Akhtar, N., Sher, F., Navarrete, A.A., Américo-Pinheiro, J.H.P., Microbial adaptation to different environmental conditions: Molecular perspective of evolved genetic and cellular systems. *Arch. Microbiol.*, 204, 144, 2022, https://doi.org/10.1007/s00203-022-02757-5.
29. Ashraf, S., Chaudhry, U., Raza, A., Ghosh, D., Zhao, X., *In vitro* activity of ivermectin against Staphylococcus aureus clinical isolates. *Antimicrob. Resist. Infect. Control*, 7, 27, 2018, https://doi.org/10.1186/s13756-018-0314-4.
30. Lim, L.E., Vilchèze, C., Ng, C., Jacobs, W.R., Ramón-García, S., Thompson, C.J., Anthelmintic avermectins kill Mycobacterium tuberculosis, including multidrug-resistant clinical strains. *Antimicrob. Agents Chemother.*, 57, 1040–1046, 2013, https://doi.org/10.1128/AAC.01696-12.
31. Tan, X., Xie, H., Zhang, B., Zhou, J., Dou, Z., Wang, X., Wang, N., A novel ivermectin-derived compound D4 and its antimicrobial/biofilm properties against MRSA. *Antibiotics*, 10, 208, 2021, https://doi.org/10.3390/antibiotics10020208.
32. Pettengill, M.A., Lam, V.W., Ollawa, I., Marques-da-Silva, C., Ojcius, D.M., Ivermectin inhibits growth of Chlamydia trachomatis in epithelial cells. *PLoS One*, 7, e48456, 2012, https://doi.org/10.1371/journal.pone.0048456.
33. Sia, D.K., Mensah, K.B., Opoku-Agyemang, T., Folitse, R.D., Darko, D.O., Mechanisms of ivermectin-induced wound healing. *BMC Vet. Res.*, 16, 397, 2020, https://doi.org/10.1186/s12917-020-02612-z.
34. Zhang, X., Song, Y., Ci, X., An, N., Ju, Y., Li, H., Wang, X., Han, C., Cui, J., Deng, X., Ivermectin inhibits LPS-induced production of inflammatory cytokines and improves LPS-induced survival in mice. *Inflamm. Res. Off. J. Eur. Histamine Res. Soc. Al.*, 57, 524–529, 2008, https://doi.org/10.1007/s00011-008-8007-8.
35. Atakisi, E., Atakisi, O., Topcu, B., Uzun, M., Effects of therapeutic dose of ivermectin on plasma nitric oxide and total antioxidant capacity in rabbits. *Eur. Rev. Med. Pharmacol. Sci.*, 13, 425–429, 2009.
36. Zhang, Y., Sun, T., Li, M., Lin, Y., Liu, Y., Tang, S., Dai, C., Ivermectin-induced apoptotic cell death in human SH-SY5Y cells involves the activation of oxidative stress and mitochondrial pathway and Akt/mTOR-pathway-mediated autophagy. *Antioxidants*, 11, 908, 2022, https://doi.org/10.3390/antiox11050908.

37. Noël, H.L., Hopkin, S.P., Hutchinson, T.H., Williams, T.D., Sibly, R.M., Population growth rate and carrying capacity for springtails Folsomia candida exposed to ivermectin. *Ecol. Appl. Publ. Ecol. Soc. Am.*, 16, 656–665, 2006, https://doi.org/10.1890/1051-0761(2006)016[0656:pgracc]2.0.co;2.
38. Mastrangelo, E., Pezzullo, M., De Burghgraeve, T., Kaptein, S., Pastorino, B., Dallmeier, K., de Lamballerie, X., Neyts, J., Hanson, A.M., Frick, D.N., Bolognesi, M., Milani, M., Ivermectin is a potent inhibitor of flavivirus replication specifically targeting NS3 helicase activity: New prospects for an old drug. *J. Antimicrob. Chemother.*, 67, 1884–1894, 2012, https://doi.org/10.1093/jac/dks147.
39. Li, N. and Zhan, X., Anti-parasite drug ivermectin can suppress ovarian cancer by regulating lncRNA-EIF4A3-mRNA axes. *EPMA J.*, 11, 289–309, 2020, https://doi.org/10.1007/s13167-020-00209-y.
40. Zhou, S., Wu, H., Ning, W., Wu, X., Xu, X., Ma, Y., Li, X., Hu, J., Wang, C., Wang, J., Ivermectin has new application in inhibiting colorectal cancer cell growth. *Front. Pharmacol.*, 12, 717529, 2021, https://doi.org/10.3389/fphar.2021.717529.
41. Slater, H.C., Foy, B.D., Kobylinski, K., Chaccour, C., Watson, O.J., Hellewell, J., Aljayyoussi, G., Bousema, T., Burrows, J., D'Alessandro, U., Alout, H., Ter Kuile, F.O., Walker, P.G.T., Ghani, A.C., Smit, M.R., Ivermectin as a novel complementary malaria control tool to reduce incidence and prevalence: A modelling study. *Lancet Infect. Dis.*, 20, 498–508, 2020, https://doi.org/10.1016/S1473-3099(19)30633-4.
42. Suputtamongkol, Y., Avirutnan, P., Mairiang, D., Angkasekwinai, N., Niwattayakul, K., Yamasmith, E., Saleh-Arong, F.A.-H., Songjaeng, A., Prommool, T., Tangthawornchaikul, N., Puttikhunt, C., Hunnangkul, S., Komoltri, C., Thammapalo, S., Malasit, P., Ivermectin accelerates circulating nonstructural protein 1 (NS1) clearance in adult dengue patients: A combined phase 2/3 randomized double-blinded placebo controlled trial. *Clin. Infect. Dis. Off. Publ. Infect. Dis. Soc. Am.*, 72, e586–e593, 2021, https://doi.org/10.1093/cid/ciaa1332.
43. Alam, M.T., Murshed, R., Bhiuyan, E., Saber, S., Alam, R.F., Robin, R.C., A case series of 100 COVID-19 positive patients treated with combination of ivermectin and doxycycline. *J. Bangladesh Coll. Phys. Surg.*, 38, 10–15, 2020.
44. Rajter, J.C., Sherman, M.S., Fatteh, N., Vogel, F., Sacks, J., Rajter, J.-J., ICON (Ivermectin in COvid Nineteen) study: Use of ivermectin is associated with lower mortality in hospitalized patients with COVID19, MedRxiv, 2020.
45. González Canga, A., Sahagún Prieto, A.M., Diez Liébana, M.J., Fernández Martínez, N., Sierra Vega, M., García Vieitez, J.J., The pharmacokinetics and interactions of ivermectin in humans—A mini-review. *AAPS J.*, 10, 42–46, 2008, https://doi.org/10.1208/s12248-007-9000-9.
46. Edwards, G., Dingsdale, A., Helsby, N., Orme, M.L.E., Breckenridge, A.M., The relative systemic availability of ivermectin after administration as

capsule, tablet, and oral solution. *Eur. J. Clin. Pharmacol.*, 35, 681–684, 1988, https://doi.org/10.1007/BF00637608.
47. Verrest, L. and Dorlo, T.P.C., Lack of clinical pharmacokinetic studies to optimize the treatment of neglected tropical diseases: A systematic review. *Clin. Pharmacokinet.*, 56, 583–606, 2017, https://doi.org/10.1007/s40262-016-0467-3.
48. Klotz, U., Ogbuokiri, J., Okonkwo, P., Ivermectin binds avidly to plasma proteins. *Eur. J. Clin. Pharmacol.*, 39, 607–608, 1990.
49. Wu, M., Fossali, T., Pandolfi, L., Carsana, L., Ottolina, D., Frangipane, V., Rech, R., Tosoni, A., Lopez, G., Agarossi, A., Hypoalbuminemia in COVID-19: Assessing the hypothesis for underlying pulmonary capillary leakage. *J. Intern. Med.*, 289, 861–872, 2021.
50. Martin, R.J. and Pennington, A.J., A patch-clamp study of effects of dihydroavermectin on Ascaris muscle. *Br. J. Pharmacol.*, 98, 747–756, 1989, https://doi.org/10.1111/j.1476-5381.1989.tb14602.x.
51. Schaeffer, J.M. and Haines, H.W., Avermectin binding in Caenorhabditis elegans: A two-state model for the avermectin binding site. *Biochem. Pharmacol.*, 38, 2329–2338, 1989, https://doi.org/10.1016/0006-2952(89)90473-5.
52. Cully, D.F. and Paress, P.S., Solubilization and characterization of a high affinity ivermectin binding site from Caenorhabditis elegans. *Mol. Pharmacol.*, 40, 326–332, 1991.
53. Ottesen, E.A. and Campbell, W.C., Ivermectin in human medicine. *J. Antimicrob. Chemother.*, 34, 195–203, 1994, https://doi.org/10.1093/jac/34.2.195.
54. Sharun, K., Dhama, K., Patel, S.K., Pathak, M., Tiwari, R., Singh, B.R., Sah, R., Bonilla-Aldana, D.K., Rodriguez-Morales, A.J., Leblebicioglu, H., Ivermectin, a new candidate therapeutic against SARS-CoV-2/COVID-19. *Ann. Clin. Microbiol. Antimicrob.*, 19, 23, 2020, https://doi.org/10.1186/s12941-020-00368-w.
55. Johnson-Arbor, K., Ivermectin: A mini-review. *Clin. Toxicol. Phila. Pa.*, 60, 571–575, 2022, https://doi.org/10.1080/15563650.2022.2043338.
56. Lammer, E., Carr, G.J., Wendler, K., Rawlings, J.M., Belanger, S.E., Braunbeck, Th., Is the fish embryo toxicity test (FET) with the zebrafish (Danio rerio) a potential alternative for the fish acute toxicity test? *Comp. Biochem. Physiol. Part C Toxicol. Pharmacol.*, 149, 196–209, 2009, https://doi.org/10.1016/j.cbpc.2008.11.006.
57. Heidary, F. and Gharebaghi, R., Ivermectin: A systematic review from antiviral effects to COVID-19 complementary regimen. *J. Antibiot., (Tokyo)*, 73, 593–602, 2020, https://doi.org/10.1038/s41429-020-0336-z.
58. Geyer, J., Döring, B., Godoy, J.R., Leidolf, R., Moritz, A., Petzinger, E., Frequency of the nt230 (del4) MDR1 mutation in Collies and related dog breeds in Germany. *J. Vet. Pharmacol. Ther.*, 28, 545–551, 2005, https://doi.org/10.1111/j.1365-2885.2005.00692.x.

59. Bowman, D.D. and Mannella, C., Macrocyclic lactones and dirofilaria immitis microfilariae, top. *Companion Anim. Med.*, 26, 160–172, 2011, https://doi.org/10.1053/j.tcam.2011.07.001.
60. Gonzalez, P., Gonzalez, F.A., Ueno, K., Ivermectin in human medicine, an overview of the current status of its clinical applications. *Curr. Pharm. Biotechnol.*, 13, 1103–1109, 2012, https://doi.org/10.2174/138920112800399248.
61. Dourmishev, A.L., Dourmishev, L.A., Schwartz, R.A., Ivermectin: Pharmacology and application in dermatology. *Int. J. Dermatol.*, 44, 981–988, 2005, https://doi.org/10.1111/j.1365-4632.2004.02253.x.
62. Zargari, O., Aghazadeh, N., Moeineddin, F., Clinical applications of topical ivermectin in dermatology. *Dermatol. Online J.*, 22, 2016, https://doi.org/10.5070/D3229032496.
63. Gokbulut, C., Cirak, V.Y., Senlik, B., Aksit, D., Durmaz, M., McKellar, Q.A., Comparative plasma disposition, bioavailability and efficacy of ivermectin following oral and pour-on administrations in horses. *Vet. Parasitol.*, 170, 120–126, 2010, https://doi.org/10.1016/j.vetpar.2010.01.041.
64. Tarr, P., Miele, P., Peregoy, K., Smith, M., Neva, F., Case report: Rectal administration of ivermectin to a patient with Strongyloides hyperinfection syndrome. *Am. J. Trop. Med. Hyg.*, 68, 453–5, 2003, https://doi.org/10.4269/ajtmh.2003.68.453.
65. C.-C. for D.C. and Prevention, CDC - Scabies - resources for health professionals - medications, 2019, https://www.cdc.gov/parasites/scabies/health_professionals/meds.html (accessed January 15, 2023).
66. Kumaraswami, V., Ottesen, E.A., Vijayasekaran, V., Devi, S.U., Swaminathan, M., Aziz, M.A., Sarma, G.R., Prabhakar, R., Tripathy, S.P., Ivermectin for the treatment of Wuchereria bancrofti filariasis: Efficacy and adverse reactions. *JAMA*, 259, 3150–3153, 1988, https://doi.org/10.1001/jama.1988.03720210040026.
67. Navarro, M., Camprubí, D., Requena-Méndez, A., Buonfrate, D., Giorli, G., Kamgno, J., Gardon, J., Boussinesq, M., Muñoz, J., Krolewiecki, A., Safety of high-dose ivermectin: A systematic review and meta-analysis. *J. Antimicrob. Chemother.*, 75, 827–834, 2020, https://doi.org/10.1093/jac/dkz524.
68. Zeng, Z., Andrew, N.W., Arison, B.H., Luffer-Atlas, D., Wang, R.W., Identification of cytochrome P4503A4 as the major enzyme responsible for the metabolism of ivermectin by human liver microsomes. *Xenobiotica*, 28, 313–321, 1998, https://doi.org/10.1080/004982598239597.
69. Campbell, W.C., Use of ivermectin in dogs and cats, in: *Ivermectin Abamectin*, W.C. Campbell (Ed.), pp. 245–259, Springer, New York, NY, 1989, https://doi.org/10.1007/978-1-4612-3626-9_18.
70. Campbell, W.C., Leaning, W.H.D., Seward, R.L., Use of ivermectin in horses, in: *Ivermectin Abamectin*, W.C. Campbell (Ed.), pp. 234–244, Springer, New York, NY, 1989, https://doi.org/10.1007/978-1-4612-3626-9_17.

71. World Health Organization. Safety evaluation of certain contaminants in food: prepared by the eighty-third meeting of the *Joint FAO/WHO Expert Committee on Food Additives (JECFA)*, 2018.
72. Lankas, G., Minsker, D., Robertson, R., Effects of ivermectin on reproduction and neonatal toxicity in rats. *Food Chem. Toxicol.*, 27, 523–529, 1989.
73. Dukuly, Z., Pacque, M., Nara, A., Taylor, H., Williams, P., Greene, B., A prospective study in high risk subjects of electrocardiographic changes with ivermectin. *Trop. Med. Parasitol. Off. Organ Dtsch. Tropenmedizinische Ges. Dtsch. Ges. Tech. Zusammenarbeit GTZ*, 41, 73–74, 1990.
74. Titlić, M., Josipović-Jelić, Z., Punda, A., Headache caused by pesticides–a review of the literature. *Acta Med. Croat. Cas. Hravatske Akad. Med. Znan.*, 62, 233–236, 2008.
75. Trailovic, S.M. and Nedeljkovic, J.T., Central and peripheral neurotoxic effects of ivermectin in rats. *J. Vet. Med. Sci.*, 73, 591–599, 2011.
76. Trailović, S. and Varagić, V.M., The effect of ivermectin on convulsions in rats produced by lidocaine and strychnine. *Vet. Res. Commun.*, 31, 863–872, 2007.
77. Blakley, B. and Rousseaux, C., Effect of ivermectin on the immune response in mice. *Am. J. Vet. Res.*, 52, 593–595, 1991.

2

Historical Background of and Synthetic Approaches to Ivermectin (IVM) and its Homologues

Shakeel Alvi[1], Mohd Kamil Hussain[2] and Rashid Ali[1*]

[1]Department of Chemistry, Jamia Millia Islamia, Jamia Nagar, Okhla, New Delhi, India
[2]Department of Chemistry, Govt. Raza P.G. College (MJP Rohilkhand University, Breilly), Rampur, India

Abstract
Notably, around 250 million people annually are using the ivermectin (IVM) to combat several parasitic diseases such as scabies, filariasis, strongyloidiasis, pediculosis and onchocerciasis, *etc.*, and many clinical trials have proven its effectiveness against these critical diseases. Moreover, the optimum doses as well as duration of the treatment have also fruitfully been reported in the scientific literature. The IVM has showed extraordinary efficacy in killing vectors of disease-causing parasites for instance tsetse flies, mosquitoes, and sandflies, and its anti-parasitic array has increased to shelter farther the parasitic infections – but entails more evaluation for myiasis as well as trichinosis. This particular chapter about IVM will not only highlights the historical perspective of IVM, but also will strongly focus on the synthetic approaches of IVM and its derivatives. Moreover, we will also shade light on the crystal structures of IVM and will conclude with the future aspects about this multipurpose Noble Prize (2015, in medicine) winning wonder drug. We are in the opinion that this much needed chapter will be very useful to the scientific community worldwide, and will also boost this still emerging field of research.

Keywords: Ivermectin, synthesis, avermection, crystal structure, history, drug, diseases

**Corresponding author*: rali1@jmi.ac.in

Rashid Ali and Shahid ul-Islam (eds.) Chemistry and Biological Activities of Ivermectin, (23–46)

2.1 Introduction

2.1.1 Historical Background of Ivermectin

Japanese soil sample's long journey to ivermectin: The story of the discovery of ivermectin (IVM) was rather incredible. In the late 1960s, a Japanese microbiologist named Satoshi Ōmura at the *Tokyo's Kitasako Institute* was hunting for the novel anti-bacterial substances and began to gather thousands of the soil samples from all around the Japan. He extracted bacteria from the soil samples (Figure 2.1). Later, he screened the cultures for medicinal application, and sent those samples 10,000 km away to from Japan to "Merck Research Laboratory – New Jersey", whereby William Campbell one of his collaborator, tested their efficacy against the parasitic worms impacting the livestock as well as other animals. After their studies, they observed that one of the cultures, obtained from a soil sample gathered near a golf-course southwest of Tokyo, was extremely effective against worms. The bacterium in that culture was a novel species, which was baptized *Streptomyces avermictilis* (*S. avermictilis*). That active component was dubbed as avermectin (AVM), and was chemically tailored to increase its potential, and undoubtedly its safety as well to produce the IVM, presently became the top-selling veterinary drug worldwide. Noticeably, after decades of hunting, *S. avermictilis* remains the sole source of IVM ever found. We believe that at present, the story

Figure 2.1 S. Ōmura collecting the historical soil sample comprising *S. avermectinius* in 1973. (Reproduced with the permission of Crump, A. J Antibiot 70, 495–505 (2017)).

IVM's success is not yet ended but still many hidden mysteries are yet to be unmarked in the years to come [1].

Notably, IVM the wonder drug discovered in 1975, is a class of macrocyclic lactones consisting of a 16-membered ring, commonly used in the treatment of several parasitic diseases for instance elephantiasis, scabies, and river blindness which belongs to the avermectins (AVMs) family. The AVMs were first discovered in the "*fermentation broth of an actinomycete culture*" in the year of 1967 in Japan [2, 3]. Later, AVMs were accomplished by means of the *S. avermitilis* in soil, by a small research team at the Kitasato Institute (Japan) in 1973, while working in collaboration with 'Merck sharpe and Dohme research laboratories' USA in the quest of novel as well as innovative antibiotics [4, 5]. An atypical and structurally distinctive organism was isolated by them from the soil sample (*each soil sample typically consists millions of different microorganisms*), and afterward its bioactivity was investigated (Figure 2.2). Next, the dispatch of a *lyophilised culture* of the microorganism to the Merck sharpe & Dohme research laboratories in 1974 to screen as a *fermentation broth* in new animal models, characterization of bacterium besides the discovery and development

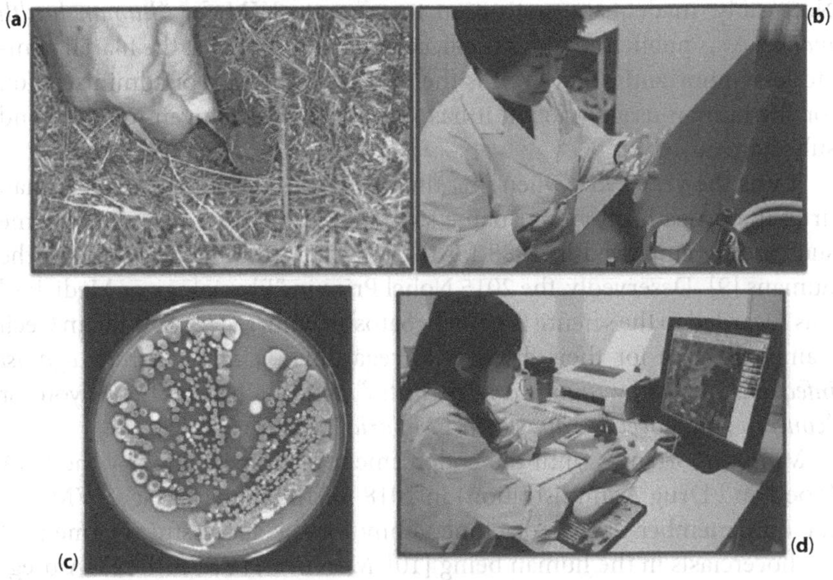

Figure 2.2 Steps in finding the *S. avermectinius*. (a) A soil sample; (b) Isolating structurally discrete organisms from a soil sample; (c) Culturing specific organism before screening for the bioactivity; (d) Elucidating the structure as well as activity of auspicious organisms. (Reproduced with the permission of Ref. [6]).

studies of the AVMs were group endeavors. Towards this remarkable mission, a multidisciplinary research team of 37 scientists, from Kitasato and Merck involving chemists, microbiologists, biologists, parasitologists, and veterinarians was composed. From the studies of *S. avermectinius,* they obtained eight natural AVMs, of which three (A_2a, B_1a, and B_2a), were formed in adequate quantity. As the whole chemical preparation of B1a – a starting material for IVM is uneconomical, thereby production on the commercial scale depends through maximizing by means of *S. avermectinius.*

Interestingly, macrocyclic lactones embracing the AVMs, are continuously receiving the invaluable therapeutic interest, as the anti-parasitic drugs for both humans as well as animals since 1980s [7]. The AVM class of drugs, incorporating avermectin (1), ivermectin (2), doramectin (3), moxidectin (4), selamectin (5), abamectin (6), eprinomectin (7), stromectol® (8) (Figure 2.3, 1-8), are structurally identical macrocyclic lactones but distinguished by the presence of A & B components. Notably, the first AVM drug, IVM (a derivative of AVM-B1 component, consisting of roughly 80% 22,23-dihydro-avermectin-B1a plus approx. 20% 22,23-dihydro-avermectin-B1b) was launched for the usage of animals, as the veterinary drug in 1981 by the Merck & Co [6, 8]. Since then, new formulations of the IVM, are being released more or less every year. Noticeably, inadequate motivation to generate IVM for "*human health market*" was publicized until its efficacy was found against the filarial nematodes. Importantly, oral IVM is the only licensed path of administration for the human usage, though it has successfully been given topically and subcutaneously [9].

It was the year 1987, when first time IVM was registered as the human drug underneath the brand name such as Mectizan®, and was first time successfully employed in 1988 for the treatment of onchocerciasis in the humans [9]. Deservedly, the 2015 Nobel Prize in "Physiology or Medicine" was Awarded to the scientists namely, Satoshi Ōmura (1/4) & William Cecil Campbell (1/4) for their discoveries regarding "*a novel therapy against infections caused by roundworm parasites*" and one half to Tu Youyou for "*concerning a novel therapy against Malaria*" (Figure 2.4) [2].

More importantly, until the endorsement of "moxidectin" by the FDA (Food and Drug Administration) in 2018 for the human usage, IVM was the only member of macrolactone family for the successful treatment of onchocerciasis in the human being [10]. Moreover, IVM is the FDA pregnancy category C. Recently, IVM has also been revealed to inhibit the proliferation of many tumor cells through regulating multiple signaling routes – suggesting that IVM may be used as an anti-cancer drug with a vast potential (Figure 2.5) [11].

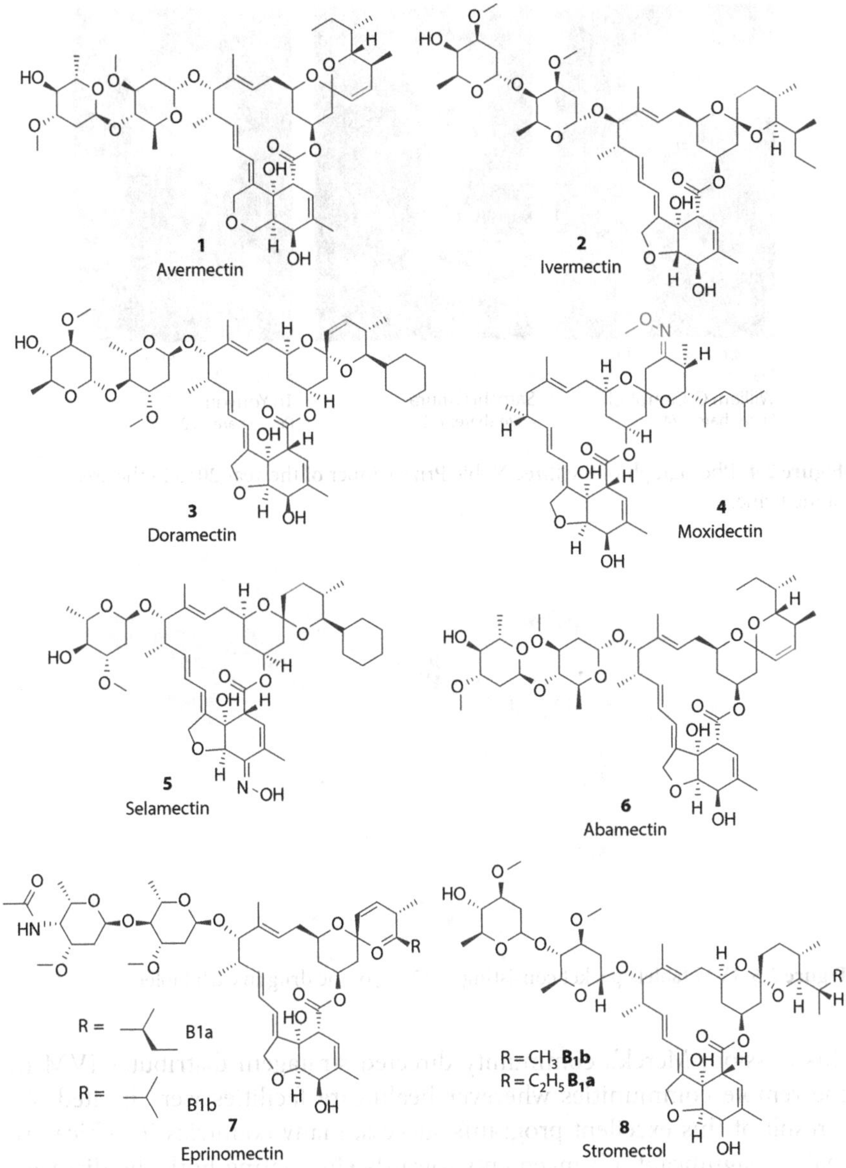

Figure 2.3 The chemical structures of IVM and other AVMs family members.

Notably, since its inception, for almost a decade, IVM was donated for the usage against the onchocerciasis (*river blindness*). In the year 1998, Merck & Co. extended the program to involve lymphatic filariasis so as to provide the IVM "for as long as required & as much as needed". Towards

Figure 2.4 Photograph of the three Noble Prize winner of the year 2015 in the area of medicine.

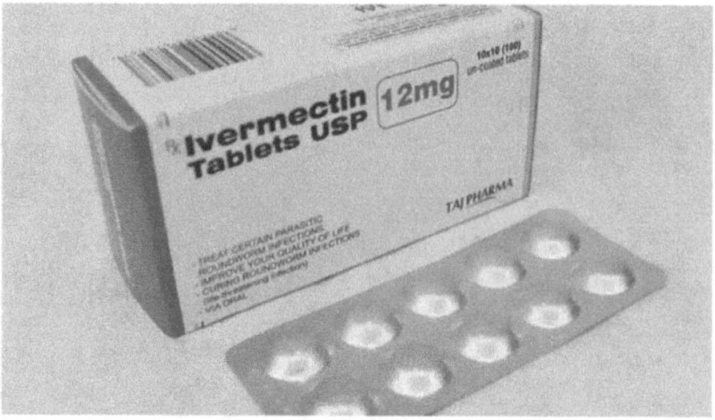

Figure 2.5 IVM tablets packet consisting of 12 mg of the drug in each tablet.

this mission, Merck's community-directed stratagem distributes IVM in the remote communities wherever healthcare facilities were limited. As a result of this excellent program's success, many countries in Africa are making significant advancements towards eliminating both the diseases. Though it has shown remarkable clinical applications, IVM was not recommended by FDA against the Covid-19 (*SARS-CoV-2*) in people due to unavailability of the validated data on the IVM's efficacy. And even the same has also been explicitly stated by Merck & Co., and presently many clinical trials involving IVM as a possible treatment of Covid-19 are under progress. We thanks, IVM for its potential applications in a varied treatment, and also wishes for its bright future for many more uses.

2.2 Synthetic Approaches Towards the Construction of IVM and Analogues

The AVMs were isolated from the culture broth of *S. avermitilis,* as vital compounds by the research team of Ōmura in Japan, and investigated their studies by W.C. Campbell. During their examination, *eight* natural AVMs (A1a, A1b, A2a, A2b, B1a, B1b, B2a, B2a,), were isolated from *S. avermectinius,* out of which *three* (A2a, B1a & B2a) were obtained in adequate quantity. It was noticed that the total synthesis of fragment B1a (*starting material for IVM*) is uneconomical, thereby a commercial scale generation mostly depends on increasing the production through the *S. avermectinius.* Because of their persuasive anthelmintic as well as insecticidal activities, in addition to their inimitable pentacyclic architecture, the AVM family has drawn a tremendous attention of the synthetic organic chemists. Basically, AVMs (*structurally related to milbemycins*), are the 16-membered macrolactones consisting of a 6,6-spiroacetal north segment directly attached to the disaccharide oleandrosyloleandrosyl, in addition to the hexahydrobenzofuran south segment. AVMs and its derivatives have attracted keen interest from synthetic organic chemists. But to our best knowledge, no report in the literature is available whereby, the synthetic portion of IVM and its derivatives has specially been focused, though a handful of research articles, books, chapters, and presentations are being published. Here, we detailed the construction of AVMs and IVM (a derivative of AVM) as shown below.

2.3 Biosynthesis

Basically, AVMs possess a worthy potency as well as broad spectrum against an array of nematode as well as insect parasites coalesced with a very low level of '*toxic-side-effects*' against the host organisms. Nonetheless, there is always a high demand to modify the systems for better performances, so is true for the AVMs – to enhance its activity spectrum and potency, besides assuming exceptionally low level of the toxicity. To this regard, researchers indented to elucidate the biochemistry, and also planned that genetics of the AVM biosynthesis may lead towards the rational tactics for improving the yields of AVMs, generation of new AVM derivatives via mutagenesis, selective production of required components, and recombinant DNA technology.

In this regard, Ikeda and Omura [12] have nicely detailed the biosynthetic pathway of the AVM which is totally based upon the evidences collected from many studies such as amalgamation of the labeled forerunners into the AVMs, transformation of the anticipated intermediates to AVMs by biosynthetically blocked mutants and/or wild type-forming strains identification of vital intermediates generated via the biosynthetically blocked mutants, and the *in vitro* estimation of enzymes included in the biosynthesis of AVMs.

As can be inspected from the Figure 2.6, the biosynthesis of the AVM could be categorized into three main stages. 1) synthesis of polyketide derived from the initial aglycon (6,8a-seco-6,8a-deoxy-5-oxoavermectin aglycons); 2) modification into the initial aglycon to produce the AVM aglycons; 3) glycosylation of the AVM aglycons to assemble AVMs. It was

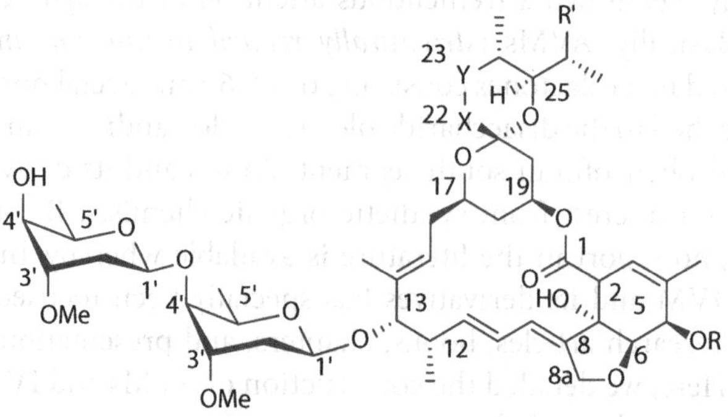

		R	R'	X-Y
Avermectin	**A1a**	CH_3	C_2H_5	CH=CH
	A1b	CH_3	CH_3	CH=CH
	A2a	CH_3	C_2H_5	C_2H-CH(OH)
	A2b	CH_3	CH_3	C_2H-CH(OH)
	B1a	H	C_2H_5	CH=CH
	B1b	H	CH_3	CH=CH
	B2a	H	C_2H_5	C_2H-CH(OH)
	B2b	H	CH_3	C_2H-CH(OH)
Ivermectin	**B1a**	H	C_2H_5	CH_2-CH_2
	B1b	H	CH_3	CH_2-CH_2

Figure 2.6 The chemical structures of AVMs as well as IVM.

also confirmed by the labeling experiments that the aglycon moiety of the AVM is derived from numerous fatty acids. Finally, a series of adjustments involving cyclization, reduction, and methylation occurred to assemble the AVM aglycons. Afterward the reasonable aliphatic polyketide originated precursor is lactonized to produce the initial aglycon. In the final biosynthetic strides, *O*-glycosylation at C13 & C4′ is being accomplished by means of the dTDP-oleandrose to generate AVMs.

As discussed above (Figure 2.5), eight allied components of the AVMs were obtained from *S. avermitilis* out of which only two, namely B1a and B1b – are generally used in the agricultural field besides their applications for medical/veterinary purposes. Despondently, separation of both these vital components at the industrial level is really a dauting task. On the other hand, organism also yields an undesirable toxic compound, dubbed as "oligomycin". Hence, undeniable issues still remain to be resolved for the industrial manufacture of avermectin (B1a). Therefore, deep understanding of the AVMs' biosynthetic pathway could mark it achievable to eliminate the unwanted products or to fabricate new AVMs through selective inactivation of the vital biosynthetic genes.

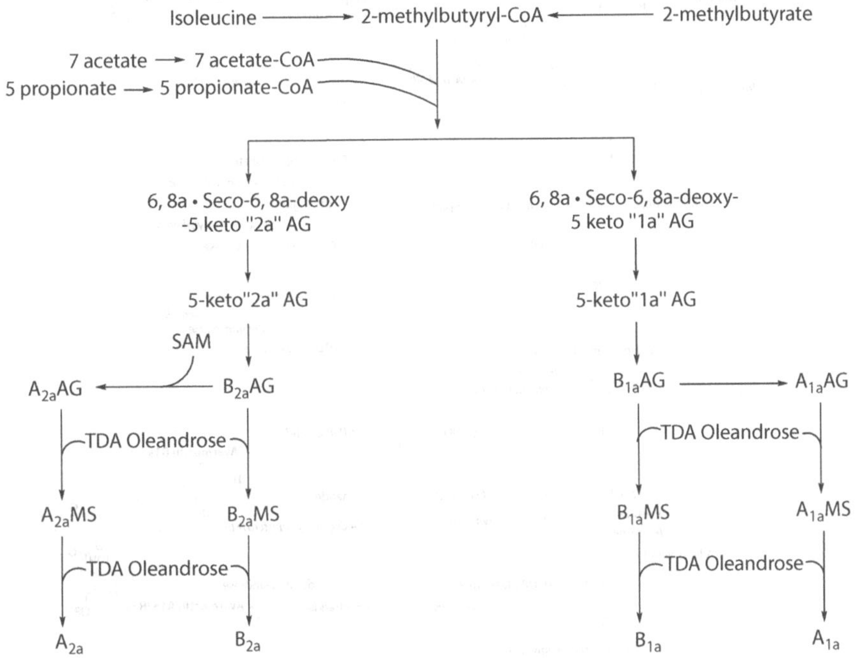

Scheme 2.1 Proposed route for the biosynthesis of AVMs from the *S. avermitilis*.

On the other hand, in the year 2003, Leadlay and co-workers have reported IVM-like drugs involving biosynthetic sequence starting from the AVM aglycone after domain exchange in the AVM polyketide synthase of the *S. avermitilis* ATCC31272 (Scheme 2.1) [13].

Generally IVM is produced by the selective chemical reduction of AVMs, which are assembled utilizing the biosynthetic route from *S. avermitilis* (Scheme 2.2). As detailed above, the biosynthesis of AVM includes extension of the starter subunit – either 2-methylbutyrate (A1) or isobutyrate (B1), with seven acetate and five propionate derived units. These 12 consecutive cycles of the extension form the aglycone 6,8a-seco-6,8a-deoxy-5-oxoavermectin – which

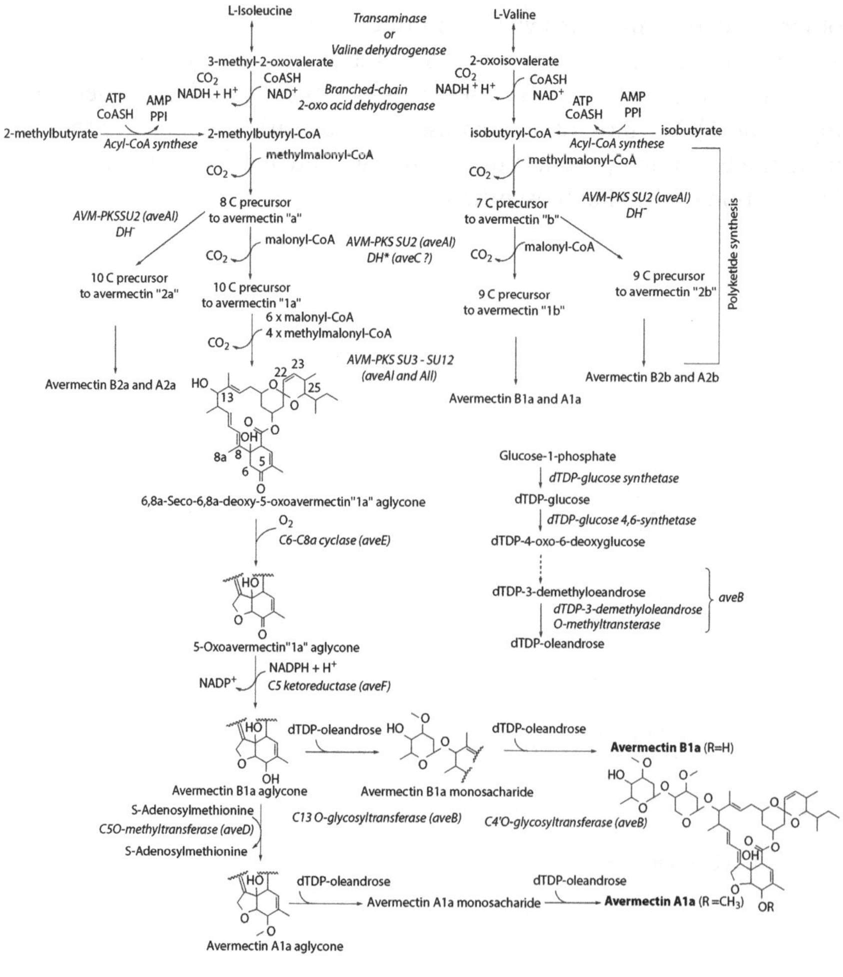

Scheme 2.2 Biosynthesis roadmap for the construction of AVMs (A1a and B1a).

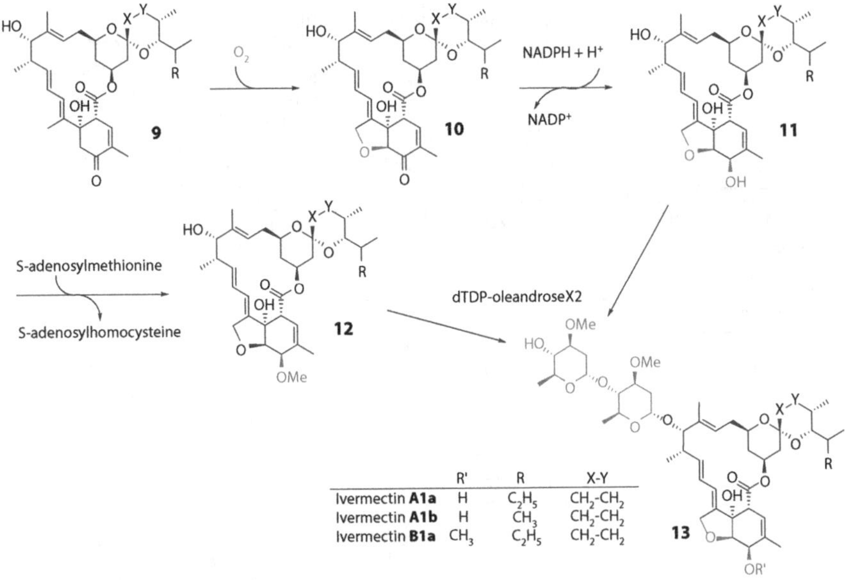

Scheme 2.3 Direct biosynthetic reaction sequence from the AVM aglycone to AVM molecules.

is consequently modified to produce the AVMs through oxidative-cyclisation, reduction, and an optional C5-*O*-methylation, delivering A/B-components, followed by the glycosylation at C13 position using addition of two *L*-oleandrosyl residues (Scheme 2.3).

Interestingly, authors obtained sufficient amounts of the anticipated 22,23-dihydroavermectins in the *cultures* of *S. avermitilis*, and the southern analysis of the recombinant strain of BIOT-1916 was copiously consistent with the estimated genetic organization From their experimentations, two of 22,23-dihydroavermectins were successfully isolated after purification, and completely characterized by means of NMR as well as high-resolution mass spectrometric analysis.

2.4 Chemical Synthetic Pathway

Remarkably, in 2016, Yamashita *et al.* revisited the total synthesis of avermectin-B$_{1a}$ as shown in the Schemes 2.4–2.6 [14]. Towards this journey, they first prepared the "*south segment*", commencing with the acid-mediated cyclization of an easily accessible *D*-sorbitol (14) – which on further acetonide formation in the presence of acetone/TsOH, provided

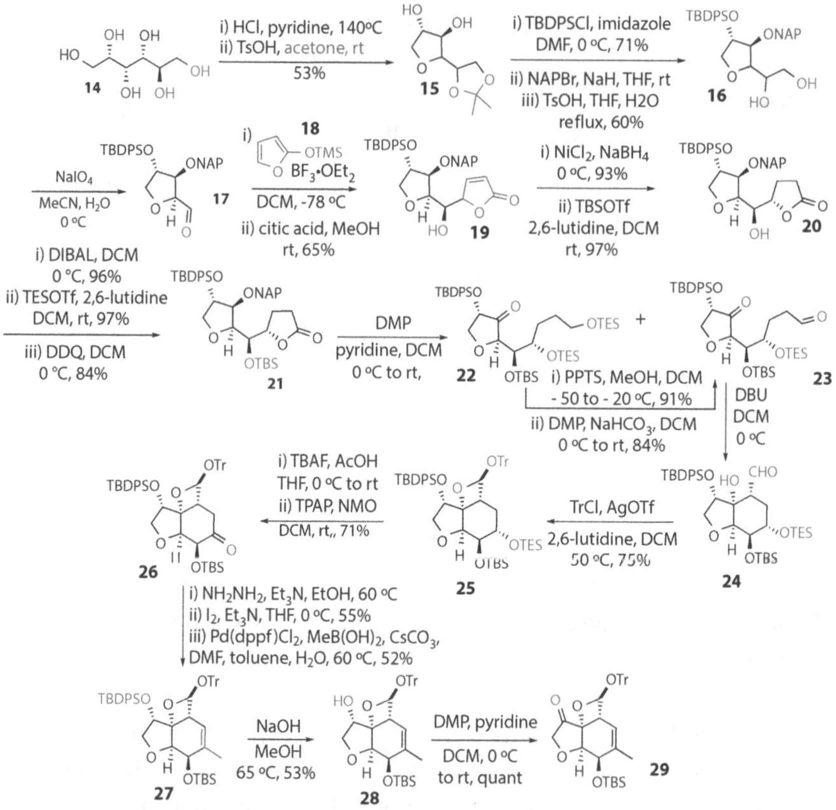

Scheme 2.4 Preparation of the component 29 from the commercially available *D*-sorbitol (14).

the tetrahydrofuran derivative (15) in overall 53% yield (Scheme 2.4). Afterward, protection of the residual diols of the compound 15 with ^{t}butyldiphenylchlorosilane (TBDPSCl) as well as 2-naphtylmethyl (NAP) bromide, furnished the fully protected compound 16. Furthermore, selective hydrolysis, followed by the oxidative cleavage using $NaIO_4$, delivered the aldehyde 17, which was directly used without further purification – for the vinylogous *Mukaiyama aldol reaction* involving 2-(trimethylsiloxy)furan (18) in the presence of $BF_3.OEt_2/CH_2Cl_2$ at −78°C to yield the required adduct 19 as a C_4-diastereomeric mixture in the ratio of 5:1 (*assumed on the basis of non-chelated Felkin–Anh model*). Next, C-C double bond hydrogenation of the compound 19 using $NiCl_2/NaBH_4$, and a subsequent ^{t}butyldimethylsilyl (TBS) protection – delivered the lactone 20 in overall excellent yield. Afterward, reduction of the 20 involving

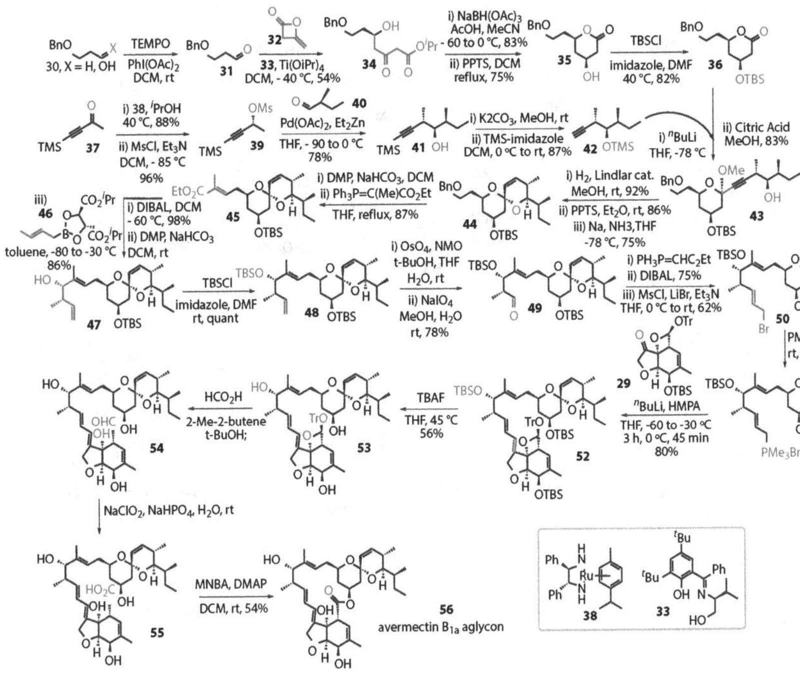

Scheme 2.5 Synthesis of avermectim ($B1_a$) aglycon **56** casting several crucial steps.

diisobutylaluminum hydride (DIBAL-H), and protection of the resulting diol with *bis*-triethylsilyl (TES) group, followed by the removal of NAP-group with 2,3-dichloro-5,6-dicyano-p-benzoquinone(DDQ) afforded the compound 21. The Dess–Martin periodinane (DMP) mediated oxidation of 21 in pyridine furnished the *bis*-carbonyl compound 23 as a major product along with the ketone 22, which was also transformed into 23 by means of conventional hydrolysis followed by oxidation reactions.

Notably, after several unsuccessful experimentations, the compound 23 was cleanly cyclized to 24 (*formed as a 3.9:1 mixture of C_2-diastereomers*) through intramolecular aldol reaction involving 1,8-diazabicyclo[5.4.0] undec-7-ene (DBU)/CH_2Cl_2 at 0°C.

Next, the crude aldehyde 24 was reacted with trityl trifluoromethanesulfonate [*in situ* prepared from trityl chloride and silver trifluoromethanesulfonate (AgOTf)], to furnish the trityl oxetane acetal 25 as a sole isomer in overall good yield. Furthermore, the selective removal of the TES-group using AcOH-buffered/tetrabutylammonium fluoride (TBAF), and subsequent oxidation with tetrapropylammonium perruthenate oxidation (TPAP) provided the ketone 26 in decent yield. Later, C_{4a}-methyl

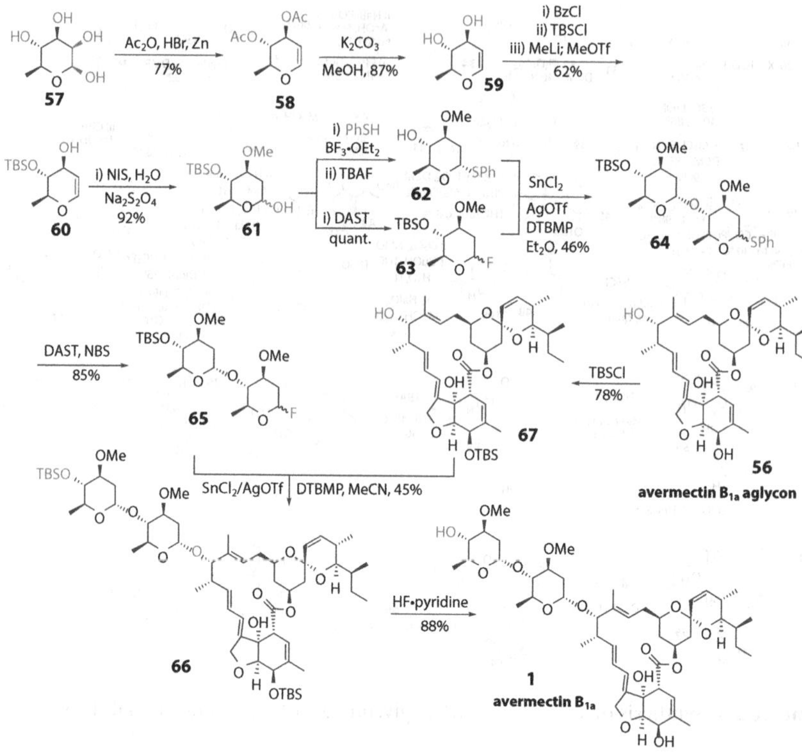

Scheme 2.6 Final total synthesis of AVM (B1a) combining fragments prepared in Scheme 4 and 5.

substituent and the C_3–C_4 double bond was installed in three steps by means of: 1) hydrazone formation; 2) iodination using I_2/Et_3N; 3) Suzuki-Miyaura cross-coupling involving methylboronic acid to deliver the desired compound 27 which was confirmed by means of X-ray crystallography analysis. After that the compound 27 was treated with NaOH to selectively remove the TBDPS-group without disturbing the TBS-group, producing a secondary alcohol 28. Finally, the DMP oxidation of 28 furnished the south segment 29 in the quantitative yield (Scheme 2.4).

After successful synthesis of 29, they next prepared the aglycon 56 in several steps (Scheme 2.5). Towards this mission, they first accomplished the enantioselective reduction of ynone 37 through '*asymmetric hydrogen transfer reaction*' using the Noyori catalyst (38) to afford the chiral secondary alcohol with enantiomeric excess (ee) of 98%, which upon mesylation with $MsCl/Et_3N$ generated the alkyne derivative 39 in overall good yield. Next, by using Marshall method, the diastereoselective cross-coupling between 39 and aldehyde 40 was accomplished using $Et_2Zn/Pd(OAc)_2$ in the presence of PPh_3 to deliver the required coupling product 41 in 78% yield with a good diastereoselectivity (dr = 14:1). Finally, trimethylsilyl (TMS)-group removal from the acetylene part using K_2CO_3/MeOH, and subsequent TMS-protection of the 2° alcoholic group provided the C_{22}–C_{28} unit 42 in excellent yield. On the other front, the C_{15}–C_{21} subunit 36 was prepared starting from the propanediol monobenzyl ether (30). To this line, first oxidation of the 30 using 2,2,6,6-tetramethylpiperidine 1-oxyl (TEMPO)/$PhI(OAc)_2$ afforded the aldehyde 31, which was then directly employed for the aldol reaction with 32 in the presence of $Ti(O^iPr)_4$ involving the chiral Schiff base (33) to yield the δ-hydroxy-β-ketoester 34 in respectable yield. Later, anti-selective Evans reduction of 34 and acid-mediated lactonization, followed by TBS-protection furnished the C_{15}–C_{21} unit 36 in overall good yield (Scheme 2.5).

Later, lactone 36 was coupled with the lithiated terminal alkyne 42 to provide the ynone – which on further treatment with citric acid in MeOH exhibited the selective removal of TMS-group to give methyl acetal 43. The acetal 43 was then transformed into the spiroacetal 44 through Lindlar reduction, acid-treatment and subsequent Birch reduction. As can be seen from the Scheme 2.5, afterward, the compound 44 was transformed into the spiroacetal 51 in several steps, which was then treated with 29 under Wittig condition to give 52, which on TBS deprotection with TBAF furnished the diol 53 in 56% yield. Interestingly, *de novo* synthesis of aglycon 56 was accomplished, first by removal of the trityl group from 53 using HCO_2H in degassed ^{t}BuOH/2-methyl-2-butene, then the degassed reaction mixture was directly reacted with aq. $NaClO_2$ and NaH_2PO_4 (Krause–Pinnick oxidation) to afford the seco-acid 55 *via* 54. Finally, Shiina macrolactonization of 55 produced the avermectin B1a aglycon 56 in 54% overall yield (Scheme 2.5).

As displayed in the Scheme 2.6, to conclude the total synthesis of AVM (B1a), these authors first prepared a disaccharide 65 in several steps – according to modified Nicolaou's protocol [15]. On the other hand, the C_5-hydroxy group of 56 was protected selectively with TBSCl, to give 67, which was then coupled with glycosyl fluoride 65 using $SnCl_2$/AgOTf in the presence of DTBMP in MeCN to furnish the anticipated α-glycoside 66 in 45% yield. Finally, removal of the TBS-groups *via* buffered HF-pyridine in MeCN deliver avermectin B1a (1) in 88% yield. Importantly, each and every compound was confirmed by spectroscopic means including the final target natural product. Notably, Omura and Crump have reported the synthesis of IVM from AVM by using Wilkinson's catalyst $[RhCl(PPh_3)_3]$ based selective hydrogenation (Scheme 2.7) [6].

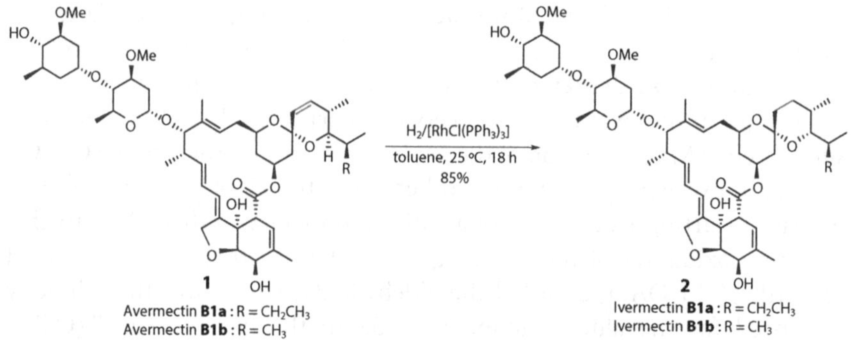

Scheme 2.7 Hydrogenation of AVM-B1, utilizing the Wilkinson's catalyst – selective reduction of the double bond at the position C22–C23 (in red), to generate the more effective IVM.

On the other hand, recently, the research group of Singh has revealed the structural modification of IVM to generate dual action molecular hybrids with worthy structure reliant *in vitro* activity against the hepatic as well as erythrocytic stages of *P. berghei* and *P. falciparum* infection – signifying

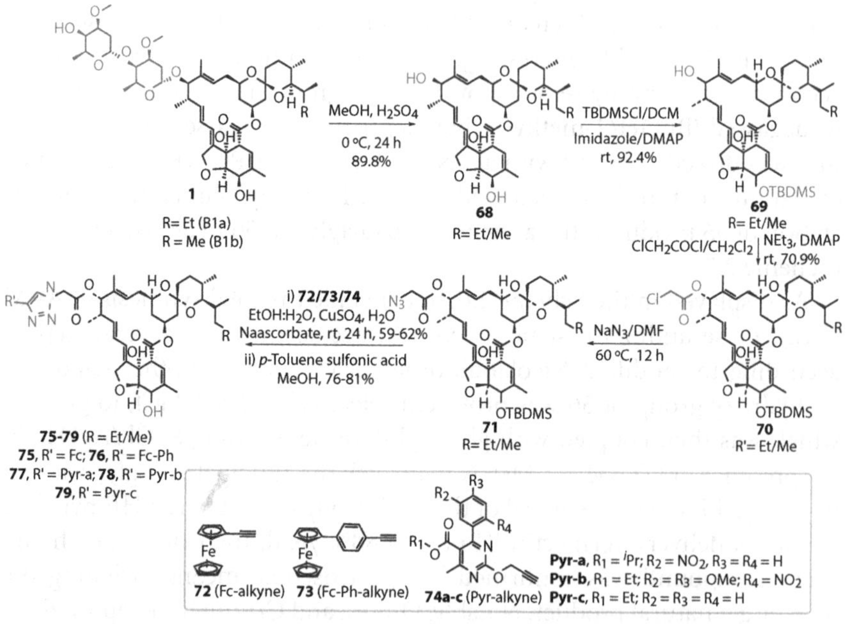

Scheme 2.8 The chemical structure of IVM (1), and synthesis of diverse IVM hybrids (75-79).

inclusion of the IVM anti-malarial hybrids in malaria control stratagems (Scheme 2.8) [16]. From their studies, they observed that the most active hybrid system exhibited over threefold and 10-fold higher *in vitro* activities as compared to the IVM against the hepatic and blood stage infections, respectively. Moreover, *in silico* docking studies were also reported by these workers, supporting the allosteric binding onto the '*glutamate-gated chloride channels*' alike to the IVM. All the synthesized compounds were completely characterized by means of spectroscopic techniques by these authors. The ^{1}H-NMR spectral allocations for all the intermediates as well as compounds were ended on the basis of the coupling connectivity, observed in the ^{1}H–^{1}H homonuclear correlation spectroscopy (COSY) spectrum. Moreover, the structures were additionally corroborated through HRMS

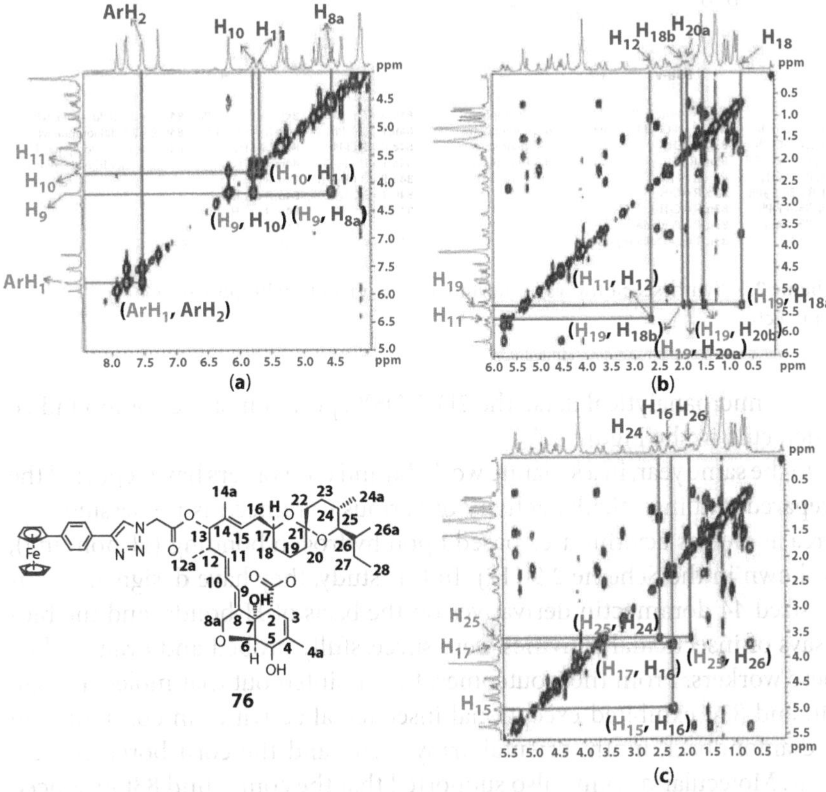

Figure 2.7 (a-c) Extended scale contour plots of the 2D-COSY spectrum (500 MHz, $CDCl_3$ at 300 K) of diverse resonances of the hybrid 76. Coupling linkages for the crucial protons are highlighted. (Reproduced with permission of Ref. [16]).

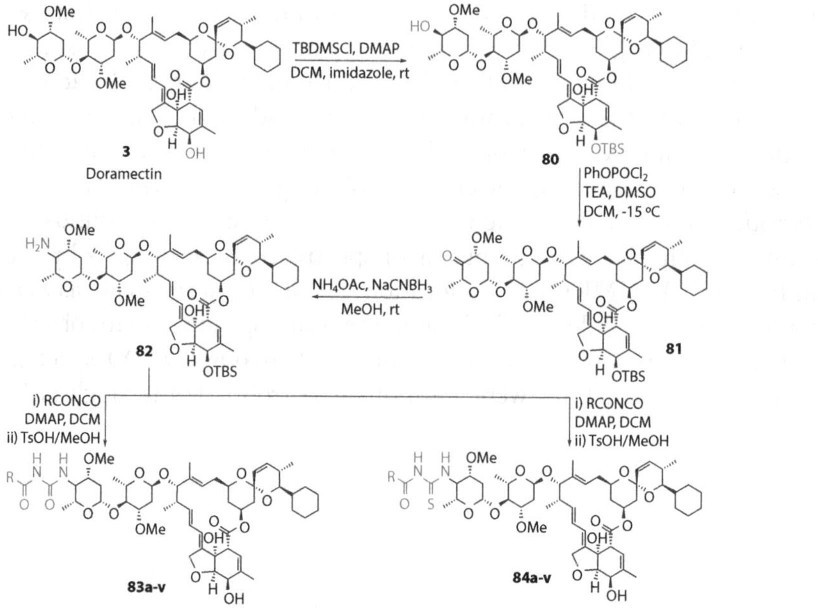

83a, R = Ph; **83b**, R = 2-F-Ph; **83c**, R = 2-Cl-Ph; **83d**, R = 2-CF$_3$-Ph; **83e**, R = 3-CH$_3$-Ph; **83f**, R = 3-CF$_3$-Ph; **83g**, R = 4-F-Ph; **83h**, R = 4-Cl-Ph; **83i**, R = 4-Br-Ph; **83j**, R = 4-OCH$_3$-Ph; **83k**, R = 4-CF$_3$-Ph; **83l**, R = CH$_3$; **83m**, R = Et; **83n**, R = Pr; **83o**, R = C$_3$H$_5$; **83p**, R = C$_6$H$_{11}$; **83q**, R = Pentane; **83r**, R = 2,4-difluorophenyl; **83s**, R = 2,5-difluorophenyl; **83t**, R = 2,6-difluorophenyl; **83u**, R = 3,5-difluorophenyl; **83v**, R = 3,4,5-Trimethoxyphenyl

84a, R = Ph; **84b**, R = 2-F-Ph; **84c**, R = 2-Cl-Ph; **84d**, R = 2-CF$_3$-Ph; **84e**, R = 3-CH$_3$-Ph; **84f**, R = 3-CF$_3$-Ph; **84g**, R = 4-F-Ph; **84h**, R = 4-Cl-Ph; **84i**, R = 4-Br-Ph; **84j**, R = 4-OCH$_3$-Ph; **84k**, R = 4-CF$_3$-Ph; **84l**, R = CH$_3$; **84m**, R = Et; **84n**, R = Pr; **84o**, R = C$_3$H$_5$; **84p**, R = C$_6$H$_{11}$; **84q**, R = Pentane; **84r**, R = 2,4-difluorophenyl; **84s**, R = 2,5-difluorophenyl; **84t**, R = 2,6-difluorophenyl; **84u**, R = 3,5-difluorophenyl; **84v**, R = 3,4,5-Trimethoxyphenyl

Scheme 2.9 Synthesis pathways for acylureas (83a-v) and acylthioureas (84a-v) of doramectin.

and/or microanalytical data. The 2D-COSY spectrum of the compound 76 is depicted in the Figure 2.7.

In the same year, in a separate work, Lu and co-workers have reported the prepared and insecticidal activity of various doramectins possessing acylurea as well as acylthiourea based upon hydrogen bonding (H-bonding), as shown in the Scheme 2.9 [17]. In this study, they have design and synthesized 44 doramectin derivatives on the basis of H-bonds, and the bioassays of insecticidal activities were successfully probed and evaluated by these workers. From their outcomes, they pointed out that molecules 83t, 84t, and 83o exhibited exceptional insecticidal activities in contradiction of diamondback moth, oriental armyworm, and the corn borer, respectively. Molecular docking also supported that the compound 83t produced stronger H-bonds with the object protein, and the docking outcome was reliable with the analogous insecticidal activity test effect.

On the other hand, Ramsay and co-workers have reported the synthesis of moxidectin (4) as shown in the Scheme 2.10 [18]. The standard mouse/*N. dubius* model was employed by them to evaluate the drug for its anti-parasitic activity in comparison to the IVM drug. And moxidectin was found to be as effective as IVM drug, according to the preliminary findings. Moreover, the alternative routes for the construction of moxidectin (4) are also displayed in the Schemes 2.11–2.13 [19, 20].

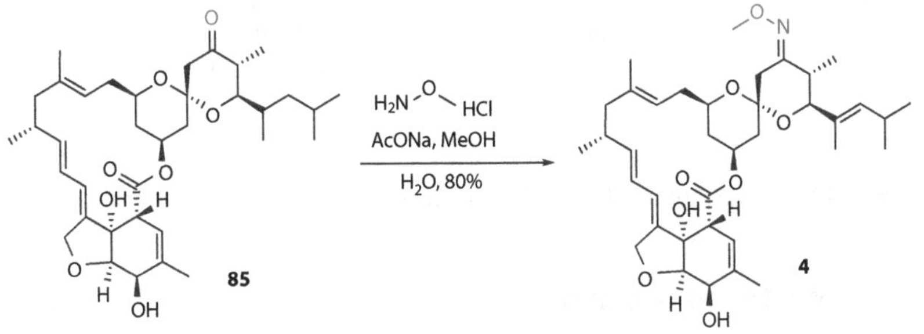

Scheme 2.10 Synthesis of moxidectin 4 starting from 85.

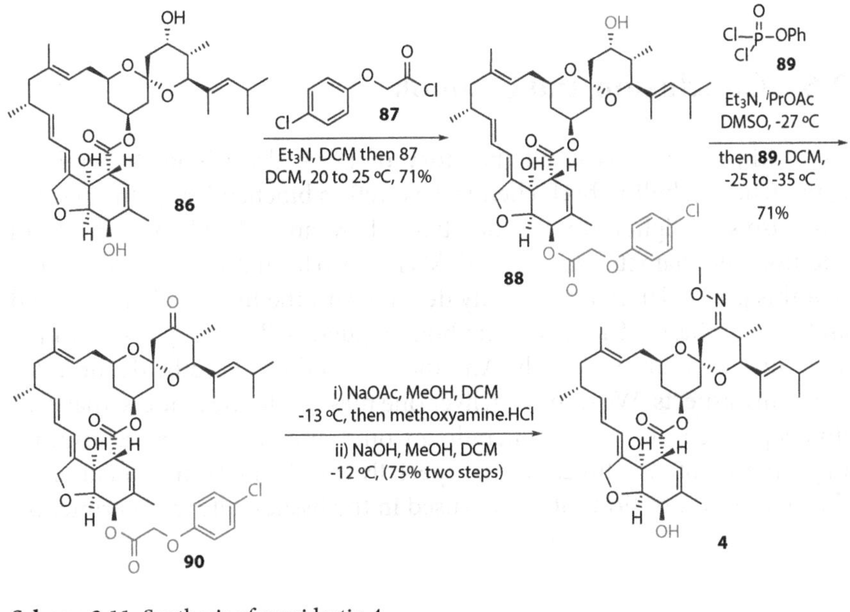

Scheme 2.11 Synthesis of moxidectin 4.

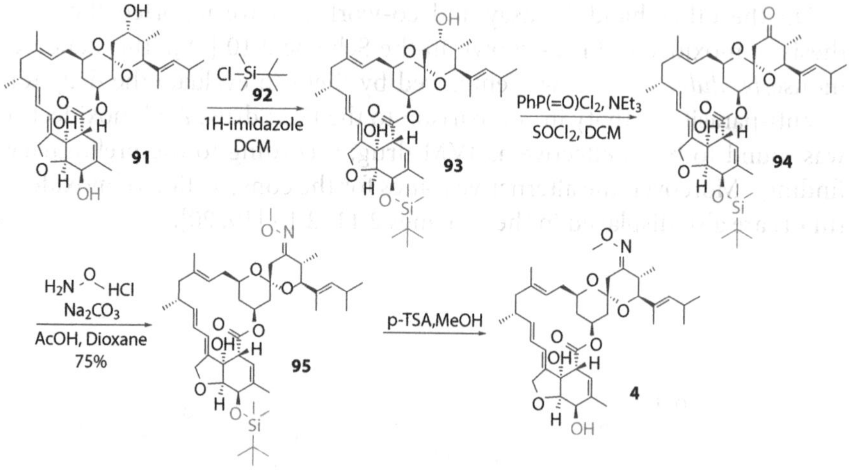

Scheme 2.12 Synthesis of moxidectin in four steps from 91.

2.5 Crystal Structure

The ^{1}H-NMR study revealed that the conformation of avermectin Bla in solution phase is closely interrelated to that of conformation in the solid state. In 2021, Belyakov and co-workers have reported the crystal structure of novel IVM pseudopolymorphs as shown in the Figure 2.8 [21].

2.6 Conclusion and Outlook

Herewith, we have covered the story of IVM, describing its synthetic approaches including both chemical as well as biochemical pathways. We have also shed light onto the potential of IVM and related compounds. In addition, the characterization of AVMs has also been discussed. We believe that this particular chapter mostly dealing with the historical background and the synthesis of IVM and its homologues, will be very useful to the scientific community globally. And the new finding could be deduced for the future aspects. We want to deliver one message to the readers, that caution is prerequisite before concluding from these preliminary repositioning experiments to clinical uses, especially for Covid-19 treatment, as of the high concentrations of IVM is used in the tissue-culture experiments.

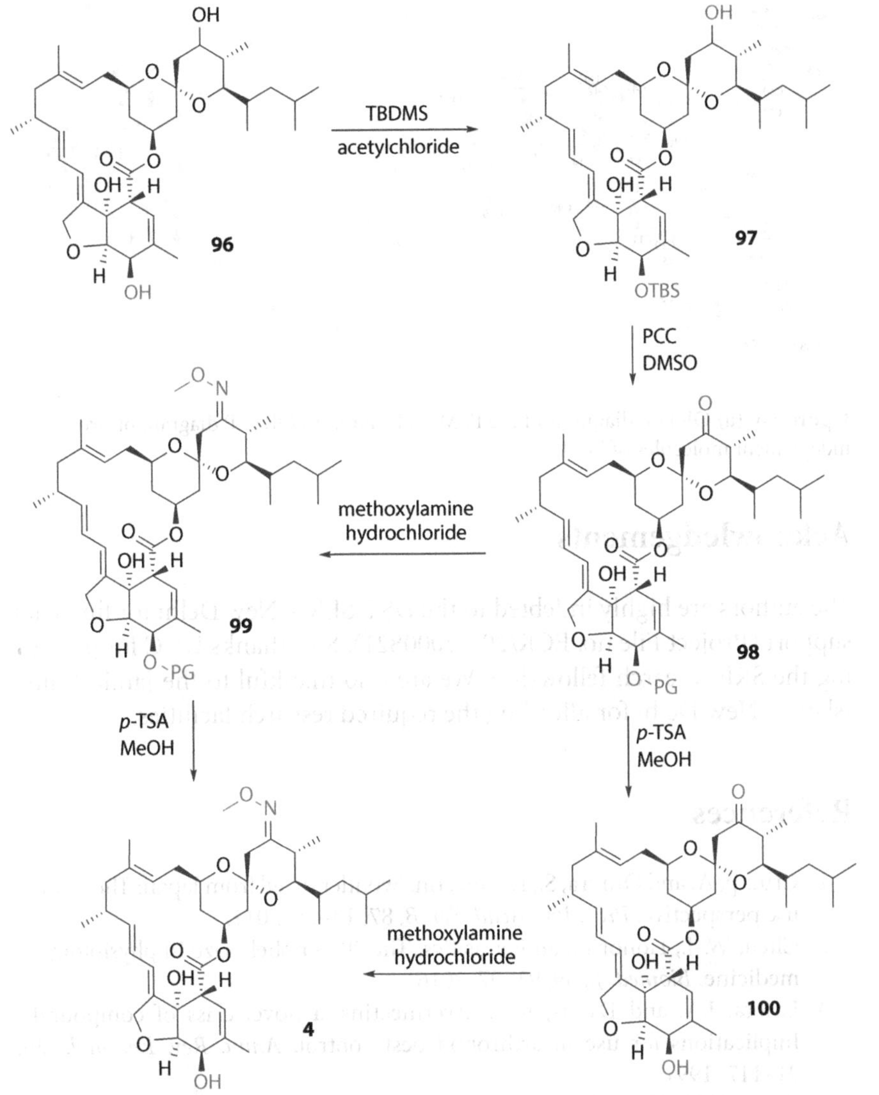

Scheme 2.13 Schematic representation of the conversion of nemadectin into the moxidectin.

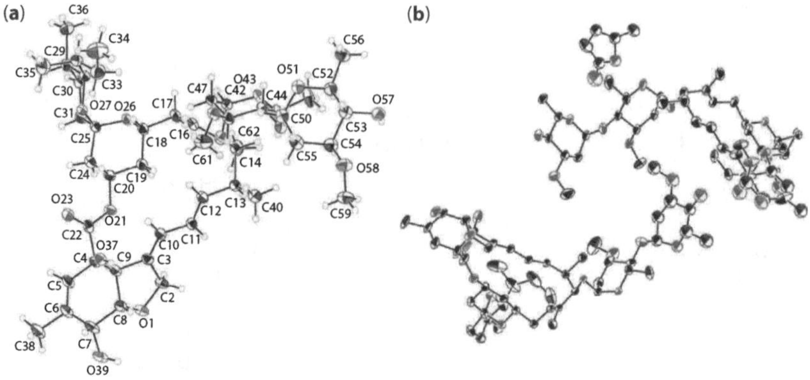

Figure 2.8 (a) ORTEP diagrams of the IVM molecules. (b) ORTEP diagram of two independent molecules of IVM.

Acknowledgements

The authors are highly indebted to the DST-SERB, New Delhi for financial support (Project File no. ECR/2017/000821). S.A. thanks UGC for providing the SRF research fellowship. We are also thankful to the Jamia Millia Islamia, New Delhi for affording the required research facilities.

References

1. Crump, A. and Omura, S., Ivermectin, 'wonder drug' from Japan: The human use perspective. *Proc. Jpn. Acad. Ser. B*, 87, 13–28, 2011.
2. Chen, W.-J., Honoring antiparasitics: The 2015 nobel prize in physiology or medicine. *Biomed. J.*, 39, 93–97, 2016.
3. Lasota, J.A. and Dybas, R.A., Avermectins, a novel class of compounds: Implications for use in arthropod pest control. *Annu. Rev. Entomol.*, 36, 91–117, 1991.
4. Kwon, D.H., Yoon, K.S., Clark, J.M., Lee, S.H., A point mutation in a glutamate-gated chloride channel confers abamectin resistance in the two-spotted spider mite, Tetranychus urticae Koch. *Insect Mol. Biol.*, 19, 583–591, 2010.
5. Hüter, O.F., Use of natural products in the crop protection industry. *Phytochem. Rev.*, 10, 185–194, 2011.
6. Ōmura, S. and Crump, A., Ivermectin: Panacea for resource-poor communities? *Trends Parasitol.*, 30, 445–455, 2014.

7. Canga, A.G., Prieto, A.M.S., Liébana, M.J.D., Martínez, N.F., Vega, M.S., Vieitez, J.J.G., The pharmacokinetics and interactions of ivermectin in humans-a mini-review. *AAPS J.*, 10, 42–46, 2008.
8. Campbell, W.C., Fisher, M.H., Stapley, E.O., Albers-Schönberg, G., Jacob, T.A., Ivermectin: A potent new antiparasitic agent. *Science*, 221, 823–828, 1983.
9. Laing, R., Gillan, V., Devaney, E., Ivermectin–old drug, new tricks? *Trends Parasitol.*, 33, 463–472, 2017.
10. Moxidectin tablets, for oral use Melbourne, Medicines Development for Global Health, Victoria, Australia, 2018, Https://www.accessdata.fda.gov/, Development for Global Health; 2018 Drugsatfda_docs/label/2018/210867lbl.pdf.
11. Tang, M., Hu, X., Wang, Y., Yao, X., Zhang, W., Yu, C., Cheng, F., Li, J., Fang, Q., Ivermectin, a potential anticancer drug derived from an antiparasitic drug. *Pharmacol. Res.*, 163, 105207, 2021.
12. Ikeda, H. and Ōmura, S., Avermectin biosynthesis. *Chem. Rev.*, 97, 2591–2610, 1997.
13. Gaisser, S., Kellenberger, L., Kaja, A.L., Weston, A.J., Lill, R.E., Wirtz, G., Kendrew, S.G., Low, L., Sheridan, R.M., Wilkinson, B., Galloway, I.S., Stutzman-Engwall, K., McArthur, H.A.I., Staunton, J., Leadlay, P.F., Direct production of ivermectin-like drugs after domain exchange in the avermectin polyketide synthase of Streptomyces avermitilis ATCC31272. *Org. Biomol. Chem.*, 1, 2840, 2003.
14. Yamashita, S., Hayashi, D., Nakano, A., Hayashi, Y., Hirama, M., Total synthesis of avermectin B1a revisited. *J. Antibiot. (Tokyo)*, 69, 31–50, 2016.
15. Nicolaou, K.C., Dolle, R.E., Papahatjis, D.P., Randall, J.L., Practical synthesis of oligosaccharides. Partial synthesis of avermectin Bla. *J. Am. Chem. Soc.*, 106, 4189–4192, 1984.
16. Singh, L., Fontinha, D., Francisco, D., Mendes, A.M., Prudencio, M., Singh, K., Molecular design and synthesis of ivermectin hybrids targeting hepatic and erythrocytic stages of plasmodium parasites. *J. Med. Chem.*, 63, 1750–1762, 2020.
17. Zhang, Q., Cheng, Y., Zheng, C., Bai, P., Yang, J., Lu, X., Design, synthesis, and insecticidal activity of novel doramectin derivatives containing acylurea and acylthiourea based on hydrogen bonding. *J. Agric. Food Chem.*, 68, 5806–5815, 2020.
18. Beddall, N.E., Howes, P.D., Ramsay, M.V.J., Roberts, S.M., Slawin, A.M.Z., Sutherland, D.R., Tiley, E.P., Williams, D.J., Chemical transformations of S541 factors (A)-(D): Preparation and reactions of the 23-ketones. *Tetrahedron Lett.*, 29, 2595–2598, 1988.

19. Flick, A.C., Leverett, C.A., Ding, H.X., McInturff, E., Fink, S.J., Helal, C.J., DeForest, J.C., Morse, P.D., Mahapatra, S., O'Donnell, C.J., Synthetic approaches to new drugs approved during 2018. *J. Med. Chem.*, 63, 10652–10704, 2020.
20. Awasthi, A., Razzak, M., Al-Kassas, R., Harvey, J., Garg, S., Analytical profile of moxidectin, in: *Profiles of Drug Substances, Excipients and Related Methodology*, vol. 38, pp. 315–366, 2013.
21. Shubin, K., Bērziņš, A., Belyakov, S., Crystal structures of new ivermectin pseudopolymorphs. *Crystals*, 11, 172, 2021.

3

Ivermectin as a Repurposed Drug for COVID-19

Hari Madhav[1], Mohammad Rehan[2], Nasimul Hoda[1] and Ehtesham Jameel[1*]

[1]*Drug Design and Synthesis Laboratory, Department of Chemistry, Jamia Millia Islamia (A Central University), New Delhi, India*
[2]*Max-Planck-Institute für Molekulare Physiologie, Abteilung Chemische Biologie, Dortmund, Germany*

Abstract
Since the COVID-19 coronavirus 2 (SARS-CoV-2) pandemic was identified, many therapeutic drugs have been re-profiled to fight the disease and used empirically before satisfactory results of any clinical trial. Ivermectin (IVM) was one of the most controversial reprofiled drugs that have been anticipated as possibly effective medication in the initial phase of COVID-19. FDA-approved anti-parasitic drug IVM showed some interesting anti-viral properties against COVID-19 *in vitro*. However, early studies also reported some clinical benefits but the quality of the results varying study to study. Furthermore, some published articles related to IVM, and COVID-19 were retracted due to lack of scientific evidence and some trials have now been recognized as potentially fraudulent. In this chapter, we have tried to summarize the proposed potential MAO of understanding on *in vitro* activity, efficacy and outcomes of the use of IVF to treat COVID-19 from SARS based on some clinical trials.

Keywords: Ivermectin, COVID-19, SARS-CoV-2, clinical results, drug repurposing

3.1 Introduction

The pandemic COVID-19 has significantly affected daily life across the globe and infected over 518 million with over 6 million causalities as of

**Corresponding author*: ehteshamjameel@gmail.com

Rashid Ali and Shahid ul-Islam (eds.) Chemistry and Biological Activities of Ivermectin, (47–68)

May 15, 2022 (WHO, 2022) [1]. Before the introduction of COVID-19 vaccines or any effective medication, the non-pharmaceutical interventions (NPIs) were rolled out to control the pandemic across the World [2]. With the development of COVID-19 vaccines, vigorous vaccination programs were used, which helped to control the spread and mortality rate of the COVID-19 pandemic and eased the NPIs. However, with the delayed production and allocations of vaccines, and absence of the effective medication, different health regulatory agencies are advised to maintain NPIs to avoid further resurgences before herd immunity can be achieved [3–7]. Along with the development of effective vaccines, the researchers worldwide tried to identify the effective medicine to treat the infected persons. Drug repurposing or drug reprofiling was one of the ways to find a quick solution to treat the pandemic.

The coronavirus disease (COVID-19) is caused by the RNA-based severe acute respiratory syndrome 2 (SARS-CoV-2) coronavirus, which is genetically referred to as β-coronavirus. In December 2019, there were reports of pneumonia of unknown foundation in Wuhan, Hubei Province, China. The novel coronavirus was isolated from the injection of bronchoalveolar lavage fluid from infected patients into human respiratory epithelial cells and cell lines Vero E6 and Huh7 cell lines and marked as 2019-nCov that was later named as SARS-CoV-2 [8]. Coronaviruses are positive single-stranded RNA viruses that are enveloped and belong to the Coronaviridae family. Currently, in total seven human coronaviruses (HCov) have been identified - belonging to the four general notations i.e., α-, β-, γ-, and δ. HCoV-229E and HCoVNL63 were labeled as α-coronaviruses, while SARS-CoV, the novel SARS-CoV-2, the Middle East Respiratory Syndrome Coronavirus (MERSCoV), HCoV-HKU1 and HCoV-OC43 have been identified as β-coronaviruses [9]. The α-coronavirus and the β-coronavirus are usually responsible for frequent colds, but also for serious infections of the lower respiratory tract [10], while zoonotic coronaviruses, SARS-CoV and MERS-CoV were frequently fatal and causes severe respiratory syndrome. Since its inception, SARS-CoV-2 has spread worldwide and has become a pandemic.

The SARS-CoV-2 virus is about 125 nm in diameter and binds to the angiotensin 2 conversion enzyme (ACE2) receptor using a glycoprotein called spike protein that enable the entry of virus into the cell through the serine protease TGRBSS2 [11, 12] as shown in the Figure 3.1.

The literature evidences established that SARS-CoV-2 virus enters through ACE2/RBD binding but recently, a new path of invasion has been induced *via* CD147 that also needs to be further insight [14].

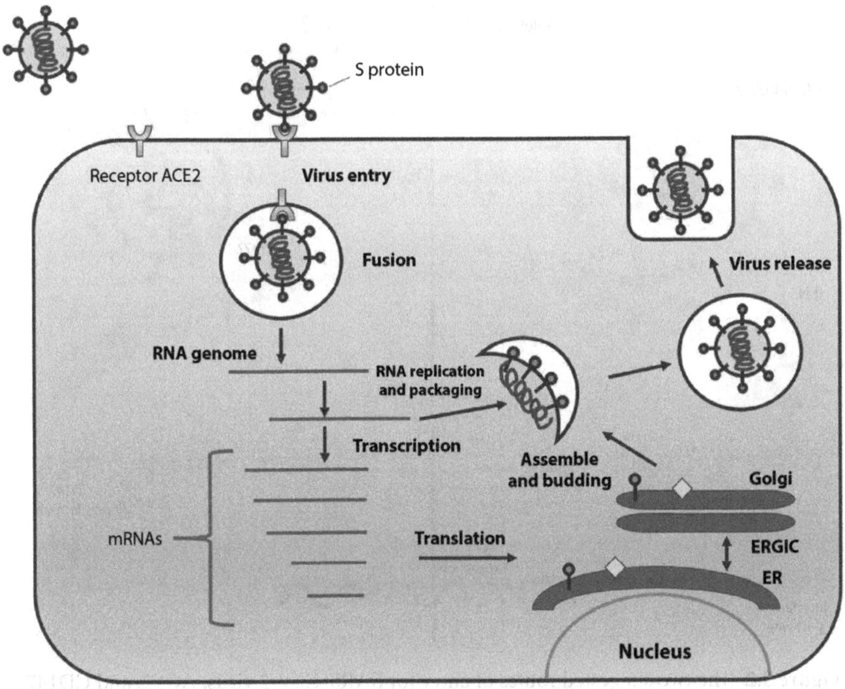

Figure 3.1 Model of SARS-CoV-2 life cycle (Adapted from [13] and licensed under CC-BY 4.0).

CD147 is a surface transmembrane glycoprotein that was suggested as a co-receptor and may play synergistic role for viral entry with ACE2 (Figure 3.2).

The CD147 can interact with different proteins *i.e.* cyclophilins or caveolin-1, monocarboxylate transporter proteins, and integrin and it was also found to be involved in the expression of matrix metallopeptidases [16]. The CD147 also regulate inflammatory responses and the pathogenesis of various inflammatory-mediated diseases by making interaction with cyclophilin [17]. A review of Behl *et al.* [18] was also published to explore the CD147 as a co-receptor of the viral entrance of SARS-CoV-2, whereas the study of Ragotte *et al.* [19] indicated that there is no direct interaction between spike protein and the CD147. A recent study of Kalejaiye *et al.* [20] identified that BSG/CD147 and ACE2 receptors plays a significant role for spike binding activity in human pod cells derived from iPS cells; although the role of CD147 for SARS-CoV-2 is still debatable.

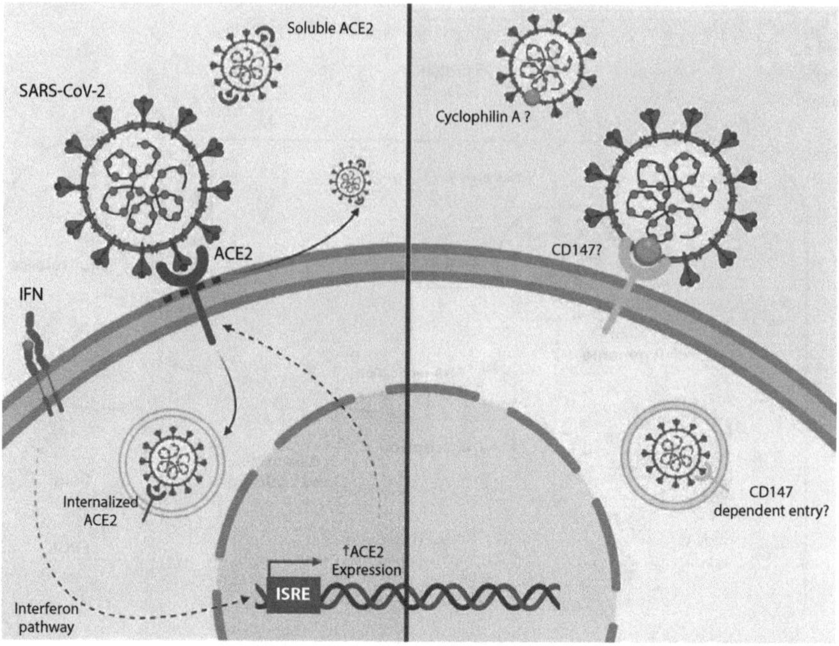

Figure 3.2 The two suggested routes of entry for SARS-CoV-2 virus, ACE2, and CD147 (Adapted from [15] and licensed under CC-BY 4.0).

3.2 Symptoms of COVID-19

The U.S. National health agency, the Centers for Disease Control and Prevention [21] has identified as a wide range of symptoms in people

Table 3.1 Symptoms of the people infected by COVID-19.

Infection type	Symptoms
Mild to moderate	Fever or chills, cough, muscle or body aches, nonpneumonia or mild pneumonia, breathing difficulty, fatigue, headache, loss of taste or smell, runny nose, sore throat, nausea, vomiting, etc.
Severe	Dyspnea, respiratory frequency ≥ 30/min, blood oxygen saturation ≤ 93%, arterial oxygen partial pressure at the fraction of inhaled oxygen ratio < 300, and/or lung infiltrates > 50% within 24/48 h
Critical	Breathing failure, septic shock, and/or multifunction or failure of several organs.

suffering from COVID-19 infected from mild range to severe illness. According to the CDC, the COVID-19 symptoms may develop 2–14 days after the exposure to the virus. The possible suggested symptoms by CDC and the review of He *et al.* [13] are listed in the Table 3.1.

In several cases, the virus SARS-CoV-2 caused to multiorgan failure response in the infected host [22, 23] which caused to the death. The possible effect of the COVID-19 infection on different organs of the human is described in the Figure 3.3.

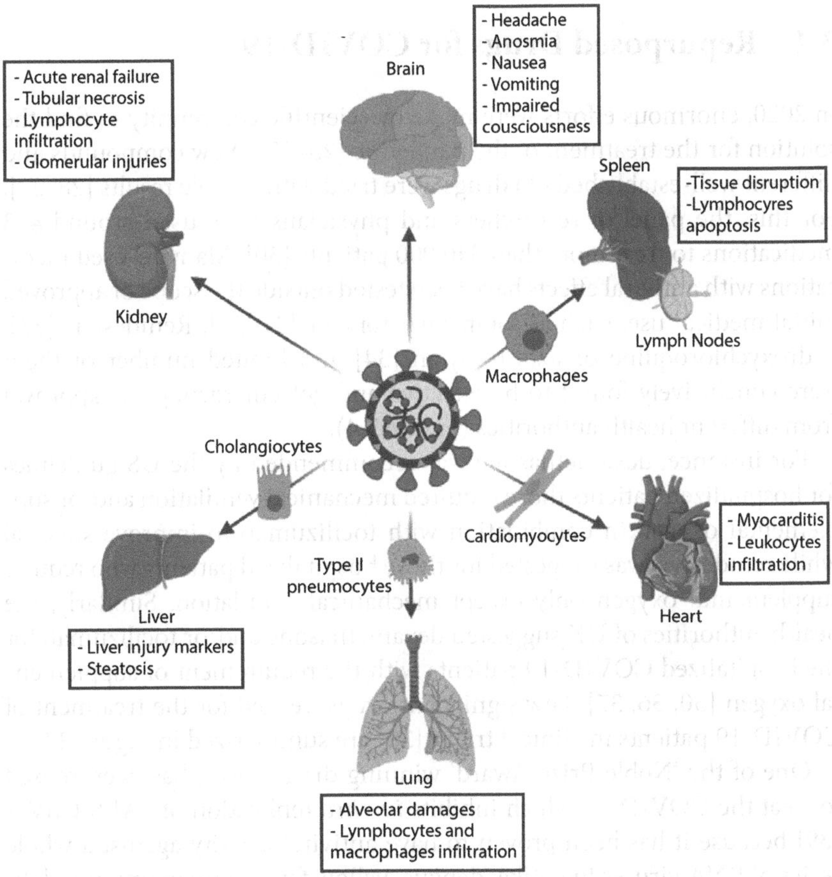

Figure 3.3 Multiple organ failure and tropism in SARS-CoV-2 infection (Adapted from [15] is licensed under CC-BY 4.0).

3.3 Repurposing of the Drugs

Drug repurposing or drug reprofiling may be identified as the novel use of the existing drugs. The drug repurposing reduces time, safety-related failure, cost, and whole development process because the existing drug is associated with the development of an established formulation and with pharmacokinetic and pharmacodynamic patterns. In addition, the first phase of the clinical study could be a bypass for many of the drugs which resulted in the reduction of several years of the development. Therefore, reprofiling or repurposing the existing drugs has the potential to minimize the whole drug development process by up to 3-12 years [24].

3.4 Repurposed Drugs for COVID-19

In 2020, enormous efforts were made by scientific community to find the solution for the treatment of the pandemic [25–27]. New compounds and different well-established old drugs were tried with flexible results [28, 29]. For this, the panel of researchers and physicians were used around 443 medications to treat more than 340,000 patients [30]. Many re-used medications with antiviral effects have been tested outside the scope of approved initial medical use, such as lopinavir/ritonavir [31, 32], Remdesivir [33], hydroxychloroquine or azithromycin [34] and limited number of them were conclusively found to be effective and got emergency use approval from different health authorities (Figure 3.4).

For instance, dexamethasone was recommended by the US guidelines for hospitalized patients those required mechanical ventilation and/or supplemental oxygen in combination with tocilizumab to improve survival while remdesivir was suggested for those hospitalized patients who require supplemental oxygen only except mechanical ventilation. Similarly, the health authorities of UK suggested dexamethasone and/or tocilizumab for the hospitalized COVID-19 patients with the requirement of supplemental oxygen [30, 36, 37]. Few significant drugs, reused for the treatment of COVID-19 patients in clinical trials, [38] are summarized in Figure 3.5.

One of the 'Noble Prize Award' winning drugs, IVM, has been reused to treat the COVID-19 which inhibits *in vitro* replication of SARS-CoV-2 [39] because it has been proven to have antiviral activity against a whole series of RNA viruses like Zika, dengue, yellow fever, human immunodeficiency virus type 1, etc. [40, 41] that was also found effective for COVID-19 *in vitro* [42].

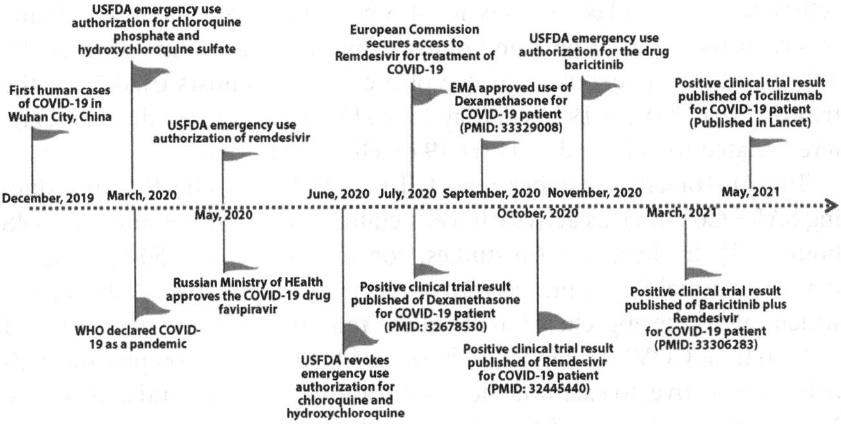

Figure 3.4 Timeline for emergency approval of different repurposed drugs for COVID-19 by different regulatory authorities (Adapted from [35] is licensed under CC-BY 4.0).

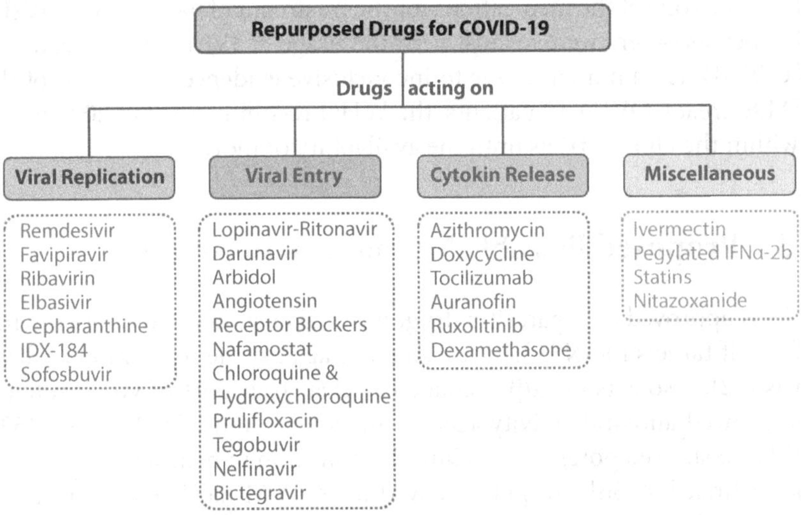

Figure 3.5 Key medications reused for in-patient COVID-19 treatment in clinical trials.

3.5 Repurposing of Ivermectin for COVID-19

The IVM is a broad-spectrum antiparasitic agent which also shows remarkable vermicidal and ectoparasiticidal properties [43–45]. It has high liposolubility, rapid oral absorption, is metabolized by the cytochrome P450 system, wide distribution profile in the body, and is excreted majorly in faces [46]. Using a standard oral dose for healthy humans, the IVM achieves the peak plasma levels at 3.4-5 h, and the half-life of the plasma was reported to be 12-66 h and binds strongly to the plasma protein [47, 48]. Based on the excellent profile of the IVM, scientists tried it for the treatment of COVID-19. As of May 13, 2022, a total of 87 studies are registered related to IVM and COVID-19 on clinicaltrials.gov.

The Australian researcher reported that IVM was effective in reducing SARS CoV-2 virus activity in cells culture by significant amount in 48 hours [39]. In these *in vitro* studies, concentrations were 50-fold higher than the normal C_{max} achieved with only one dose of ivermectin 200 μg/kg, which raised the apprehension about the required dose and tolerability of IVM to treat COVID infection in humans [49, 50]. Based on pharmacokinetic perspective, to examine the *in vitro* antiviral activity and end-points, Momekov *et al.* proposed that available pharmacokinetic data from clinically relevant and excessive dose-related studies, the inhibitory concentrations of SARS-CoV-2 are unlikely to be achievable in humans [51].

Therefore, it was observed that some of the researchers were advised to use IVM as one of the medications for the treatment of patients of COVID-19, whereas others were against with the usage of IVM for the treatment of COVID-19 in humans. Due to inconclusive evidence on the use of the IVM to treat COVID-19 patients, the WHO recommended in 2021 to use it within the clinical trials until the availability of more data.

3.6 Proposed Possible Mechanism of Action

IVM is approved as a parasitic drug, but it also shows potential antiviral effects. It targets the NS3 helicase and acts against the replication of flavivirus [52], also acts on α/β-mediated nuclear transport of viral proteins, and showed antiviral activity against the dengue and HIV-1 viruses [53]. IVM also showed potential for clinical application in neurological diseases and asthma [54–56]. In quite a few studies, IVM has been proposed as having different mechanisms of action against SARS-CoV-2, among them, some articles have been retracted due to lack of scientific evidences [57]. Researchers proposed that the activity of IVM may be due to the dependency of most RNA-based viruses on IMPα/β1 during infection. IVM is proposed to act both intracellularly and extracellularly. IVM provokes the ionophores along the virus lipoprotein nucleocapsid outside the host cell, which allows the appearance of oxhidril compounds which mutilate the virus structure. Further, if the virus passes to the cytosol, IVM inhibits its conveyance to the cell nucleus using importins α and β1 to do so [42]. It was reported from *in vitro* studies that IVM reduced the viral load 5000

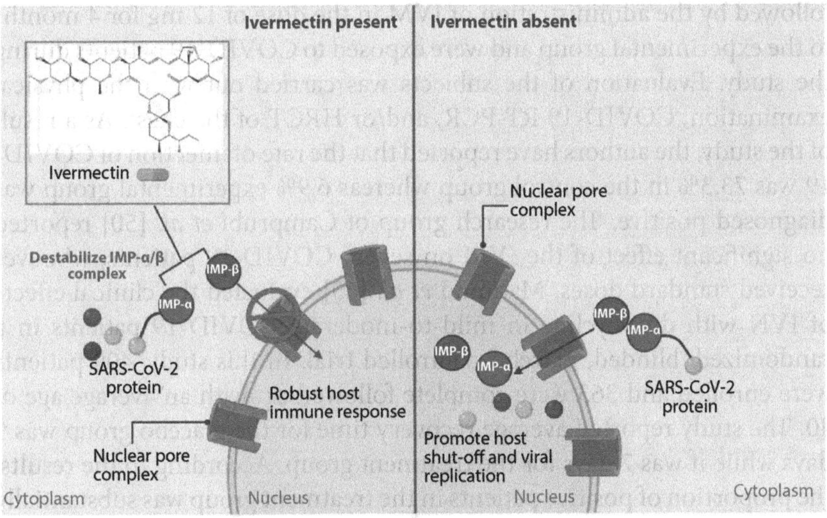

Figure 3.6 Probable proposed mechanism of action of IVM against SARS-CoV-2 (Adapted from [61] is licensed under CC-BY).

times in laboratory specimens [42, 58, 59]. The SARS-CoV proteins have been examined for the potential role of the signal-dependent nucleocytoplasmic shuttling of the SARS-CoV nucleocapsid protein, which might affect the host cell division [41, 60, 61]. A schematic figure based on the reported probable proposed mechanism of action of IVM against SARS-CoV-2 is shown in Figure 3.6.

3.7 SARS COVID-19 Clinical Studies with Ivermectin

IVM is widely used as an anti-parasitic clinical drug and its safety as well as efficacy has been nicely proved. As IVM has already shown diversified activity on different viruses such as West Nile Encephalitis, dengue fever, etc., and good results in *in vitro* against SARS-CoV-2. Hence, scientists tried IVM as a repurposed drug in clinical trials to combat the pandemic COVID-19, while some studies also addressed potential bias and medical fraud due to not following ethics and/or lack of scientific shreds of evidences [62, 63]. Furthermore, this chapter summarizes some clinical studies regarding IVM's therapeutic potential in treating COVID-19. An observational study was fruitfully conducted by Alam *et al.* [64] in a tertiary hospital in Dhaka from May-to-August 2020 with 118 healthcare providers. The authors divided all subjects into control and experimental groups followed by the administration of IVM in the dose of 12 mg for 4 months to the experimental group and were exposed to COVID-19 patients during the study. Evaluation of the subjects was carried out with the physical examination, COVID-19 RT-PCR, and/or HRCT of the chest. As a result of the study, the authors have reported that the rate of infection of COVID-19 was 73.3% in the control group whereas 6.9% experimental group was diagnosed positive. The research group of Camprubí *et al.* [50] reported no significant effect of the IVM on severe COVID-19 patients, whoever received standard doses. Mahmud *et al.* [65] evaluated the clinical effects of IVN with doxycycline in mild-to-moderate COVID-19 patients in a randomized, blinded, placebo-controlled trial. In this study, 400 patients were enrolled and 363 were complete followed-up with an average age of 40. The study reported average recovery time for the placebo group was 9 days while it was 7 days for the treatment group. According to the results, the proportion of positive patients in the treatment group was substantially lower than that of the placebo group on day 14. The study further recommended the combination of IVM and doxycycline for mild-to-moderate COVID-19 infected patients. Another study by the Aref *et al.*, [66] 114 patients diagnosed with mild COVID-19 in a clinical trial. Two groups of patients were treated for mild COVID-19, group A, which include 57 patients received the Egyptian protocol of treatment for COVID-19 only, while group B, which included 57 patients, received IVM nanosuspension nasal spray twice daily along with the Egyptian protocol. The presented results indicated that 54 (94.7%) patients of group B got negative in comparison to 43 (75.4%) of group A. The authors also highlighted that the group B experienced fever, cough, dyspnea, and anosmia for noticeably less time than group A. On the other hand, a separate clinical trial on 72 hospitalized patients in Dhaka, Bangladesh was also reported.

To ascertain the viral clearance and safety of IVM among the adults, SARS-CoV-2 patients, Ahmed *et al.* [67] undertook a randomized, double-blind, placebo-controlled experiments. The authors divided the patients into three groups in which one group received oral IVM of 12 mg once in a day for 5 days, whereas the second one received IVM with a combination of doxycycline (12 mg IVM once daily and 200 mg doxycycline on day 1, along with 100 mg every 12 h for the next 4 days) orally with the placebo-controlled third group. The reported consequences indicated that viral clearance for the IVM group was 5 days, IVM + doxycycline for 11.5 days, and 12.7 days for the placebo group. The authors did not found any negative results of the drug with mild COVID-19 and recommended the larger trials. The study of Krolewiecki *et al.* [68] evaluated the antiviral effect of IVM on viral load of respiratory secretions and its connection with drug concentrations in plasma as a precise, pilot, proof-of-concept, randomized, outcome-assessor blinded trial of high-dose of IVM in 45 COVID-19 patients. In four hospitals in Argentina, the patients were into groups of two to receive standard care with IVM at 0.6 mg/kg/day for 5 days and standard care in 4 hospitals in Argentina. The study concluded with no significant difference in viral load in both categories notable difference was reported in plasma level, 59–77 IQR for IVM category and 31–73 IQR for standard care category. The purpose of the study [69] was to assess the effectiveness of IVM in treating 164 COVID-19 patients with mild-to-moderate symptoms. Patients were randomly split into the treatment group and the control group. The control group received only standard care while IVM was given standard care with a dose of 12 mg once a day for three days. The results indicated that the patients of the IVM group stayed in the hospital for 8.82 ± 4.94 days as compared to the control group for 10.97 ± 5.28 days. The mechanical ventilation requirement for each group was 3.7% while the mortality rate for the IVM group was 3.7% versus 4.9% for the control group. The authors didn't conclude any significant difference except reduction in hospital stay rate. In a separate report, Shahbaznejad *et al.* [70] conducted a randomized, double-blind clinical trial of 69 patients with COVID-19 into 2 groups: intervention and control. The authors used 0.2 mg/kg of IVM with standard care while the control group was treated with only standard care. The study declared that 51.4% of patients in the intervention group and 52.9% of patients in the control group were males. The results of the study are clearly summarized in the Table 3.2.

The results indicated that the single dose of IVM improves significant clinical features, and the authors recommend further trials with larger samples size and different dosages to understand potential clinical benefits.

To analyze the efficacy of the IVM for mild COVID-19, López-Medina *et al.* [71] also conducted a study in Colombia. The authors conducted a double-blind, randomized experiment on 400 COVID-19 laboratory-confirmed adult patients with mild symptoms between July 15, and

Table 3.2 The clinical results of the trial study of Shahbaznejad *et al.* [70].

Group	Dyspnea (days)	Cough (days)	Mean hospital stay (days)	Lymphopenia decreased
Intervention	2.6	3.1	7.1	14.3%
Control	3.8	4.8	8.4	NA

November 30, 2020, that were followed up until December 21, 2020. IVM 300 μg/kg of body weight was given to the intervention patients every day for 5 days. The authors reported that the average time to resolve the symptoms of intervention patients was 10 days versus 12 days for the placebo group. It was observed on the day 21, that the symptoms of 82% of the intervention group's participants were resolved as compared to 79% of the placebo group. The significant adverse events were the same in both groups, thus the authors concluded that their research does not support using IVM to treat mild COVID-19. Another large-scale study highlighted the significance of IVM in COVID-19 treatment [72]. 159, 561 patients in total were enrolled in the study, of which 113, 845 (71.3%) were regular IVM users and 45, 716 (23.3%) were not. The study indicated that a total of 4,311 IVM users got infected while 3,034 non-IVM users were infected. The rate of infection was recorded at 3.7% and 6.6% for IVM and non-IVM users i.e., 44% reduction in the infection rate. The authors reported the mortality rate for IVM users and non-IVM users was 0.8% and 2.6%, respectively. Furthermore, the hospitalization rate was 67% lower for IVM users as compared with non-IVM users which indicated that using IVM as a preventive measure was linked to considerably lower rates of COVID-19 infection, hospitalization, and mortality.

The effectiveness of IVM in reducing the advancement of severe disease among high-risk COVID-19 patients was investigated in a clinical experiment by Lim *et al.* [73]. The authors reported that the study was an open-label randomized clinical trial, and it was carried out in Malaysia between May 31 and October 25, 2021, at 20 public hospitals and a COVID-19 quarantine facility. The patients were separated in a 1:1 ratio and got oral IVM (0.4 mg/kg body weight daily for 5 days with standard care, according to the authors. The authors carefully investigated the percentage of patients who developed serious illness, which was determined by the hypoxic stage of the level of supplemental oxygen in the patients i.e., ≥95%, and requirement of mechanical ventilation, admission to an intensive care unit, hospital mortality, and adverse events. The results of different parameters of the study are reported in the Table 3.3.

On analyzing the results, the authors came to the conclusion that IVM treatment during early stages of illness did not stop the development of severe disease, and they did not advise IVM to treat COVID-19. Later, Reis *et al.* [74] reported a symptomatic SARS-CoV-2–positive adult randomized, placebo-controlled, double-blind experiment that participants were recruited from 12 public health clinics in Brazil. The authors randomly assigned 3515 patients who had COVID-19 symptoms for up to 7 days to administer IVM (679 patients) of 400 μg/kg dose/day for 3 days

Table 3.3 Different reported results of the study of Lim *et al.* [73].

Group	Patients progressed to severe disease (%)	Mechanical ventilation occurred (%)	Admission to an intensive care unit (%)	28-day in-hospital mortality (%)	Adverse event: diarrhea (%)
Intervention	21.6	1.7	2.4	1.2	5.8
Control	17.3	4.0	3.2	4.0	1.6

with placebo-controlled (679 patients) and the rest were treated based on other interventions (2157 patients). Based upon the results, these workers concluded that IVM was not found effective to minimize the headway of COVID-19 and was not found useful to reduce the incidence of medical admission to a hospital.

3.8 Conclusions

In conclusion, many drugs were repurposed to find the effective treatment of COVID-19, and Ivermectin was one of them. IVM is widely used as an anti-parasitic clinical drug; its safety and efficacy have been proved. As IVM is already shown diversified activity on different viruses such as West Nile Encephalitis, dengue fever, etc., and good results in *in vitro* against SARS-CoV-2. Therefore, scientists tried IVM as a repurposed drug in clinical trials to combat the pandemic COVID-19. After showing initial hopeful *in vitro* results by IVM, different clinical studies were conducted that showed inconclusive results. The number of published articles based on IVM and COVID-19 has been retracted while some studies also have been addressed as potential bias and medical fraud due to not following ethics and/or lack of scientific pieces of evidence. Scientific papers were also published to define the IVM's inhibitory effect against SARS-CoV-2 required a mechanism of action, but those investigations likewise had a dearth of scientific evidence. Therefore, in 2021, WHO advised using IVM only in clinical trials. This chapter summarized the published and *in vitro* findings that may be related to SARS-CoV-2 replication and the illness COVID-19. The study's findings showed that none of the outcomes for which IVM was used were significant, such as progression to severe disease, mortality, prevention of ICU, etc. and the results do not support the use of IVM for COVID-19 patients. Although, there were some observed trends in reducing hospital stay in IVM intervention groups, a bibber clinical investigation is still required to corroborate these findings.

3.9 Acknowledgments

The author HM wishes to acknowledge the Indian Council of Medical Research (Government of India), New Delhi, India for providing financial support during the study as a Senior Research Fellowship (Award No. 45/02/2020-Nan/BMS).

References

1. WHO, Weekly epidemiological update on COVID-19 - 18 May 2022, 2022, https://www.who.int/publications/m/item/weekly-epidemiological-update-on-covid-19—18-may-2022 (accessed May 25, 2022).
2. Perra, N., Non-pharmaceutical interventions during the COVID-19 pandemic: A review. *Phys. Rep.*, 913, 1–52, 2021, https://doi.org/10.1016/j.physrep.2021.02.001.
3. Ge, Y., Zhang, W.-B., Liu, H., Ruktanonchai, C.W., Hu, M., Wu, X., Song, Y., Ruktanonchai, N.W., Yan, W., Cleary, E., Feng, L., Li, Z., Yang, W., Liu, M., Tatem, A.J., Wang, J.-F., Lai, S., Impacts of worldwide individual non-pharmaceutical interventions on COVID-19 transmission across waves and space. *Int. J. Appl. Earth Obs. Geoinf.*, 10, 102649, 2022, https://doi.org/10.1016/j.jag.2021.102649.
4. Kupferschmidt, K., Viral mutations may cause another 'very, very bad' COVID-19 wave, scientists warn. 05 January 2021, 2021, https://www.science.org/content/article/viral-mutations-may-cause-another-very-very-bad-covid-19-wave-scientists-warn (accessed May 25, 2022).
5. Zhou, Q., International collaboration for global accessibility of COVID-19 vaccines. *Natl. Sci. Rev.*, 7, 1269–1269, 2020, https://doi.org/10.1093/nsr/nwaa147.
6. Pang, X., Ren, L., Wu, S., Ma, W., Yang, J., Di, L., Li, J., Xiao, Y., Kang, L., Du, S., Du, J., Wang, J., Li, G., Zhai, S., Chen, L., Zhou, W., Lai, S., Gao, L., Pan, Y., Wang, Q., Li, M., Wang, J., Huang, Y., Wang, J., Cold-chain food contamination as the possible origin of COVID-19 resurgence in Beijing. *Natl. Sci. Rev.*, 7, 1861–1864, 2020, https://doi.org/10.1093/nsr/nwaa264.
7. Yang, J., Marziano, V., Deng, X., Guzzetta, G., Zhang, J., Trentini, F., Cai, J., Poletti, P., Zheng, W., Wang, W., Wu, Q., Zhao, Z., Dong, K., Zhong, G., Viboud, C., Merler, S., Ajelli, M., Yu, H., Despite vaccination, China needs non-pharmaceutical interventions to prevent widespread outbreaks

of COVID-19 in 2021. *Nat. Hum. Behav.*, 5, 1009–1020, 2021, https://doi.org/10.1038/s41562-021-01155-z.

8. Zhu, N., Zhang, D., Wang, W., Li, X., Yang, B., Song, J., Zhao, X., Huang, B., Shi, W., Lu, R., Niu, P., Zhan, F., Ma, X., Wang, D., Xu, W., Wu, G., Gao, G.F., Tan, W., A novel coronavirus from patients with pneumonia in China, 2019. *N. Engl. J. Med.*, 382, 727–733, 2020, https://doi.org/10.1056/NEJMoa2001017.
9. Ciotti, M., Angeletti, S., Minieri, M., Giovannetti, M., Benvenuto, D., Pascarella, S., Sagnelli, C., Bianchi, M., Bernardini, S., Ciccozzi, M., COVID-19 outbreak: An overview. *Chemotherapy*, 64, 215–223, 2019, https://doi.org/10.1159/000507423.
10. Wevers, B.A. and van der Hoek, L., Recently discovered human coronaviruses. *Clin. Lab. Med.*, 29, 715–724, 2009, https://doi.org/10.1016/j.cll.2009.07.007.
11. Stasi, C., Fallani, S., Voller, F., Silvestri, C., Treatment for COVID-19: An overview. *Eur. J. Pharmacol.*, 889, 173644, 2020, https://doi.org/10.1016/j.ejphar.2020.173644.
12. Matricardi, P.M., Dal Negro, R.W., Nisini, R., The first, holistic immunological model of COVID-19: Implications for prevention, diagnosis, and public health measures. *Pediatr. Allergy Immunol.*, 31, 454–470, 2020, https://doi.org/10.1111/pai.13271.
13. He, F., Deng, Y., Li, W., Coronavirus disease 2019: What we know? *J. Med. Virol.*, 92, 719–725, 2020, https://doi.org/10.1002/jmv.25766.
14. Wang, K., Chen, W., Zhang, Z., Deng, Y., Lian, J.-Q., Du, P., Wei, D., Zhang, Y., Sun, X.-X., Gong, L., Yang, X., He, L., Zhang, L., Yang, Z., Geng, J.-J., Chen, R., Zhang, H., Wang, B., Zhu, Y.-M., Nan, G., Jiang, J.-L., Li, L., Wu, J., Lin, P., Huang, W., Xie, L., Zheng, Z.-H., Zhang, K., Miao, J.-L., Cui, H.-Y., Huang, M., Zhang, J., Fu, L., Yang, X.-M., Zhao, Z., Sun, S., Gu, H., Wang, Z., Wang, C.-F., Lu, Y., Liu, Y.-Y., Wang, Q.-Y., Bian, H., Zhu, P., Chen, Z.-N., CD147-spike protein is a novel route for SARS-CoV-2 infection to host cells. *Signal Transduction Targeting Ther.*, 5, 283, 2020, https://doi.org/10.1038/s41392-020-00426-x.
15. Lebeau, G., Vagner, D., Frumence, É., Ah-Pine, F., Guillot, X., Nobécourt, E., Raffray, L., Gasque, P., Deciphering SARS-CoV-2 virologic and immunologic features. *Int. J. Mol. Sci.*, 21, 5932, 2020, https://doi.org/10.3390/ijms21165932.
16. Xiong, L., Edwards, C., Zhou, L., The biological function and clinical utilization of CD147 in human diseases: A review of the current scientific literature. *Int. J. Mol. Sci.*, 15, 17411–17441, 2014, https://doi.org/10.3390/ijms151017411.
17. Yurchenko, V., Constant, S., Eisenmesser, E., Bukrinsky, M., Cyclophilin–CD147 interactions: A new target for anti-inflammatory therapeutics. *Clin. Exp. Immunol.*, 160, 305–317, 2010, https://doi.org/10.1111/j.1365-2249.2010.04115.x.

18. Behl, T., Kaur, I., Aleya, L., Sehgal, A., Singh, S., Sharma, N., Bhatia, S., Al-Harrasi, A., Bungau, S., CD147-spike protein interaction in COVID-19: Get the ball rolling with a novel receptor and therapeutic target. *Sci. Total Environ.*, 808, 152072, 2022, https://doi.org/10.1016/j.scitotenv.2021.152072.
19. Ragotte, R.J., Pulido, D., Donnellan, F.R., Hill, M.L., Gorini, G., Davies, H., Brun, J., McHugh, K., King, L.D.W., Skinner, K., Miura, K., Long, C.A., Zitzmann, N., Draper, S.J., Human basigin (CD147) does not directly interact with SARS-CoV-2 spike glycoprotein. *MSphere*, 6, e00647–21, 2021, https://doi.org/10.1128/mSphere.00647-21.
20. Kalejaiye, T.D., Bhattacharya, R., Burt, M.A., Travieso, T., Okafor, A.E., Mou, X., Blasi, M., Musah, S., SARS-CoV-2 employ BSG/CD147 and ACE2 receptors to directly infect human induced pluripotent stem cell-derived kidney podocytes. *Front. Cell Dev. Biol.*, 10, 855340, 2022, https://doi.org/10.3389/fcell.2022.855340.
21. CDC, Symptoms of COVID-19 | CDC, 2022, https://www.cdc.gov/coronavirus/2019-ncov/symptoms-testing/symptoms.html (accessed July 12, 2022).
22. Bhalerao, A., Raut, S., Noorani, B., Mancuso, S., Cucullo, L., Molecular mechanisms of multi-organ failure in COVID-19 and potential of stem cell therapy. *Cells*, 10, 2878, 2021, https://doi.org/10.3390/cells10112878.
23. Mokhtari, T., Hassani, F., Ghaffari, N., Ebrahimi, B., Yarahmadi, A., Hassanzadeh, G., COVID-19 and multiorgan failure: A narrative review on potential mechanisms. *J. Mol. Histol.*, 51, 613–628, 2020, https://doi.org/10.1007/s10735-020-09915-3.
24. Novac, N., Challenges and opportunities of drug repositioning. *Trends Pharmacol. Sci.*, 34, 267–272, 2013, https://doi.org/10.1016/j.tips.2013.03.004.
25. Siminea, N., Popescu, V., Sanchez Martin, J.A., Florea, D., Gavril, G., Gheorghe, A.-M., Iţcuş, C., Kanhaiya, K., Pacioglu, O., Popa, L.I., Trandafir, R., Tusa, M.I., Sidoroff, M., Păun, M., Czeizler, E., Păun, A., Petre, I., Network analytics for drug repurposing in COVID-19. *Brief. Bioinform.*, 23, bbab490, 2022, https://doi.org/10.1093/bib/bbab490.
26. Bakowski, M.A., Beutler, N., Wolff, K.C., Kirkpatrick, M.G., Chen, E., Nguyen, T.-T.H., Riva, L., Shaabani, N., Parren, M., Ricketts, J., Gupta, A.K., Pan, K., Kuo, P., Fuller, M., Garcia, E., Teijaro, J.R., Yang, L., Sahoo, D., Chi, V., Huang, E., Vargas, N., Roberts, A.J., Das, S., Ghosh, P., Woods, A.K., Joseph, S.B., Hull, M.V., Schultz, P.G., Burton, D.R., Chatterjee, A.K., McNamara, C.W., Rogers, T.F., Drug repurposing screens identify chemical entities for the development of COVID-19 interventions. *Nat. Commun.*, 12, 3309, 2021, https://doi.org/10.1038/s41467-021-23328-0.
27. Jang, W.D., Jeon, S., Kim, S., Lee, S.Y., Drugs repurposed for COVID-19 by virtual screening of 6,218 drugs and cell-based assay. *Proc. Natl. Acad. Sci.*, 118, e2024302118, 2021, https://doi.org/10.1073/pnas.2024302118.
28. Guy, R.K., DiPaola, R.S., Romanelli, F., Dutch, R.E., Rapid repurposing of drugs for COVID-19. *Sci. (80-.)*, 368, 829–830, 2020, https://doi.org/10.1126/science.abb9332.

29. Srivastava, K. and Singh, M.K., Drug repurposing in COVID-19: A review with past, present and future. *Metab. Open*, 12, 100121, 2021, https://doi.org/10.1016/j.metop.2021.100121.
30. Venkatesan, P., Repurposing drugs for treatment of COVID-19. *Lancet Respir. Med.*, 9, e63, 2021, https://doi.org/10.1016/S2213-2600(21)00270-8.
31. Horby, P.W., Mafham, M., Bell, J.L., Linsell, L., Staplin, N., Emberson, J., Palfreeman, A., Raw, J., Elmahi, E., Prudon, B., Green, C., Carley, S., Chadwick, D., Davies, M., Wise, M.P., Baillie, J.K., Chappell, L.C., Faust, S.N., Jaki, T., Jefferey, K., Lim, W.S., Montgomery, A., Rowan, K., Juszczak, E., Haynes, R., Landray, M.J., Lopinavir–ritonavir in patients admitted to hospital with COVID-19 (RECOVERY): A randomised, controlled, open-label, platform trial. *Lancet*, 396, 1345–1352, 2020, https://doi.org/10.1016/S0140-6736(20)32013-4.
32. Cao, B., Wang, Y., Wen, D., Liu, W., Wang, J., Fan, G., Ruan, L., Song, B., Cai, Y., Wei, M., Li, X., Xia, J., Chen, N., Xiang, J., Yu, T., Bai, T., Xie, X., Zhang, L., Li, C., Yuan, Y., Chen, H., Li, H., Huang, H., Tu, S., Gong, F., Liu, Y., Wei, Y., Dong, C., Zhou, F., Gu, X., Xu, J., Liu, Z., Zhang, Y., Li, H., Shang, L., Wang, K., Li, K., Zhou, X., Dong, X., Qu, Z., Lu, S., Hu, X., Ruan, S., Luo, S., Wu, J., Peng, L., Cheng, F., Pan, L., Zou, J., Jia, C., Wang, J., Liu, X., Wang, S., Wu, X., Ge, Q., He, J., Zhan, H., Qiu, F., Guo, L., Huang, C., Jaki, T., Hayden, F.G., Horby, P.W., Zhang, D., Wang, C., A trial of lopinavir–ritonavir in adults hospitalized with severe COVID-19. *N. Engl. J. Med.*, 382, 1787–1799, 2020, https://doi.org/10.1056/NEJMoa2001282.
33. Beigel, J.H., Tomashek, K.M., Dodd, L.E., Mehta, A.K., Zingman, B.S., Kalil, A.C., Hohmann, E., Chu, H.Y., Luetkemeyer, A., Kline, S., Lopez de Castilla, D., Finberg, R.W., Dierberg, K., Tapson, V., Hsieh, L., Patterson, T.F., Paredes, R., Sweeney, D.A., Short, W.R., Touloumi, G., Lye, D.C., Ohmagari, N., Oh, M., Ruiz-Palacios, G.M., Benfield, T., Fätkenheuer, G., Kortepeter, M.G., Atmar, R.L., Creech, C.B., Lundgren, J., Babiker, A.G., Pett, S., Neaton, J.D., Burgess, T.H., Bonnett, T., Green, M., Makowski, M., Osinusi, A., Nayak, S., Lane, H.C., Remdesivir for the treatment of COVID-19 — Final report. *N. Engl. J. Med.*, 383, 1813–1826, 2020, https://doi.org/10.1056/NEJMoa2007764.
34. Gautret, P., Lagier, J.-C., Parola, P., Hoang, V.T., Meddeb, L., Mailhe, M., Doudier, B., Courjon, J., Giordanengo, V., Vieira, V.E., Tissot Dupont, H., Honoré, S., Colson, P., Chabrière, E., La Scola, B., Rolain, J.-M., Brouqui, P., Raoult, D., Hydroxychloroquine and azithromycin as a treatment of COVID-19: Results of an open-label non-randomized clinical trial. *Int. J. Antimicrob. Agents*, 56, 105949, 2020, https://doi.org/10.1016/j.ijantimicag.2020.105949.
35. Chakraborty, C., Sharma, A.R., Bhattacharya, M., Agoramoorthy, G., Lee, S.-S., The drug repurposing for COVID-19 clinical trials provide very effective therapeutic combinations: Lessons learned from major clinical studies. *Front. Pharmacol.*, 12, 704205, 2021, https://doi.org/10.3389/fphar.2021.704205.

36. van de Veerdonk, F.L., Giamarellos-Bourboulis, E., Pickkers, P., Derde, L., Leavis, H., van Crevel, R., Engel, J.J., Wiersinga, W.J., Vlaar, A.P.J., Shankar-Hari, M., van der Poll, T., Bonten, M., Angus, D.C., van der Meer, J.W.M., Netea, M.G., A guide to immunotherapy for COVID-19. *Nat. Med.*, 28, 39–50, 2022, https://doi.org/10.1038/s41591-021-01643-9.
37. Dexamethasone in hospitalized patients with COVID-19. *N. Engl. J. Med.*, 384, 693–704, 2021, https://doi.org/10.1056/NEJMoa2021436.
38. Singh, T.U., Parida, S., Lingaraju, M.C., Kesavan, M., Kumar, D., Singh, R.K., Drug repurposing approach to fight COVID-19. *Pharmacol. Rep.*, 72, 1479–1508, 2020, https://doi.org/10.1007/s43440-020-00155-6.
39. Caly, L., Druce, J.D., Catton, M.G., Jans, D.A., Wagstaff, K.M., The FDA-approved drug ivermectin inhibits the replication of SARS-CoV-2 *in vitro*. *Antivir. Res.*, 178, 104787, 2020, https://doi.org/10.1016/j.antiviral.2020.104787.
40. Formiga, F.R., Leblanc, R., de Souza Rebouças, J., Farias, L.P., de Oliveira, R.N., Pena, L., Ivermectin: An award-winning drug with expected antiviral activity against COVID-19. *J. Control. Release*, 329, 758–761, 2021, https://doi.org/10.1016/j.jconrel.2020.10.009.
41. Yang, S.N.Y., Atkinson, S.C., Wang, C., Lee, A., Bogoyevitch, M.A., Borg, N.A., Jans, D.A., The broad spectrum antiviral ivermectin targets the host nuclear transport importin α/β1 heterodimer. *Antivir. Res.*, 177, 104760, 2020, https://doi.org/10.1016/j.antiviral.2020.104760.
42. Héctor, C., Roberto, H., Psaltis, A., Veronica, C., Study of the efficacy and safety of topical ivermectin + iota-carrageenan in the prophylaxis against COVID-19 in health personnel. *J. Biomed. Res. Clin. Investig.*, 2, 1007, 2020, https://doi.org/10.31546/2633-8653.1007.
43. Kar, S.K., Mania, J., Patnaik, S., The use of ivermectin for scabies. *Natl. Med. J. India*, 7, 15–6, 1994, http://www.ncbi.nlm.nih.gov/pubmed/8156025.
44. Chabala, J.C., Mrozik, H., Tolman, R.L., Eskola, P., Lusi, A., Peterson, L.H., Woods, M.F., Fisher, M.H., Campbell, W.C., Ivermectin, a new broad-spectrum antiparasitic agent. *J. Med. Chem.*, 23, 1134–1136, 1980, https://doi.org/10.1021/jm00184a014.
45. Wang, C., Huang, X.X., Zhang, Y.Q., Yen, Q.Y., Wen, Y., Efficacy of ivermectin in hookworms as examined in ancylostoma caninum infections. *J. Parasitol.*, 75, 373, 1989, https://doi.org/10.2307/3282591.
46. González Canga, A., Sahagún Prieto, A.M., Diez Liébana, M.J., Fernández Martínez, N., Sierra Vega, M., García Vieitez, J.J., The pharmacokinetics and interactions of ivermectin in humans—A mini-review. *AAPS J.*, 10, 42–46, 2008, https://doi.org/10.1208/s12248-007-9000-9.
47. Edwards, G., Dingsdale, A., Helsby, N., Orme, M.L.E., Breckenridge, A.M., The relative systemic availability of ivermectin after administration as capsule, tablet, and oral solution. *Eur. J. Clin. Pharmacol.*, 35, 681–684, 1988, https://doi.org/10.1007/BF00637608.

48. Klotz, U., Ogbuokiri, J.E., Okonkwo, P.O., Ivermectin binds avidly to plasma proteins. *Eur. J. Clin. Pharmacol.*, 39, 607–608, 1990, https://doi.org/10.1007/BF00316107.
49. Chaccour, C., Hammann, F., Ramón-García, S., Rabinovich, N.R., Ivermectin and COVID-19: Keeping rigor in times of urgency. *Am. J. Trop. Med. Hyg.*, 102, 1156–1157, 2020, https://doi.org/10.4269/ajtmh.20-0271.
50. Camprubí, D., Almuedo-Riera, A., Martí-Soler, H., Soriano, A., Hurtado, J.C., Subirà, C., Grau-Pujol, B., Krolewiecki, A., Muñoz, J., Lack of efficacy of standard doses of ivermectin in severe COVID-19 patients. *PLoS One*, 15, e0242184, 2020, https://doi.org/10.1371/journal.pone.0242184.
51. Momekov, G. and Momekova, D., Ivermectin as a potential COVID-19 treatment from the pharmacokinetic point of view: Antiviral levels are not likely attainable with known dosing regimens. *Biotechnol. Biotechnol. Equip.*, 34, 469–474, 2020, https://doi.org/10.1080/13102818.2020.1775118.
52. Mastrangelo, E., Pezzullo, M., De Burghgraeve, T., Kaptein, S., Pastorino, B., Dallmeier, K., de Lamballerie, X., Neyts, J., Hanson, A.M., Frick, D.N., Bolognesi, M., Milani, M., Ivermectin is a potent inhibitor of flavivirus replication specifically targeting NS3 helicase activity: New prospects for an old drug. *J. Antimicrob. Chemother.*, 67, 1884–1894, 2012, https://doi.org/10.1093/jac/dks147.
53. Wagstaff, K.M., Sivakumaran, H., Heaton, S.M., Harrich, D., Jans, D.A., Ivermectin is a specific inhibitor of importin α/β-mediated nuclear import able to inhibit replication of HIV-1 and dengue virus. *Biochem. J.*, 443, 851–856, 2012, https://doi.org/10.1042/BJ20120150.
54. Yan, S., Ci, X., Chen, N., Chen, C., Li, X., Chu, X., Li, J., Deng, X., Anti-inflammatory effects of ivermectin in mouse model of allergic asthma. *Inflamm. Res.*, 60, 589–596, 2011, https://doi.org/10.1007/s00011-011-0307-8.
55. Franklin, K.M., Asatryan, L., Jakowec, M.W., Trudell, J.R., Bell, R.L., Davies, D.L., P2X4 receptors (P2X4Rs) represent a novel target for the development of drugs to prevent and/or treat alcohol use disorders. *Front. Neurosci.*, 8, 176, 2014, https://doi.org/10.3389/fnins.2014.00176.
56. Tang, M., Hu, X., Wang, Y., Yao, X., Zhang, W., Yu, C., Cheng, F., Li, J., Fang, Q., Ivermectin, a potential anticancer drug derived from an antiparasitic drug. *Pharmacol. Res.*, 163, 105207, 2021, https://doi.org/10.1016/j.phrs.2020.105207.
57. Zaidi, A.K. and Dehgani-Mobaraki, P., Retracted article: The mechanisms of action of ivermectin against SARS-CoV-2: An evidence-based clinical review article. *J. Antibiot. (Tokyo)*, 75, 122–122, 2022, https://doi.org/10.1038/s41429-021-00430-5.
58. Ordóñez, L.E. and Angulo, E.S., [Efficacy of ivermectin in the treatment of children parasitized by Strongyloides stercoralis]. *Biomedica*, 24, 33–41, 2004, http://www.ncbi.nlm.nih.gov/pubmed/15239599.
59. To, K.K.-W., Tsang, O.T.-Y., Leung, W.-S., Tam, A.R., Wu, T.-C., Lung, D.C., Yip, C.C.-Y., Cai, J.-P., Chan, J.M.-C., Chik, T.S.-H., Lau, D.P.-L., Choi,

C.Y.-C., Chen, L.-L., Chan, W.-M., Chan, K.-H., Ip, J.D., Ng, A.C.-K., Poon, R.W.-S., Luo, C.-T., Cheng, V.C.-C., Chan, J.F.-W., Hung, I.F.-N., Chen, Z., Chen, H., Yuen, K.-Y., Temporal profiles of viral load in posterior oropharyngeal saliva samples and serum antibody responses during infection by SARS-CoV-2: An observational cohort study. *Lancet Infect. Dis.*, 20, 565–574, 2020, https://doi.org/10.1016/S1473-3099(20)30196-1.

60. Timani, K.A., Liao, Q., Ye, L., Zeng, Y., Liu, J., Zheng, Y., Ye, L., Yang, X., Lingbao, K., Gao, J., Zhu, Y., Nuclear/nucleolar localization properties of C-terminal nucleocapsid protein of SARS coronavirus. *Virus Res.*, 114, 23–34, 2005, https://doi.org/10.1016/j.virusres.2005.05.007.
61. Mudatsir, M., Yufika, A., Nainu, F., Frediansyah, A., Megawati, D., Pranata, A., Mahdani, W., Ichsan, I., Dhama, K., Harapan, H., Antiviral activity of ivermectin against SARS-CoV-2: An old-fashioned dog with a new trick—A literature review. *Sci. Pharm.*, 88, 36, 2020, https://doi.org/10.3390/scipharm88030036.
62. Hill, A., Mirchandani, M., Pilkington, V., Ivermectin for COVID-19: Addressing potential bias and medical fraud. *Open Forum Infect. Dis.*, 9, ofab645, 2022, https://doi.org/10.1093/ofid/ofab645.
63. Taccone, F.S., Hites, M., Dauby, N., From hydroxychloroquine to ivermectin: How unproven "cures" can go viral. *Clin. Microbiol. Infect.*, 28, 472–474, 2022, https://doi.org/10.1016/j.cmi.2022.01.008.
64. Alam, M.T., Murshed, R., Gomes, P.F., Masud, Z.M., Saber, S., Chaklader, M.A., Khanam, F., Hossain, M., Momen, A.B.I.M., Yasmin, N., Alam, R.F., Sultana, A., Robin, R.C., Ivermectin as pre-exposure prophylaxis for COVID-19 among healthcare providers in a selected tertiary hospital in Dhaka – an observational study. *Eur. J. Med. Health Sci.*, 2, 1–5, 2020, https://doi.org/10.24018/ejmed.2020.2.6.599.
65. Mahmud, R., Rahman, M.M., Alam, I., Ahmed, K.G.U., Kabir, A.K.M.H., Sayeed, S.K.J.B., Rassel, M.A., Monayem, F.B., Islam, M.S., Islam, M.M., Das Barshan, A., Hoque, M.M., Mallik, M.U., Yusuf, M.A., Hossain, M.Z., Ivermectin in combination with doxycycline for treating COVID-19 symptoms: A randomized trial. *J. Int. Med. Res.*, 49, 030006052110135, 2021, https://doi.org/10.1177/03000605211013550.
66. Aref, Z.F., Bazeed, S.E.E.S., Hassan, M.H., Hassan, A.S., Rashad, A., Hassan, R.G., Abdelmaksoud, A.A., Clinical, Biochemical and Molecular Evaluations of Ivermectin Mucoadhesive Nanosuspension Nasal Spray in Reducing Upper Respiratory Symptoms of Mild COVID-19. *Int. J. Nanomed.*, 16, 4063–4072, 2021, https://doi.org/10.2147/IJN.S313093.
67. Ahmed, S., Karim, M.M., Ross, A.G., Hossain, M.S., Clemens, J.D., Sumiya, M.K., Phru, C.S., Rahman, M., Zaman, K., Somani, J., Yasmin, R., Hasnat, M.A., Kabir, A., Aziz, A.B., Khan, W.A., A five-day course of ivermectin for the treatment of COVID-19 may reduce the duration of illness. *Int. J. Infect. Dis.*, 103, 214–216, 2021, https://doi.org/10.1016/j.ijid.2020.11.191.

68. Krolewiecki, A., Lifschitz, A., Moragas, M., Travacio, M., Valentini, R., Alonso, D.F., Solari, R., Tinelli, M.A., Cimino, R.O., Álvarez, L., Fleitas, P.E., Ceballos, L., Golemba, M., Fernández, F., Fernández de Oliveira, D., Astudillo, G., Baeck, I., Farina, J., Cardama, G.A., Mangano, A., Spitzer, E., Gold, S., Lanusse, C., Antiviral effect of high-dose ivermectin in adults with COVID-19: A proof-of-concept randomized trial. *EClinicalMedicine*, 37, 100959, 2021, https://doi.org/10.1016/j.eclinm.2021.100959.
69. Abd-Elsalam, S., Noor, R.A., Badawi, R., Khalaf, M., Esmail, E.S., Soliman, S., Abd El Ghafar, M.S., Elbahnasawy, M., Moustafa, E.F., Hassany, S.M., Medhat, M.A., Ramadan, H.K., Eldeen, M.A.S., Alboraie, M., Cordie, A., Esmat, G., Clinical study evaluating the efficacy of ivermectin in COVID-19 treatment: A randomized controlled study. *J. Med. Virol.*, 93, 5833–5838, 2021, https://doi.org/10.1002/jmv.27122.
70. Shahbaznejad, L., Davoudi, A., Eslami, G., Markowitz, J.S., Navaeifar, M.R., Hosseinzadeh, F., Movahedi, F.S., Rezai, M.S., Effects of ivermectin in patients with COVID-19: A multicenter, double-blind, randomized, controlled clinical trial. *Clin. Ther.*, 43, 1007–1019, 2021, https://doi.org/10.1016/j.clinthera.2021.04.007.
71. López-Medina, E., López, P., Hurtado, I.C., Dávalos, D.M., Ramirez, O., Martínez, E., Díazgranados, J.A., Oñate, J.M., Chavarriaga, H., Herrera, S., Parra, B., Libreros, G., Jaramillo, R., Avendaño, A.C., Toro, D.F., Torres, M., Lesmes, M.C., Rios, C.A., Caicedo, I., Effect of ivermectin on time to resolution of symptoms among adults with mild COVID-19. *JAMA*, 325, 1426, 2021, https://doi.org/10.1001/jama.2021.3071.
72. Kerr, L., Cadegiani, F.A., Baldi, F., Lobo, R.B., Assagra, W.L.O., Proença, F.C., Kory, P., Hibberd, J.A., Chamie-Quintero, J.J., Ivermectin prophylaxis used for COVID-19: A citywide, prospective, observational study of 223,128 subjects using propensity score matching. *Cureus*, 14, 1, e21272, 2022, https://doi.org/10.7759/cureus.21272.
73. Lim, S.C.L., Hor, C.P., Tay, K.H., Mat Jelani, A., Tan, W.H., Ker, H.B., Chow, T.S., Zaid, M., Cheah, W.K., Lim, H.H., Khalid, K.E., Cheng, J.T., Mohd Unit, H., An, N., Nasruddin, A.B., Low, L.L., Khoo, S.W.R., Loh, J.H., Zaidan, N.Z., Ab Wahab, S., Song, L.H., Koh, H.M., King, T.L., Lai, N.M., Chidambaram, S.K., Peariasamy, K.M., Hwong, W.Y., Low, E.V., Pathmanathan, M.D., Hamzah, M.L., Chan, Y.C., Voo, J.Y.H., Yap, C.F., Chan, Y.Q., Vun, L.K., Kong, K.K.K., Lim, Y.F., Teoh, Y.J., Abdullah, A.R., Ramadas, A., Leong, C.L., Wahab, N.H., Ismail, N., Ismail, I., Lee, T.M., Khoo, P.J., Phua, S.H., Gopalakrishnan, P.P., Jaya Selan, S., Ampalakan, I., Khuan, J.F., Abdul Rashid, W.N.F., Zakaria, S.S., Gemini, K., Burahan, H., Santokh Singh, T.S., Jaafar, N., Mohd Shukri, N.A., Izhar Hisham, S., Teow, S.H., Lim, C.Y., Rajantran, S., Kamaruddin, S.I.A., Izhar, I.I., Mohd Mustapha, N.S., Mohamad, Z., Abu Salim, S.R.N., Othman Andu, D.F., Kamarudin, N., Sarban Singh, K.K., Tay, E.P., Md Tahir, S.H.H., Vijayasingham, S., Kum, Y.Z., Natarajan, P.A., Soh, Y.H., Syed Alwi, S.O.F., Murugan, H., Chuah, C.H., Tan, S.W., Leong, K.N., Wong, P.S., Chen, W.T.T.,

Ng, R.S., Lim, Y.L., Bidin, F.N., Chin, M.L., Guan, H.L., Mohamad Rasli, M.H., Abdullah, R., Jamaludin, M.A., Mohd Shohaime, N., Mohd Mansor, S., Rasliza, R., Mohamed Nor, L., Thong, K.M., Muniandy, B., Saw, P.V.T., Thong, K.S., Wong, K.C., Muthi, V.R.N., Ahmad Shanizza, Q.I., Jeevaraj, L., Chew, E.L., Huang, P.C., Retnasamy, J., Devesahayam, P.R., Lim, M.S., Viswanathan, T.T., Mahamad Azazis, M.S., Domnic, G., Tengku, M.F., Wong, J.Q.Y., Choo, X.H., Prabhaharan, A.N., Zaharudin, N.S., Abu Bakar Sayuti, A.U., Abdul Wahid, N., Saat, N.H., Othman, N.H., Ahmad Zubaidi, A., Abdul Mutalib, N.M.S., Lekh Raj Sharma, V.D., Gunaraj, D., Hanafi, M.N., Embok Ungah, N.A., Mohd Zahari, M.A., Chaw, C.L., Arokisamy, J., Mohd Hassan, P.A., Ayub, A.J., Zainal Abidin, A.N., Choong, K.S., Teoh, L.R., Kang, H.Y., Krishnan, K., Purusothman, P., Zainol, M.I., Tew, M.M., Bahrudin, M.F., Lim, K.C., Mohd Nadzir, S., Narayanan, L., Shamsuddin, A.N., Tan, K.T., Kamaludeen, S., Ibrahim, N.M., Sim, P.K.A., Aminuddi, I., Raja Nahar Putra, R.N., Yah, L.Y., Liew, B.S., Ravi, T., Syed Badaruddin, S.N.A., Mah Hassan, N.S., Roslan, Z., Nadarajan, R., Ang, J.-G., Arumugam, M., Chua, K.W., Ooi, C.G.S., Lee, S.H., Ooi, W.T., Tang, X.Y., Perumalu, K., Razali, M.H., Murat, M.S.A., Hamdan, N.S., Hamidi, M.S., Anuar, A., Ang, W.C., Wong, C.K., Mushaddik, I.L., Mohamed, S.H., Raja Lope Ahmad, R.A.R., Wan Zainudin, W.M.K., Mohd Zin, A.F., Teoh, S.K., Yusoff, M.Y., Abdul Rani, S.N., Ab Rahman, M., Mohd Noor, M.A., Tuan Mat, T.N., Othman, M.K., Sayed Sahul Hamid Gani, M., Ngua, C.Z., Chang, A.K.W., Wong, Z.H., Ko, A.T.Y., Thung, S.F., Tiong, X.T., Chua, H.H., Goh, K.S., Muthusamy, S., Loo, W.Y., Supramaniam, T., Lingam, R., Chandra Kumar, L., Chun, S.T., Selvarajah, D.R., Mohan Raja, D., Low, O.L., Supparmaniam, P., Ad Suhadak, H., Beh, B.C., Lee, Y.L., Ooi, C.L., Ishak, K.N., Harun, R., Lee, S.L., Lee, K.S., Ow, J.K., Kaisbain, N., Leong, C.J.W., Chee, Y.L., Teh, K.L., Chan, K.V., Lee, K.T., Wong, E.J., Ismail, I., Mohd Suan, M.A., Mohamed Yusoff, A.L., Tuan Ismail, T.M.F., Tuan Manah, Ibrahim, K.A., Mohd Unit, H., Sidek, N.N., Seman, N., Efficacy of ivermectin treatment on disease progression among adults with mild to moderate COVID-19 and comorbidities. *JAMA Intern. Med.*, 182, 426, 2022, https://doi.org/10.1001/jamainternmed.2022.0189.

74. Reis, G., Silva, E.A.S.M., Silva, D.C.M., Thabane, L., Milagres, A.C., Ferreira, T.S., dos Santos, C.V.Q., Campos, V.H.S., Nogueira, A.M.R., de Almeida, A.P.F.G., Callegari, E.D., Neto, A.D.F., Savassi, L.C.M., Simplicio, M.I.C., Ribeiro, L.B., Oliveira, R., Harari, O., Forrest, J.I., Ruton, H., Sprague, S., McKay, P., Guo, C.M., Rowland-Yeo, K., Guyatt, G.H., Boulware, D.R., Rayner, C.R., Mills, E.J., Effect of early treatment with ivermectin among patients with COVID-19. *N. Engl. J. Med.*, 386, 1721–1731, 2022, https://doi.org/10.1056/NEJMoa2115869.

4

Ivermectin as an Anti-Parasitic Agent

Areeba Rahman[1], Abdulaziz S. Alouffi[2], Mohammad Husain[1] and Abdur Rub[1*]

[1]Department of Biotechnology, Jamia Millia Islamia (A Central University), New Delhi, India
[2]King Abdulaziz City for Science and Technology, Riyadh, Saudi Arabia

Abstract
Ivermectin is the world's first 'endectocide' derived from avermectin B, a metabolite obtained from *Streptomyces avermectinius*. It is a macrocyclic lactone, originally used as a veterinary drug. Later, it was used more extensively to treat human parasitic diseases. Ivermectin is used by more than 250 million people each year to treat a variety of parasitic disorders, such as filariasis, onchocerciasis, strongyloidiasis, scabies, and pediculosis despite its recently reported resistance in some of the infections. It is well documented that it selectively activates glutamate-gated chloride channels which causes hyperpolarization of affected parasitic cells leading to muscle paralysis and death. Here, in this chapter, we reviewed its clinical uses in human parasitic infections, mechanism of action, mode of administration, side effects, and reported resistance in detail.

***Keywords*:** Ivermectin, anti-parasitic, parasites, strongyloidiasis, onchocerciasis, scabies, pediculosis, wonder drug

Abbreviations

WHO	World health organization
DEC	diethylcarbamazine
mf	microfilariae
IVM	Ivermectin

**Corresponding author*: arub@jmi.ac.in

Rashid Ali and Shahid ul-Islam (eds.) Chemistry and Biological Activities of Ivermectin, (69–92)

ROS	Reactive oxygen species
SRPs	Signal recognition particles
RNA	Ribonucleic acid
NPC	Nuclear pore complex
MMP	Mitochondrial membrane potential
Cl	Chloride ion
Kapα	Karyopherin α
Kap β	Karyopherin β

4.1 Introduction

Parasites are the organisms that live on or inside the host organisms. These are responsible for a plethora of infections in various groups of organisms, particularly human beings, mostly prevalent in tropical areas. Scientists have been working to combat parasitic diseases for decades by discovering various anti-parasite drugs. These drugs are a class of medications employed in treating and managing infections caused by parasites. One of the breakthroughs in the discovery of potent anti-parasitic drugs is the synthesis of a broad spectrum, avermectin-derived drug - Ivermectin. It is a booming drug with an outstanding tolerance and sterling safety profile. Since 1987, this drug has been widely used in veterinary medicine (Table 4.1), and its use has been extended to treat parasitic infections in humans caused by both ectoparasites as well as endoparasites (Figure 4.1). Ivermectin is used by over 250 million people worldwide each year to treat a variety of parasitic disorders, such as filariasis, onchocerciasis, strongyloidiasis, scabies, and pediculosis [1]. Moreover, it continues to astound scientists because of its significant role in the treatment of several other tropical diseases (Table 4.2) (such as Chagas, leishmaniasis, worm infections, etc.). The possible mechanism of action lies in its interference with glutamate-gated chloride channels in the cell membrane, present in invertebrates and absent in vertebrates. Primarily, it causes the inflow of negatively charged ions, resulting in hyperpolarization of the afflicted cells, ultimately leading to muscle paralysis (Figure 4.2) [2]. The adaptability of Ivermectin might be a potent weapon for combating the numerous overlapping endemic and emerging illnesses that now impact people worldwide. Further, repurposing this medicine would be a crucial therapeutic move to lessen the enormous burden of infections caused by parasites across the globe.

The World Health Organization (WHO) has led several Ivermectin-based control projects aimed at eradicating onchocerciasis and lymphatic filariasis and reducing malaria transmission. Like many other drugs used

Table 4.1 Use of Ivermectin in the treatment of parasitic infections in non-human hosts.

Effective Ivermectin treatment in animals	Major family of infecting parasites	Name of the parasite	References
Cattles	Gastrointestinal Nematodes	*Bunostomum phlebotomum, C. pectinata, C. punctata, Haemonchus placei, Oesophagostomum radiatum, C. oncophora, Ostertagia ostertagi, Mecistocirrus digitatus, Nematodirus spathiger, Nematodirus helvetianus, O. lyrat, Strongyloides papillosus, Toxocara vitulorum* *T. colubriformi, Trichostrongylus axei*	[4–11]
	Lungworm	*Dictyocaulus viviparus*	[5, 8]
	Extraintestinal Nematode	*Parafilaria bovicola*	[12]
	Grubs	*Dermatobia hominis, H.lineatum* *Hypoderma bovis*	[8, 13]
	Sucking lice	*Haematopinus eurysternus, Linognathus vituli, Solenopotes capillatus*	[14, 15]
	Biting Lice	*Damalinia bovis*	[8]
	Mites	*Sarcoptes scabiei var. bovis, Chorioptes bovis, Psoroptes ovis*	[16, 17]
	Ticks	*Boophilus microplus*	[18]

(*Continued*)

Table 4.1 Use of Ivermectin in the treatment of parasitic infections in non-human hosts. (*Continued*)

Effective Ivermectin treatment in animals	Major family of infecting parasites	Name of the parasite	References
	Screwworms	*B.decoloratus, Ornithodoros savignyi, Chrysomya bezziana*	[19, 20]
Sheeps	Gastrointestinal nematodes	*Bunostomum trigonocephalum, Cooperia oncophora, Cooperia surnabada, Cooperia curticei, Haemonchus contortus, Marshallagia marshalli, Nematodirus filicollis, Nematodirus spathiger, Nematodirus roscidus, Ostertagia leptospicularis, Ostertagia Ostertagi, Ostertagia circumcincta, Strongyloides papillosus, Trichostrongylus colubriformis, Trichostrongylus vitrinus, Trichostrongylus axei, Trichuris ovis*	[21, 25, 26]
	Lungworms	*Dictyocaulus filaria, Protostrongylus rufescens*	[21, 22]
	Mange mite Itch mite	*Sarcoptes scabiei* *Psorergates ovis*	[23, 24]

(*Continued*)

Table 4.1 Use of Ivermectin in the treatment of parasitic infections in non-human hosts. (*Continued*)

Effective Ivermectin treatment in animals	Major family of infecting parasites	Name of the parasite	References
Goats	Gastrointestinal Nematodes	*Bunostomum spp., Haemonchus spp., Oesophagostomum spp, Strongyloides spp., Trichostrongylus colubriformis*	[27, 28]
	Mite	*Sarcoptes scabiei*	[29]
	Sucking lice	*Linognathus stenopsis*	[30]
	Chewing lice	*Bovicola caprae*	[31]
Pigs	Gastrointestinal Nematodes	*Ascaris suum, Hyostrongylus rubidus, Oesophagostomum spp., Strongyloides ransomi, Macracanthorhynchus hirudinaceus, Ascarops strongylina*	[32, 35]
	Lungworm	*Metastrongylus spp.*	[33]
	Kidney worms	*Stephanurus dentatus*	[36]
	Sucking lice	*Haematopinus suis*	[35]
	Mites	*Sarcoptes scabiei var. suis*	[33]
Horses	Gastrointestinal nematodes	*Cyanthostomum spp., Cyclicocyclus spp., Cyclicostephanus spp, Habronema spp., Oxyuris equi, Parascaris equorum, S. edentatus, S. equinus, Strongylus vulgaris, Trichostrongylus axei, Tridontophorus spp*	[34, 37, 38]

(*Continued*)

Table 4.1 Use of Ivermectin in the treatment of parasitic infections in non-human hosts. (*Continued*)

Effective Ivermectin treatment in animals	Major family of infecting parasites	Name of the parasite	References
	Other nematodes	*Onchocerca sp.*	[37]
	Bots	*Gastrophilus spp.*	[38]
	Lungworm	*Dictyocaulus arnfieli*	[39]
Chickens	Nematodes	*Ascaridia galli, Heterakis gallinarum*	[40, 42]
	Mites	*Dermanyssus gallinae*	[41]
	Lice	*Menacanthus stramineus, Menopon gallinae*	[41]
Cats	Gastrointestinal nematodes	*A. tubaeforme, Ancylostoma braziliense, Toxocara cati*	[43, 46]
	Mite	*Nodoedres cati Otodectes cynotis*	[44, 45]
	Lungworm	*Aelurostrongylus abstrusus*	[46]
Dogs	Gastrointestinal nematodes	*A. braziliense, Ancylostoma caninum, Toxascaris leonina, Toxocara canis, Trichuris vulpis, Uncinaria stenocephala*	[47–49]
	Tick	*Rhipicephalus sanguineous*	[50]
	Mites	*Demodex canis, Otodectes cynotis, Sarcoptes scabiei*	[51, 52]
	Heartworm	*Dirofilaria immitis Dipetalonema recondition*	[53, 54]

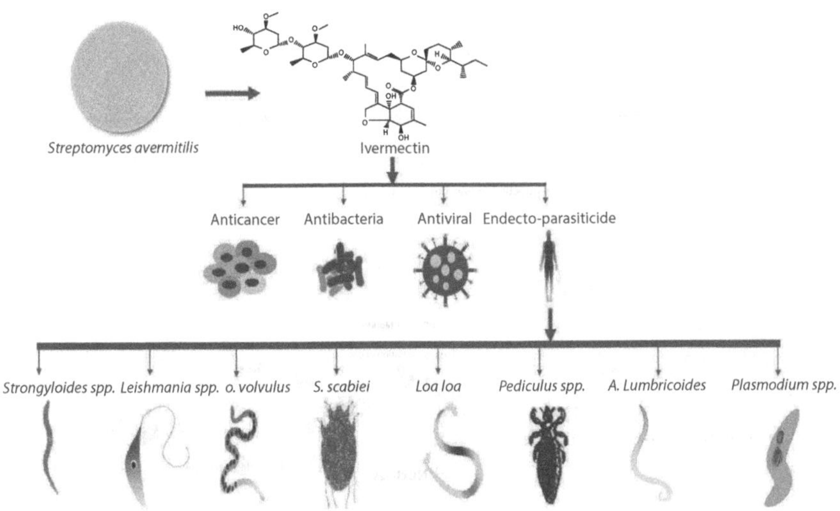

Figure 4.1 Ivermectin as an anticancer, antibacterial, antiviral and endecto-parasiticide.

Table 4.2 List of other human-parasitic infections treated by Ivermectin.

S. no.	Disease	% Efficacy	References
1	Trichuriasis	84.1	[115]
2	Enterobiasis	100	
3	Hymenolopiases	50	
4	Tungiasis	64	
5	Hookworm disease	67.7	
6	Bedbug infection	86	[116]
7	Toxoplasmosis	-	[117]
8	Myiasis	100	[118]
9	African trypanosomiasis	-	[119]
10	Cutaneous larval migran infection	94	[120]

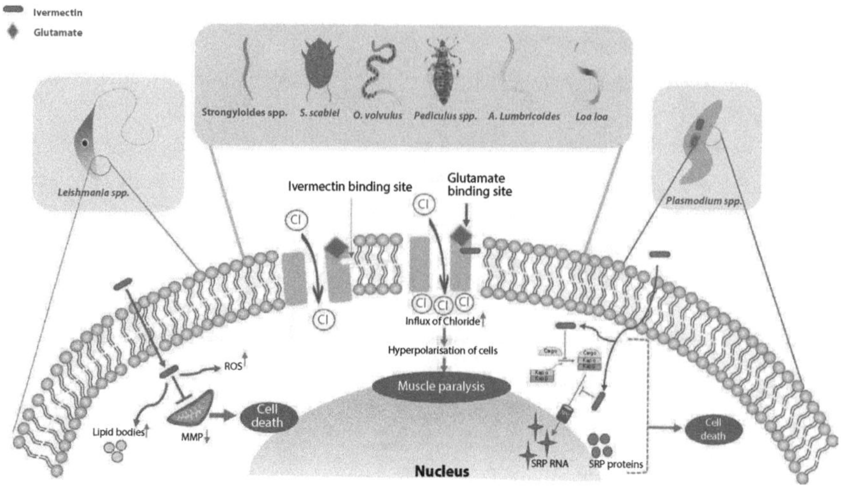

Figure 4.2 A schematic mechanism of anti-parasitic action of Ivermectin on different parasites.

in treating infection, Ivermectin too develops resistance due to its widespread uses [3]. Recent studies showed the possibilities for new drug delivery methods for Ivermectin that might expand its ability to treat a wider variety of disorders and prevent drug resistance. In this chapter, the authors highlighted the clinical uses of Ivermectin in human parasitic infections, its mechanism of antiparasitic action, efficacy alone, and in combination with other drugs, mode of administration, side effects, and recently reported resistance in some cases.

4.2 Use of Ivermectin Against Various Human Parasitic Infections

Ivermectin has been extensively used for the treatment of various parasitic infections including filariasis, onchocerciasis, strongyloidiasis, scabies, and pediculosis.

4.2.1 Strongyloidiasis

Strongyloidiasis is a neglected tropical disease caused by the organism-*Strongyloides stercoralis*. It has been estimated to affect over 600 million people worldwide [55]. The causative organism is a nematode, majorly prevalent in tropical and subtropical climates, especially in areas with improper sanitary conditions. The clinical manifestation of this disease includes chronic infection that may be life-threatening in an immunocompromised host. Hence, it is imperative to conduct various clinical trials of drugs for the management of strongyloidiasis. Currently, Ivermectin has emerged as the treatment of choice for this disease. It has efficacy similar to the previously prescribed drug, thiabendazole but has higher safety standards [56]. Further, it has been observed that in the case of larval clearance, Ivermectin has better performance than albendazole [57]. Ivermectin is administered orally (200 $\mu g\ kg^{-1}$) for two days as the preferred medicament for treating strongyloidiasis. Besides this, a single dose of Ivermectin is sufficient for its treatment [58]. It is also recommended by experts to repeat the course after 2 weeks to combat with parasite's autoinfection cycle [59]. Other methods of Ivermectin administration are also used, specifically in the cases of paralytic ileus where patients cannot be prescribed oral medication. These involve per rectal and parenteral formulations [60]. The Parenteral formulation of Ivermectin is used specifically in animals; however, it could be an option for extreme situations when no other options are left for clearing Strongyloides infection.

4.2.2 Onchocerciasis

Onchocerciasis is the vector-borne infection caused by the tissue residing filaria - *Onchocerca volvulus* and is transmitted by blackflies. The Global Burden of Disease survey, conducted in 2017, estimated that 20.9 million people are infected with *O. volvulus* worldwide [61]. It is endemic in the regions of Africa and the Arabian Peninsula. Most importantly, according to the WHO, it is the second leading cause of blindness after trachoma. The initial manifestation of this infection includes itching of the skin. Further, symptoms like lichenification, depigmentation, papules development, and atrophy have also been reported in onchocerciasis patients [62]. The historical treatments for this disease were suramin and diethylcarbamazine (DEC), but due to their limited efficacy and safety, these drugs were discontinued [63–65]. Moreover, Ivermectin was introduced and licensed in 1987 for the treatment of Onchocerciasis and became the first choice of drug. Since then, the standard dose of Ivermectin 150μg/kg has been employed and is reported to clear the skin-residing microfilariae [66]. In the case of adult worms, its activity is deemed limited, however, it is efficacious in sterilizing them [67]. In addition, the drug dose is repeated annually or biannually to prevent the reoccurrence of the infection in patients.[68].

4.2.3 Lymphatic Filariasis

Lymphatic filariasis is a major health concern and is endemic in more than 72 countries across the world [69]. Transmission of lymphatic filariasis has been confirmed in the Caribbean, South America, the Middle East, Southeast Asia, and Africa [70]. In Addition, it is a mosquito-borne disease, caused by three nematodes *Wuchereria bancrofti*, *Brugia Malawi*, and *Brugia timori* [71]. It can permanently disable a person by interfering with their lymphatic system, and is characterized by symptoms like Lymphedema, hydrocele, and Edema. For the treatment and management of this disease, various drugs are employed and Ivermectin is one of them. Although it is not effective against adult larvae, it has reasonable microfilaricidal properties. Moreover, the Global Programme for the Elimination of Lymphatic Filariasis in sub-Saharan Africa has adopted the combination of Ivermectin and Albendazole for mass therapy [72, 73]. It has also been noted that even at very low dosages (21.3±0.7 μg/kg), a single dose of Ivermectin has substantially cleared microfilariae from the blood within 36 h, and 100% was cleared in 5–12 days [74]. Most importantly, due to its comparatively favorable efficacy and side effects, Ivermectin therapy has an advantage over standard anti-filarial drug diethylcarbamazine citrate. Hence it makes Ivermectin an excellent candidate to replace DEC as the treatment of choice for filariasis [75].

4.2.4 Loiasis

Loiasis is a vector-borne disease caused by the parasite *Loa loa*, transmitted by the flies of the genus Chrysops. It is reported that 3 to 13 million people are infected by *Loa loa*, which is indigenous to Central and West Africa. The diagnostic stage of the disease is marked by the presence of sheathed microfilariae in spinal fluid, urine, sputum, peripheral blood, and the lungs. In addition, the adult worm can occasionally be seen moving through the outer layer (sclera) of the eye; hence, the Loiasis-causing parasite is also known as the African eye worm. The most common clinical manifestations due to adult worms are "Calabar swelling" (localized angioedema) and pruritus. The current treatment of Loiasis includes Ivermectin. It has been seen, that within the first three days following a single dosage of Ivermectin (150 μg/ml), Loa microfilaremia drops by 70%–80% [76]. Moreover, Ivermectin not only kills the parasite but also improves the clinical symptoms of loiasis and tends to stop the recurrence of Calabar swellings for several months [77], albeit, Ivermectin has been reported to induce potentially lethal encephalopathies in people harboring

>30,000–50,000 microfilariae [78]. However, Ivermectin can be prescribed in cases with microfilariae densities (8000 and 30,000 mf/mL), but close surveillance is needed along with hospitalization for the first 3–4 days [79].

4.2.5 Scabies

The skin condition known as scabies is an ectoparasitic infection, caused by *Sarcoptes scabiei* mite infestation. Each year, there are about 455 million new cases of scabies globally [80]. The mite penetrates the skin and produces a popular rash and excruciating itching. Secondary skin infections with bacteria like *Streptococcus pyogenes* and *Staphylococcus aureus*, which can result in more serious skin infections, invasive illness, and immune-mediated disease, further aggravate the infestations [81]. The treatment of scabies infection includes topical and oral drugs. According to observational studies, Ivermectin is successful in treating individuals with scabies. Moreover, treatment of classic scabies includes two doses of Ivermectin (200 μg /kg) along with food as ingestion of food increases the bioavailability of Ivermectin by a factor of two. In addition, taking the drug with food will enhance the penetration of the drug into the epidermis. Ivermectin is not ovicidal, hence two doses should be given for the treatment of classic scabies, one to two weeks apart [82]. Ivermectin can also be used as a topical medicine (cream 1%) [83]. Nonetheless, Ivermectin's usage in the treatment of scabies hasn't been conclusively linked to any severe side effects as of now, but it is advisable to avoid the usage of Ivermectin during pregnancy [82].

4.2.6 Pediculosis

Pediculosis is caused by blood-sucking lice. The three main types of lice that infest individuals are *Pediculus humanus humanus, Pthirus pubis, and Pediculus humanus capitis* (body louse). Symptoms of a louse infestation include conjunctivitis, excoriations, and itching on the scalp. Body lice also transmit harmful microorganisms like trench fever caused by *Bartonella quintana*, relapsing fever by *Borrelia recurentis*, and epidemic typhus by *Rickettsia prowazekii* [84]. While body lice mostly affect the destitute and dispossessed, head lice infection affects humans from all economic or social backgrounds. Further, due to the mode of transmission of lice and the ineffectiveness of the treatment options, managing and controlling infestations of lice is still a challenge. Ivermectin, however, is a possible medicine to cure a lice infestation in various observational studies [85]. Additionally, scientists have identified its mode of action and noted that it selectively disrupts glutamate channels (Figure 4.2) [86]. Most importantly, Ivermectin is the only systemic drug against lice. Nevertheless, despite these outstanding benefits, the rise of lice with resistance is now compromising Ivermectin's effectiveness [87].

4.2.7 Mansonellosis

Mansonellosis is a vector-borne parasitic disease caused by three species of parasite nematodes (*Mansonella streptocerca, Mansonella ozzardi,* and *Mansonella pertans*) that harbor the human body as the definitive host. These are transmitted by blood-feeding female flies belonging to two separate families, Ceratopogonidae (Biting Midges) and Simuliidae (blackflies). The majority of the time, mansonellosis infections are believed to be asymptomatic and have low pathogenicity, but they can occasionally cause itching, joint pain, swollen lymph nodes, and stomach symptoms. Although currently, there is no standard treatment for mansonella infection, Ivermectin has been employed to treat mansnellosis. In the case of the most challenging *M. pertans* infection, Ivermectin has proven to be ineffective [88]. However, it has been very effective in the treatment of *M. ozzardi* (150 μg/kg) and *M .streptocerca* (200μg/kg) infections [89–92]. A single dose of Ivermectin administered to infected individuals was found to suppress *M. ozzardi* microfilaremia. Additionally, Ivermectin has also been used as a mansonellosis control measure in the Brazilian Amazon due to its success in treating *M. ozzardi* [93, 94]. The drawback of Ivermectin includes its ineffectiveness against adult worms.

4.2.8 Ascariasis

Ascariasis is caused by *Ascaris lumbricoides* and infects more than 1.2 billion people worldwide. It is a soil-transmitted helminth transmitted through the fecal-oral route. These parasites reside in the intestine and prevail in areas with poor sanitation, poor hygiene conditions, and inappropriate sewage disposal. People ingesting *Ascaris* usually show no symptoms, however, if symptoms do persist, they may include discomfort and abdominal pain. In severe cases, it may block the intestines and slow down the growth of children. Current treatments for Ascariasis include drugs such as Albendazole, Mebendazole, and Ivermectin. Furthermore, studies have demonstrated that Albendazole and Ivermectin have similar anti-Ascariasis efficacies, however the combinational therapy showed greater efficacy [95]. A single dose of 200 μg/kg of Ivermectin resulted in cure rates of about 78% for Ascariasis, further demonstrated the drug's improved efficacy and safety in several trials. Despite improved efficacy, Ivermectin mass treatments given over an extended period have not been found to have any impact on reducing the transmission or prevalence of *A. lumbricoides* [96].

4.2.9 Gnathostomiases

Gnathostomiasis is caused by the parasite *Gnathostoma spp*., mostly prevalent in the tropical and subtropical regions. It is endemic in regions of Thailand, Mexico, Southeast Asia, and Japan. Humans are generally infected with third-stage larvae by eating raw or uncooked snakes, frogs, pigs, and fish infested with *Gnathostoma spp*. The clinical manifestations of this disease comprise visceral and cutaneous forms. The cutaneous form is characterized by nodular migratory panniculitis and creeping eruption while the visceral form includes a wide range of symptoms affecting pulmonary, gastrointestinal and genitourinary, and ocular parts of the body [97]. For several years there was no potent treatment for Gnathostomiasis, surgical excision of the larvae persisted as the only efficacious management. However, after a trial of variety of drugs in animal models and humans, a few drugs were proved to be successful in the treatment of Gnathostomiases. These included Ivermectin and Albendazole. The efficacy of Ivermectin is reported to be similar to that of Albendazole and found to be well tolerated too [98]. The further study elaborated that treating 15 patients with a single dosage of 0.2 mg/kg of Ivermectin administered twice in a row may result in cure rates of 100% [99].

4.2.10 Leishmaniasis

Leishmaniasis is a neglected tropical disease, endemic in 98 counties, affecting 12–15 million people worldwide. An estimated 1.5 to 2 million new cases are reported per year, which results in 70,000 fatalities [100, 101]. It has a wide range of clinical symptoms that varies based on the species of *Leishmania*, disseminated by sandflies. The three main types of leishmaniasis are cutaneous, mucocutaneous, and visceral, with visceral being the most severe. Treatment of these clinical forms comprises pentavalent antimonial, miltefosine, and Amphotericin B. While these drugs are effective against Leishmaniasis, they possess various side effects like anorexia, fever, renal and cardiac alterations, liver damage, teratogenicity, and parasite resistance [102, 103]. Therefore, efforts to find new, efficient, and affordable antileishmanial agents are instrumental to combat the disease. Various studies have demonstrated the effect of Ivermectin on the

Leishmania parasite. Results obtained from these studies conferred that Ivermectin is highly effective against promastigotes and amastigotes while being less toxic to human cells [104–106]. It has also been observed that hydrogels containing 1.5% w/w of Ivermectin can be used as an alternative to boost the treatment of cutaneous leishmaniasis [107]. Apart from these studies, sero analogs of Ivermectin were also prepared from commercial Ivermectin and their antileishmanial activity was assayed. It has been concluded that Conjugated Delta (2,3)-IVM shows the higher antileishmanial activity as compared to the parental compound [108]. Moreover, it is highly recommended as a safer treatment option. The mode of action that causes stress and eventually leads to the death of leishmania is suggested by the depolarization of the mitochondrial membrane potential and rise in reactive oxygen species levels in the parasite, together with the production of lipid bodies (Figure 4.2) [105]. Although, Ivermectin showed promising results, it remains a subject for further investigations to explore its antileishmanial property in humans.

4.2.11 Malaria

Malaria is a tropical parasitic disease transmitted by female anopheles mosquitoes. According to the WHO estimates, there were 241 million global incidences of malaria, with a fatality rate of 627,000 cases in the year 2020 [109]. It is mostly driven by five Plasmodium species in humans: *P. falciparum*, *P. malariae*, *P. ovale*, *P. vivax*, and *P. knowlesi*. Clinical manifestations include sweating, anorexia, anemia, headaches, and recurrent fever. Random diagnostic tests and a microscopic review of blood are used to diagnose malaria. In addition, an array of strategies, such as avoiding mosquito bites, receiving intermittent prophylactic care, and administering large doses of medication, are used to control this disease. Moreover, it can be treated normally with antimalarial chemotherapies. The best treatment available is artemisinin-based therapy. However, there are various ongoing projects aimed to produce pipelines of robust approaches to combat this disease. One of them is the use of long-lasting Ivermectin in the treatment of malaria. In a study conducted by scientists, it was demonstrated that Ivermectin reduces the liver plasmodium density *in vivo*, further it also affects mosquito survival [110]. Similarly, other findings reveal the successful inhibition of plasmodium sporogony [111, 112]. Moreover, Ivermectin leads to the inhibition of the development of asexual parasites and mature gametocytes even at nanomolar concentrations and arrests the parasite cycle at the trophozoite stage [113]. The proposed mechanism of action of Ivermectin against malaria is shown in Figure 4.2 [114].

4.3 Mode of Action Against Various Parasites

Although Ivermectin has established itself as a robust broad-spectrum anti-parasitic drug, its mechanism of action is still not very clear. However, investigations have been conducted to envisage its mechanism of action against various parasites. The general mode of action against invertebrates includes ligand-gated ion channels as shown in Figure 4.2. Ivermectin affects the glutamate-activated chloride channel of nerve or muscle cells in invertebrates. It binds to the channel, causing an influx of chloride ions that causes cells to become hyperpolarized, which ultimately results in the paralysis of parasites' muscles and death (Figure 4.2). In the case of *Leishmania*, Ivermectin causes a change in the mitochondrial membrane potential (MMP). It also produces reactive oxygen species (ROS). Furthermore, it triggers the production of lipid bodies, which has been linked to the consequences of cell death in several investigations [105]. Ivermectin inhibits the growth of the Plasmodium parasite by preventing the nucleo-cytoplasmic transfer of signal recognition particles (SRPs), which are essential for the parasite's survival. It does this by changing the stage of karyopherin dimerization or the point at which SRPs connect to karyopherin dimers. This ultimately has the effect of preventing SRP from reaching its location of RNA assembly inside the nucleus. Additionally, the buildup of proteins in the cytoplasm causes cell death as illustrated in (Figure 4.2) [114].

4.4 Conclusions

Since its discovery in the late 1970s, Ivermectin has benefitted public health worldwide and contributed a lot to the improvement of socioeconomic wellbeing globally. For polyparasitised, underprivileged communities, this medicine has proven to be a panacea, further, its unique structure has enthralled scientists across the globe, and as a consequence, significant research has been conducted to combat a variety of diseases. Moreover, this has led to unmatched repurposing of the drug as a potential antiparasitic drug. Furthermore, in the current scenario, Ivermectin is acknowledged as a relatively safe and practical anti-parasitic medication for humans. It is a robust drug that has been used immensely as a treatment option for various illnesses including strongyloidiasis, onchocerciasis, and lymphatic filariasis. Moreover, current studies have identified it as a promising drug against malaria and coronavirus. Despite new efforts to maximize Ivermectin's advantage, more human trials need to be conducted. The global community needs to be more attentive and innovative in attempting to eliminate the world's most invasive, disabling, and stigmatizing parasitic infections in the world.

4.5 Acknowledgements

The authors are thankful to the Ministry of AYUSH for funding [Z.28015/252/2016-HPC (EMR)AYUSH-C].

References

1. Ashour, D.S., Ivermectin: From theory to clinical application. *Int. J. Antimicrob. Agents*, *54*, 2, 134–142, 2019, https://doi.org/10.1016/j.ijantimicag.2019.05.003.
2. Ottesen, E.A. and Campbell, W.C., Ivermectin in human medicine. *J. Antimicrob. Chemother.*, 34, 2, 195–203, 1994, http://jac.oxfordjournals.org/.
3. Prichard, R.K., Ivermectin resistance and overview of the Consortium for Anthelmintic Resistance SNPs. *Expert Opin. Drug Discovery*, *2*, s1, S41–S52, 2007, https://doi.org/10.1517/17460441.2.S1.S41.
4. Tassi, P., Barth, D., Gross, S.J., The efficacy of ivermectin against Strongyloides papillosus in cattle. *Parassitologia*, *32*, 3, 347–352, 1990.
5. Armour, J., Bairden, K., Batty, A.F., Davison, C.C., Ross, D.B., Persistent anthelmintic activity of ivermectin in cattle. *Vet. Rec.*, *116*, 6, 151–153, 1985.
6. Mendoza-de Gives, P., Herrera-Rodríguez, D., Roncalli, R.A., Efficacy of ivermectin against the abomasal nematode mecistocirrus digitatus in naturally infected zebu calves. *Am. J. Vet. Res.*, *48*, 11, 1611–1612, 1987.
7. Whelan, N.C., Charleston, W.A.G., Alexander, A.M., Lees, S., Efficacy of ivermectin against benzimidazole-resistant Nematodirus spathiger in sheep. *N Z Vet. J.*, 43, 3, 99–100, 1995.
8. Alva-Valdes, R., Wallace, D.H., Holste, J.E., Egerton, J.R., Cox, J.L., Wooden, J.W., Barrick, R.A., Efficacy of ivermectin in a topical formulation against induced gastrointestinal and pulmonary nematode infections, and naturally acquired grubs and lice in cattle. *Am. J. Vet. Res.*, *47*, 11, 2389–2392, 1986.
9. Yazwinski, T.A., Williams, M., Greenway, T., Tilley, W., Anthelmintic activities of ivermectin against gastrointestinal nematodes of cattle. *Am. J. Vet. Res.*, *42*, 3, 481–482, 1981.
10. Tassi, P., Barth, D., Gross, S.J., The efficacy of ivermectin against Strongyloides papillosus in cattle. *Parassitologia*, *32*, 3, 347–352, 1990.
11. Avcioglu, H. and Balkaya, I., A comparison of the efficacy of subcutaneously administered ivermectin, doramectin, and moxidectin against naturally

infected Toxocara vitulorum in calves. *Trop. Anim. Health Prod.*, *43*, 6, 1097–1099, 2011.

12. Swan, G.E., Soll, M.D., Carmichael, I.H., Schroder, J., Efficacy of ivermectin against Parafilaria bovicola and lesion resolution in cattle. *Vet. Parasitol.*, 40, 3–4, 267–272, 1983, doi: 10.1016/0304-4017(91)90106-6.
13. Roncalli, R.A. and Usher, C.B., Efficacy of ivermectin against Dermatobia hominis in cattle. *Vet. Parasitol.*, *28*, 4, 343–346, 1988.
14. Hussain, M.A., Khan, M.N., Iqbal, Z., Sajid, M.S., Arshad, M., Bovine pediculosis: Prevalence and chemotherapeutic control in Pakistan. *Livest. Res. Rural Dev.*, *18*, 10–17, 2006.
15. Titchener, R.N., The control of lice on domestic livestock. *Vet. Parasitol.*, *18*, 3, 281–288, 1985.
16. Barth, D. and Preston, J.M., Efficacy of topically administered ivermectin against chorioptic and sarcoptic mange of cattle. *Vet. Rec.*, *123*, 4, 101–104, 1988.
17. Guillot, F.S. and Meleney, W.P., The infectivity of surviving Psoroptes ovis (Hering) on cattle treated with ivermectin. *Vet. Parasitol.*, *10*, 1, 73–78, 1982.
18. Davey, R.B., Pound, J.M., Miller, J.A., Klavons, J.A., Therapeutic and persistent efficacy of a long-acting (LA) formulation of ivermectin against Rhipicephalus (Boophilus) microplus (Acari: Ixodidae) and sera concentration through time in treated cattle. *Vet. Parasitol.*, *169*, 1-2, 149–156, 2010.
19. Schroder, J., Swan, G.E., Soll, M.D., Hotson, I., Efficacy of ivermectin against ectoparasites of cattle in South Africa. *J. S. Afr. Vet. Assoc.*, *56*, 1, 31–35, 1985.
20. Spradbery, J.P., Tozer, R.S., Drewett, N., Lindsey, M.J., The efficacy of ivermectin against larvae of the screw-worm fly (Chrysomya bezziana). *Aust. Vet. J.*, *62*, 9, 311–314, 1985.
21. Wescott, R.B. and LeaMaster, B.R., Efficacy of ivermectin against naturally acquired and experimentally induced nematode infections in sheep. *Am. J. Vet. Res.*, *43*, 3, 531–533, 1982.
22. Rehbein, S. and Visser, M., Efficacy of ivermectin delivered via a controlled-release capsule against small lungworms (Protostrongylidae) in sheep. *J. Vet. Med., Ser. B*, *49*, 7, 313–316, 2002.
23. Iqbal, A., Baba, M.A., Shah, M., Mushtaq, I., Sakina, A., Wani, S., Treatment of mange infection in a weaner flock of sheep with ivermectin at sheep breeding farm Hardishiva of Kashmir valley. *J. Parasitol. Dis.*, *39*, 2, 171–173, 2015.
24. Soll, M.D. and Carmichael, I.H., Efficacy of injectable ivermectin against the itch mit (Psorergates ovis) of sheep. *Parasitol. Res.*, *75*, 1, 81–82, 1988.
25. Rehbein, S., Barth, D., Visser, M., Winter, R., Langholff, W.K., Efficacy of an ivermectin controlled-release capsule against some rarer nematode parasites of sheep. *Vet. Parasitol.*, *88*, 3-4, 293–298, 2000.
26. Yazwinski, T.A., Greenway, T., Presson, B.L., Pote, L.M., Featherstone, H., Williams, M., Antiparasitic efficacy of ivermectin in naturally parasitized sheep. *Am. J. Vet. Res.*, 44, 11, 2186–2187, 1983.

27. Lespine, A., Alvinerie, M., Sutra, J.F., Pors, I., Chartier, C., Influence of the route of administration on efficacy and tissue distribution of ivermectin in goat. *Vet. Parasitol.*, 128, 3-4, 251–260, 2005.
28. Godara, R., Sharma, R.L., Sodhi, S.S., Efficacy of fenbendazole, levamisole and ivermectin against gastrointestinal nematodes in Jamunapari goats. *J. Parasitol. Dis.*, 35, 2, 219–221, 2011.
29. Manurung, J., Stevenson, P., Knox, M.R., Use of ivermectin to control sarcoptic mange in goats in Indonesia. *Trop. Anim. Health Prod.*, 22, 3, 206–212, 1990.
30. Nizamov, N. and Prelezov, P., A study on efficacy of ivermectin against Linognathus stenopsis (Phthiraptera: Linognathidae) in goats. *Trad. Mod. Vet. Med.*, 4, 2, 16–20, 2019.
31. Dimri, U., Devi, G., Ajith, Y., Gopalakrishnan, A., A field study on the efficacy of ivermectin via subcutaneous route against chewing lice (Bovicola caprae) infestation in naturally infested goats. *Onderstepoort J. Vet. Res.*, 86, 1, 1–5, 2019.
32. Lichtensteiger, C.A., DiPietro, J.A., Paul, A.J., Neumann, E.J., Thompson, L., Persistent activity of doramectin and ivermectin against Ascaris suum in experimentally infected pigs. *Vet. Parasitol.*, 82, 3, 235–241, 1999.
33. Alva-Valdes, R., Wallace, D.H., Foster, A.G., Ericsson, G.F., Wooden, J.W., Efficacy of an in-feed ivermectin formulation against gastrointestinal helminths, lungworms, and sarcoptic mites in swine. *Am. J. Vet. Res.*, 50, 8, 1392–1395, 1989.
34. Barth, D., Rehbein, S., Reid, J.F.S., Barrick, R.A., Efficacy of an in-feed formulation of ivermectin against adult worms and somatic larvae of Strongyloides ransomi. *Vet. Parasitol.*, *65*, 1-2, 89–97, 1996.
35. Stewart, T.B., Marti, O.G., Hale, O.M., Efficacy of ivermectin against five genera of swine nematodes and the hog louse, Haematopinus suis. *Am. J. Vet. Res.*, 42, 8, 1425–1426, 1981.
36. Stewart, T.B., Stromberg, B.E., Lawhorn, B., *Internal parasites in swine*, Michigan State University, Cooperative Extension Service, 108 Agriculture Hall, East Lansing, MI 48824, United States, 1994.
37. Egerton, J.R., Brokken, E.S., Suhayda, D., Eary, C.H., Wooden, J.W., Kilgore, R.L., The antiparasitic activity of ivermectin in horses. *Vet. Parasitol.*, 8, 1, 83–88, 1981.
38. Schröder, J. and Swan, G.E., Ivermectin as an antiparasitic agent in horses. *J. S. Afr. Vet. Assoc.*, 53, 2, 127–128, 1982.
39. Britt, D.P. and Preston, J.M., Efficacy of ivermectin against Dictyocaulus arnfieldi in ponies. *Vet. Rec.*, 116, 13, 343–345, 1985.
40. Sharma, R.L. and Bhat, T.K., Anthelmintic activity of ivermectin against experimental Ascaridia galli infection in chickens. *Vet. Parasitol.*, 37, 3-4, 307–314, 1990.
41. Arisova, G.B., Efficacy of ivermectin-based drugs against ectoparasites in broiler chickens. *World Vet. J.*, 10, 2, 160–164, 2020.

42. Khayatnouri, M.H., Garedaghi, Y., Arbati, A.R., Khalili, H., The effect of Ivermectin pour-on administration against natural Heterakis gallinarum infestation and its prevalence in native poultry. *Am. J. Anim. Vet. Sci.*, 6, 1, 55–58, 2011.
43. Nolan, T.J., Niamatali, S., Bhopale, V., Longhofer, S.L., Schad, G.A., Efficacy of a chewable formulation of ivermectin against a mixed infection of Ancylostoma braziliense and Ancylostoma tubaeforme in cats. *Am. J. Vet. Res.*, 53, 8, 1411–1413, 1992.
44. Sivajothi, S., Sudhakara Reddy, B., Rayulu, V.C., Sreedevi, C., Notoedres cati in cats and its management. *J. Parasitol. Dis.*, 39, 2, 303–305, 2015.
45. Nunn-Brooks, L., Michael, R., Ravitz, L.B., Kordick, D., Lappin, M.R., Efficacy of a single dose of an otic ivermectin preparation or selamectin for the treatment of Otodectes cynotis infestation in naturally infected cats. *J. Feline Med. Surg.*, 13, 8, 622–624, 2011.
46. Kirkpatrick, C.E. and Megella, C., Use of ivermectin in treatment of Aelurostrongylus abstrusus and Toxocara cati infection in a cat. *J. Am. Vet. Med. Assoc.*, 190, 10, 1309–1310, 1987.
47. Daurio, C.P., Roberson, E.L., Seward, R.L., Efficacy of ivermectin in a beef-based chewable formulation against Ancylostoma caninum and Uncinaria stenocephala in dogs. *J. Parasitol.*, *79*, 5, 768–770, 1993.
48. Khayatnouri, M.H. and Garedaghi, Y., Efficacy of ivermectin pour-on administration against natural Toxascaris leonina infestation in native dogs. *World J. Zool.*, 7, 1, 12–16, 2012.
49. Anderson, D.L. and Roberson, E.L., Activity of ivermectin against canine intestinal helminths. *Am. J. Vet. Res.*, 43, 9, 1681–1683, 1982.
50. Roy, S. and Roy, M., Therapeutic evaluation of ivermectin pour on against tick infestation in dogs. *Vet. World*, 3, 3, 113–114, 2010.
51. Mueller, R.S. and Bettenay, S.V., A proposed new therapeutic protocol for the treatment of canine mange with ivermectin. *J. Am. Anim. Hosp. Assoc.*, 35, 1, 77–80, 1999.
52. Yazwinski, T.A., Pote, L., Tilley, W., Rodriguez, C., Greenway, T., Efficacy of ivermectin against Sarcoptes scabiei and Otodectes cynotis infestations of dogs. *Vet. Med. Small Anim. Clin.: VM, SAC*, 76, 12, 1749–1751, 1981.
53. Paul, A.J., Todd, K.S., Jr., Acre, K. E. Sr., Plue, R.E., Wallace, D.H., French, R.A., Wallig, M.A., Efficacy of ivermectin chewable tablets and two new ivermectin tablet formulations against Dirofilaria immitis larvae in dogs. *Am. J. Vet. Res.*, 52, 11, 1922–1923, 1991.
54. Lindemann, B.A. and McCall, J.W., Microfilaricidal activity of ivermectin against Dipetalonema reconditum. *J. Vet. Pharmacol. Ther.*, 6, 1, 75–76, 1983.
55. Buonfrate, D., Bisanzio, D., Giorli, G., Odermatt, P., Fürst, T., Greenaway, C., French, M., Reithinger, R., Gobbi, F., Montresor, A., Bisoffi, Z., The global prevalence of strongyloides stercoralis infection. *Pathogens*, 9, 6, 1–9, 2020, https://doi.org/10.3390/pathogens9060468.

56. Gann, P.H., Neva, F.A., Gam, A.A., A randomized trial of single- and two-dose ivermectin versus thiabendazole for treatment of strongyloidiasis. *J. Infect. Dis.*, *169*, 5, 1076–1079, 1994.
57. Datry, A., Hilmarsdottir, I., Mayorga-Sagastume, R., Lyagoubi, M., Gaxotte, P., Biligui, S., Gentilini, M., Treatment of Strongyloides stercoralis infection with ivermectin compared with albendazole: Results of an open study of 60 cases. *Trans. R. Soc. Trop. Med. Hyg.*, *88*, 3, 344–345, 1994.
58. Hürlimann, E. and Keiser, J., A single dose of ivermectin is sufficient for strongyloidiasis. *Lancet Infect. Dis.*, 19, 11, 1150–1151, 2019, http://doi.org/10.1016/S1473-3099(19)30519-5.
59. Greaves, D., Coggle, S., Pollard, C., Aliyu, S.H., Moore, E.M., Strongyloides stercoralis infection. *BMJ*, 347–352, 2013, https://doi.org/10.1136/bmj.f4610.
60. Grein, J.D., Mathisen, G.E., Donovan, S., Fleckenstein, L., Serum ivermectin levels after enteral and subcutaneous administration for Strongyloides hyperinfection: A case report. *Scand. J. Infect. Dis.*, 42, 3, 234–236, 2010, https://doi.org/10.3109/00365540903443165.
61. Global burden of disease, 2018, Keys Facts. [(Accessed on 24 May 2021)] Available online:https://www.who.int/news-room/fact-sheets/detail/onchocerciasis.
62. Brattig, N.W., Cheke, R.A., Garms, R., Onchocerciasis (river blindness) – more than a century of research and control. *Acta Trop.*, 218, 105677–105700, 2021, https://doi.org/10.1016/j.actatropica.2020.105677.
63. Awadzi, K. and Gilles, H.M., Diethylcarbamazine in the treatment of patients with onchocerciasis. *Br. J. Clin. Pharmacol.*, 34, 4, 281–288, 1992, https://doi.org/10.1111/j.1365-2125.1992.tb05632.x.
64. Budden, F.H., The natural history of ocular onchocerciasis over a period of 14–15 years and the efect on this of a single course of suramin therapy. *Trans. R. Soc. Trop. Med. Hyg.*, 70, 5–6, 484–491, 1976, https://doi.org/10.1016/0035-9203(76)90134-6.
65. Duke, B.O., The effects of drugs on Onchocerca volvulus trials of suramin at different dosages and a comparison of the brands Antrypol, Moranyl and Naganol. *Bull. World Health Organ.*, 39, 2, 157–67, 1968.
66. Basáñez, M.G., Pion, S.D., Boakes, E., Filipe, J.A., Churcher, T.S., Boussinesq, M., Effect of single-dose ivermectin on Onchocerca volvulus: A systematic review and meta-analysis. *Lancet Infect. Dis.*, *8*, 5, 310–322, 2008, https://doi.org/10.1016/S1473-3099(08)70099-9.
67. Walker, M., Pion, S.D.S., Fang, H., Gardon, J., Kamgno, J., Basáñez, M.-G., Boussinesq, M., Macrofilaricidal efficacy of repeated doses of ivermectin for the treatment of river blindness. *Clin. Infect. Dis.*, *2017*, 65, 2026, https://doi.org/10.1093/cid/cix616.
68. Batsa Debrah, L., Klarmann-Schulz, U., Osei-Mensah, J., Dubben, B., Fischer, K., Mubarik, Y., Hoerauf, A., Comparison of repeated doses of ivermectin versus ivermectin plus albendazole for the treatment of onchocerciasis: A

randomized, open-label, clinical trial. *Clin. Infect. Dis.*, *71*, 4, 933–943, 2020, https://doi.org/10.1093/cid/ciz889.

69. World Health Organisation, Global health observatory, 2020, [(Accessed on 14 May 2022)] Available online:https://www.who.int/data/gho/data/themes/topics/lymphatic-filariasis.
70. Global mapping of lymphatic filariasis. *Parasitol. Today*, 13, 472–476, 1997.
71. Strategies and tools for the control/elimination of lymphatic filariasis. *Bull. World Health Organ.*, 75, 491–503, 1997.
72. Ottesen, E.A., Hooper, P.J., Bradley, M., Biswas, G., The global programme to eliminate lymphatic filariasis: Health impact after 8 years. *PLoS Negl. Trop. Dis.*, *2*, 10, e317, 2008, https://doi.org/10.1371/journal.pntd.0000317.
73. Massa, K., Magnussen, P., Sheshe, A., Ntakamulenga, R., Ndawi, B., Olsen, A., The combined effect of the Lymphatic Filariasis Elimination Programme and the Schistosomiasis and Soil-transmitted Helminthiasis Control Programme on soil-transmitted helminthiasis in schoolchildren in Tanzania. *Trans. R. Soc. Trop. Med. Hyg.*, *103*, 1, 25–30, 2009, https://doi.org/10.1016/j.trstmh.2008.07.011.
74. Ottesen, E.A., Vijayasekaran, V., Kumaraswami, V., Pillai, S.P., Sadanandam, A., Frederick, S., Tripathy, S.P., A controlled trial of ivermectin and diethylcarbamazine in lymphatic filariasis. *N. Engl. J. Med.*, *322*, 16, 1113–1117, 1990, https://doi.org/10.1056/NEJM199004193221604.
75. Kumaraswami, V., Ottesen, E.A., Vijayasekaran, V., Devi, S.U., Swaminathan, M., Aziz, M.A., Tripathy, S.P., Ivermectin for the treatment of Wuchereria bancrofti filariasis: Efficacy and adverse reactions. *JAMA*, 259, 21, 3150–3153, 1988, https://doi.org/10.1001/jama.1988.03720210040026.
76. Richard-Lenoble, D., Kombila, M., Rupp, E.A., Pappayliou, E.S., Gaxotte, P., Nguiri, C., Aziz, M.A., Ivermectin in loiasis and concomitant O. volvulus and M. perstans infections. *Am. J. Trop. Med. Hyg.*, *39*, 5, 480–483, 1988, https://doi.org/10.4269/ajtmh.1988.39.480.
77. Hovette, P., Debonne, J.M., Touze, J.E., Gaxotte, P., Imbert, P., Fourcade, L., Laroche, R., Efficacy of ivermectin treatment of Loa loa filariasis patients without microfilaraemias. *Ann. Trop. Med. Parasitol.*, *88*, 1, 93–94, 1994, https://doi.org/10.1080/00034983.1994.11812842.
78. Gardon, J., Gardon-Wendel, N., Kamgno, J., Chippaux, J.P., Boussinesq, M., Serious reactions after mass treatment of onchocerciasis with ivermectin in an area endemic for Loa loa infection. *Lancet*, 350, 9070, 18–22, 1997.
79. Boussinesq, M., Loiasis: New epidemiologic insights and proposed treatment strategy. *J. Travel Med.*, 19, 3, 140–143, 2012, https://doi.org/10.1111/j.1708-8305.2012.00605.x.
80. Lake, S.J., Kaldor, J.M., Hardy, M., Engelman, D., Steer, A.C., Romani, L., Mass drug administration for the control of scabies: A systematic review and meta-analysis. *Clin. Infect. Dis.*, 75, 6, 959–967, 2022, https://doi.org/10.1093/cid/ciac042.

81. Engelman, D., Cantey, P.T., Marks, M. *et al.*, The public health control of scabies: Priorities for research and action. *Lancet*, 394, 81–92, 2019.
82. Currie, B.J. and Mccarthy, J.S., Clinical therapeutics permethrin and ivermectin for scabies. *N. Engl. J. Med.*, 362, 717–742, 2010.
83. Absil, G., Lebas, E., Libon, F., El Hayderi, L., Dezfoulian, B., Nikkels, A.F., Scabies and therapeutic resistance: Current knowledge and future perspectives, 157–164, 2022, https://doi.org/10.1002/jvc2.25.
84. Badiaga, S. and Brouqui, P., Human louse- transmitted infectious disease. *Clin. Microbiol. Infect.*, 18, 4, 332–7, 2012.
85. Martin, P.M.V., Efficacy of ivermectin for the treatment of head lice (Pediculosis capitis). *Trop. Med. Parasitol.*, 45, 3, 253–254, 1994.
86. Lamassiaude, N., Toubate, B., Dric Neveu, C., Charnet, P., Dupuy, C., Oise Debierre-Grockiego, F., Dimier-Poisson, I., Charvet, C.L., The molecular targets of ivermectin and lotilaner in the human louse Pediculus humanus humanus: New prospects for the treatment of pediculosis. *PLoS Pathog.*, 17, 2, 2021, https://doi.org/10.1371/journal.ppat.1008863.
87. Amanzougaghene, N., Fenollar, F., Nappez, C., Ben-Amara, A., Decloquement, P., Azza, S., Mediannikov, O., Complexin in ivermectin resistance in body lice. *PLoS Genet.*, *14*, 8, e1007569, 2018, https://doi.org/10.1371/journal.pgen.1007569.
88. Bregani, E.R., Rovellini, A., Mbaïdoum, N., Magnini, M.G., Comparison of different anthelminthic drug regimens against Mansonella perstans filariasis. *Trans. R. Soc. Trop. Med. Hyg.*, *100*, 5, 458–463, 2006.
89. de Almeida Basano, S., Fontes, G., Medeiros, J.F., Camargo, J. S. D. A. A., Vera, L.J.S., Araújo, M.P.P., Camargo, L.M.A., Sustained clearance of Mansonella ozzardi infection after treatment with ivermectin in the Brazilian Amazon. *Am. J. Trop. Med. Hyg.*, 90, 6, 1170, 2014.
90. Asio, S.M., Simonsen, P.E., Onapa, A.W., Mansonella perstans: Safety and efficacy of ivermectin alone, albendazole alone and the two drugs in combination. *Ann. Trop. Med. Parasitol.*, *103*, 1, 31–37, 2009.
91. Fischer, P., Bamuhiiga, J., Büttner, D.W., Treatment of human Mansonella streptocerca infection with ivermectin. *Trop. Med. Int. Health, 2*, 2, 191–199, 1997.
92. Fischer, P., Tukesiga, E., Büttner, D.W., Long-term suppression of Mansonella streptocerca microfilariae after treatment with ivermectin. *J. Infect. Dis.*, *180*, 4, 1403–1405, 1999, https://doi.org/10.1086/315014.
93. Pérez-Molina, J.A., Diaz-Menendez, M., Perez-Ayala, A., Ferrere, F., Monje, B., Norman, F., Lopez-Velez, R., Treatment of diseases caused by parasites. *Enferm. Infecc. Microbiol. Clin.*, *28*, 1, 44–59, 2009.
94. Díaz-Menéndez, M., Norman, F., Monge-Maillo, B., Pérez-Molina, J.A., López-Vélez, R., Filariasis in clinical practice. *Enferm. Infecc. Microbiol. Clin.*, *29*, 27–37, 2011, https://doi.org/10.1016/s0213-005x(11)70041-6.
95. Belizario, V.Y., Amarillo, M.E., Leon, W.D., Reyes, A.D.L., Bugayong, M.G., Macatangay, B.J.C., A comparison of the efficacy of single doses of

albendazole, ivermectin, and diethylcarbamazine alone or in combinations against Ascaris and Trichuris spp. *Bull. World Health Organ.*, *81*, 1, 35–42, 2003.

96. Moncayo, A.L., Vaca, M., Amorim, L., Rodriguez, A., Erazo, S., Oviedo, G., Cooper, P.J., Impact of long-term treatment with ivermectin on the prevalence and intensity of soil-transmitted helminth infections. *PLoS Negl. Trop. Dis.*, *2*, 9, e293, 2008, https://doi.org/10.1371/journal.pntd.0000293.
97. Herman, J.S. and Chiodini, P.L., Gnathostomiasis, another emerging imported disease. *Clin. Microbiol. Rev.*, 22, 3, 484–492, 2009, https://doi.org/10.1128/CMR.00003-09.
98. Nontasut, P., Bussaratid, V., Chullawichit, S., Charoensook, N., Visetsuk, K., Comparison of ivermectin and albendazole treatment for gnathostomiasis. *Southeast Asian J. Trop. Med. Public Health*, *31*, 2, 374–377, 2000.
99. Nontasut, P., Claesson, B.A., Dekumyoy, P., Pakdee, W., Chullawichit, S., Double-dose ivermectin vs albendazole for the treatment of gnathostomiasis. *Southeast Asian J. Trop. Med. Public Health*, *36*, 3, 650, 2005.
100. Reithinger, R., Dujardin, J.C., Louzir, H., Pirmez, C., Alexander, B., Brooker, S., Cutaneous leishmaniasis. *Lancet Infect. Dis.*, 7, 9, 581–596, 2007.
101. Bravo, F. and Sanchez, M.R., New and re-emerging cutaneous infectious diseases in Latin America and other geographic areas. *Dermatol. Clin.*, 21, 4, 655–668, 2003.
102. Oliveira, L.F., Schubach, A.O., Martins, M.M., Passos, S.L., Oliveira, R.V., Marzochi, M.C., Andrade, C.A., Systematic review of the adverse effects of cutaneous leishmaniasis treatment in the New World. *Acta Trop.*, 118, 2, 87–96, 2011, https://doi.org/10.1016/J.ACTATROPICA.2011.02.007.
103. Uliana, S.R., Trinconi, C.T., Coelho, A.C., Chemotherapy of leishmaniasis: Present challenges. *Parasitology*, 145, 4, 464–480, 2018.
104. Salazar, B.M., Pereira, V. C., Torre, Y.H.L., Khamesipour, A., Cottier, F.T., Evaluation of a new topical treatment for the control of cutaneous leishmaniasis. *Microorganisms*, 8, 11, 1803–1819, 2020, doi: 10.3390/microorganisms8111803.
105. Reis, T.A.R., Oliveira-da-Silva, J.A., Tavares, G.S.V., Mendonça, D.V.C., Freitas, C.S., Costa, R.R., Lage, D.P., Martins, V.T., Machado, A.S., Ramos, F.F., Silva, A.M., Ludolf, F., Antinarelli, L.M.R., Brito, R.C.F., Chávez-Fumagalli, M.A., Humbert, M.V., Roatt, B.M., Coimbra, E.S., Coelho, E.A.F., Ivermectin presents effective and selective antileishmanial activity *in vitro* and *in vivo* against Leishmania infantum and is therapeutic against visceral leishmaniasis. *Exp. Parasitol.*, 221, 108059–108069, 2021, https://doi.org/10.1016/j.exppara.2020.108059.
106. Morsy, T.A., Rasheid, K.A., Efficacy of ivermectin on the infectivity of Leishmania major promastigotes. *J. Egypt. Soc. Parasitol.*, 28, 1, 207–219, 1998.
107. Romero, A.I., Cid, A.G., Minetti, N.E., Briones Nieva, C.A., García Bustos, M.F., Gonzo, E.E., Bermúdez, J.M., Sustained-release hydrogels of ivermectin

as alternative systems to improve the treatment of cutaneous leishmaniasis. *Ther. Deliv.*, 11, 12, 779–790, 2020.

108. dos Santos, A.R., Falcão, C.A.B., Muzitano, M.F., Kaiser, C.R., Rossi-Bergmann, B., Férézou, J.P., Ivermectin-derived leishmanicidal compounds. *Bioorg. Med. Chem.*, 17, 2, 496–502, 2009, https://doi.org/10.1016/j.bmc.2008.12.003.
109. World Health Organisation, 2020. [(Accessed on 8 May 2022)] Available online: https://www.who.int/news-room/fact-sheets/detail/malaria.
110. Mendes, A.M., Albuquerque, I.S., Machado, M., Pissarra, J., Meireles, P., Prudêncio, M., Inhibition of Plasmodium liver infection by ivermectin. *Antimicrob. Agents Chemother.*, 61, 2, e02005-16, 2017.
111. Kobylinski, K.C., Foy, B.D., Richardson, J.H., Ivermectin inhibits the sporogony of Plasmodium falciparum in Anopheles gambiae. *Malar. J.*, 11, 1, 1–9, 2012.
112. Sylla, M., Kobylinski, K.C., Gray, M., Chapman, P.L., Sarr, M.D., Rasgon, J.L., Foy, B.D., Mass drug administration of ivermectin in south-eastern Senegal reduces the survivorship of wild-caught, blood fed malaria vectors. *Malar. J.*, 9, 1, 1–10, 2010.
113. de Carvalho, L.P., Sandri, T.L., José Tenório de Melo, E., Fendel, R., Kremsner, P.G., Mordmüller, B., Held, J., Ivermectin impairs the development of sexual and asexual stages of Plasmodium falciparum *in vitro*. *Antimicrob. Agents Chemother.*, 63, 8, e00085-19, 2019.
114. Panchal, M., Rawat, K., Kumar, G., Kibria, K.M., Singh, S., Kalamuddin, M., Tuteja, R., Plasmodium falciparum signal recognition particle components and anti-parasitic effect of ivermectin in blocking nucleo-cytoplasmic shuttling of SRP. *Cell Death Dis.*, 5, 1, e994–e994, 2014.
115. Heukelbach, J., Wilcke, T., Winter, B., de Oliveira, F.A.S., Moura, R.C.S., Harms, G., Feldmeier, H., Efficacy of ivermectin in a patient population concomitantly infected with intestinal helminths and ectoparasites. *Arzneimittelforschung*, 54, 07, 416–421, 2004.
116. Sheele, J.M., Anderson, J.F., Tran, T.D., Teng, Y.A., Byers, P.A., Ravi, B.S., Sonenshine, D.E., Ivermectin causes Cimex lectularius (bedbug) morbidity and mortality. *J. Emerg. Med.*, 45, 3, 433–440, 2013.
117. Bilgin, M., Yildirim, T., Hökelek, M., *In vitro* effects of ivermectin and sulphadiazine on Toxoplasma gondii. *Balkan Med. J.*, 30, 19–22, 2013.
118. Victoria, J., Trujillo, R., Barreto, M., Myiasis: A successful treatment with topical ivermectin. *Int. J. Dermatol.*, 38, 2, 142–144, 1999.
119. Udensi, U.K. and Fagbenro-Beyioku, A.F., Effect of ivermectin on Trypanosoma brucei brucei in experimentally infected mice. *J. Vector Borne Dis.*, 49, 3, 143, 2012.
120. Van den Enden, E., Stevens, A., Van Gompel, A., Treatment of cutaneous larva migrans. *N. Engl. J. Med.*, 339, 17, 1246–1247, 1998.

5

Emerging Paradigm of Ivermectin and its Hybrids in Elimination of Malaria

Iram Irfan[1], Diksha Shahi[1,2], Mukesh C. Joshi[3], Shailja Singh[2]* and Mohammad Abid[1]†

[1]*Medicinal Chemistry Laboratory, Department of Biosciences, Jamia Millia Islamia, New Delhi, India*

[2]*Special Centre for Molecular Medicine, Jawaharlal Nehru University, New Delhi, India*

[3]*Department of Chemistry, Kirori Mal College, University of Delhi, New Delhi, India*

Abstract

Ivermectin, a most commonly used avermectin a chemically modified derivative of the natural product obtained from *Streptomyces avermitilis*. It has been found to be effective against a number of parasitic vector-borne diseases and receiving attention to control malaria transmission. Particularly, it has been found to inhibit the hepatic stage of *Plasmodium* infection both *in vitro* and *in vivo* by impairing parasite development inside the hepatocytes. Recent reports suggested Ivermectin hybrids inhibit sporogonic as well as blood-stage development in *Plasmodium falciparum* as dual-action molecular hybrids with better potency compared to Ivermectin. Herein, we have reviewed the role of Ivermectin and its hybrids reported till date on various stages of the malaria parasite and their effectiveness in malaria control.

Keywords: Malaria, *Plasmodium*, drug-repurposing, Ivermectin, hybrids, liver-stage, blood-stage

**Corresponding author*: shailja.jnu@gmail.com
†*Corresponding author*: mabid@jmi.ac.in

Rashid Ali and Shahid ul-Islam (eds.) Chemistry and Biological Activities of Ivermectin, (93–120)

5.1 Introduction

Ivermectin is a macrocyclic lactone compound discovered in the starting of 1970s and licensed first as an antiparasitic drug for animals in 1981 [1]. It is a mixture of two semisynthetic analogues derived from avermectin B_1 which is produced from bacterium *S. avermitilis* [2]. It has been used in affected human to treat onchocerciasis around the tropical areas since 1987 with low mortality and high safety measurements [3, 4]. In present scenario, it is widely applied to treat some tropical diseases, *viz.*, strongyloidiasis (onchocerciasis) and lymphatic filariasis. However, it is also found active against lice and scabies [5–7]. Ivermectin is also very useful as broad spectrum drug against insects, acarine parasites and nematodes at very low doses [5]. Ivermectin is an effective endectocide (also known as systemic insecticides) and it could be administered to the affected community as a supportive device for vector control. It is efficient in killing anthropods that feed on treated individuals, including *Anopheles* mosquitoes. This feature makes ivermectin with mass drug administration (MSD), a most promising combination to minimize malaria transmission [8].

Malaria is caused by five prominent species of *Plasmodium* viz., *Plasmodium falciparum, P. malariae, P. vivax, P. knowlesi,* and *P. ovale.* However, *P. falciparum* is lethal to human especially in Sub-Saharan Africa and South Asia [9]. According to the World Malaria Report 2022, 247 million cases were reported in 2022, recording an increase of about 6 million cases of infection compared to 2021, whereas 619000 deaths were recorded in 2021. Recently, artimisinin-based combination therapies (ACTs) are in use to treat uncomplicated malaria which rapidly eliminates young gametocytes and asexual parasites. However, ACTs are not effective against mature gametocytes [10, 11]. Despite of development of potent anti-malarials against various stages of malaria, the parasites have got resistant against those drugs. Due to less susceptibility of the parasites to the drugs, there is an urgent need of the development of new entity.

Currently, few groups are involved in investigation of emerging capabilities of Ivermectin. They focus on its ability to block parasite transmission and sporogony as well as its insecticidal activity [12–21]. It could be used alone or in combination with partner drug to combat overlooked tropical diseases such as filariasis [22, 23]. As per the results, the role and scope of Ivermectin towards the anti-malarial activity pave the way for a totally new drug-based approach to stop and manage malaria transmission [8]. Further work needs to be explored with regard to Ivermectin for potential anti-malarial agent.

The safety and potency of the Ivermectin has been evaluated using two distinct dose regimens, either three daily doses of 300 μg/kg for three days in a row, as determined by the IVERMAL dose-ranging research, or a single dosage of 400 μg/kg, which is already used to treat lymphatic filariasis. These results were also supported by pharmacokinetics and modeling studies [24].

5.2 Malaria

Malaria is originated by protozoan parasite of the genus, eukaryotic microbe, *Plasmodium* associated to kingdom Protista and phylum *Apicomplexa.* Global malaria cases declined from 2000 (238 million cases) and deaths have steadily declined between 2000 and 2019, falling from 736,000 in year 2000 to 619,000 deaths in 2021. The proportion of total death due to malaria among children under the age of 5 years has decreased from 84% in 2000 to 77% in 2021.

Malaria deaths in the WHO African Region by 44% between 2000 and 2019, and the mortality rate of declined by 67% during the same time duration. India contributed the most significant absolute reductions in the WHO south-East Asia Region, from approximately 20 million cases in 2000 to approximately 5.6 million in 2019 [25]. Clinical manifestations of malaria are largely caused by the parasite's exponential growth in blood. Parasites invade, multiply within and evade erythrocytes to spread. There is currently no drug regimen available that can target the parasite's multiple intra-erythrocytic proliferative stages. In complex using a peptide mimic limits merozoite penetration of RBCs, suggesting a unique method for inhibiting malaria parasites' blood-stage development.

5.2.1 World Malaria Report 2021

According to new statistics from the WHO's current World malaria report 2021, Malaria services have been disrupted due to the COVID-19 pandemic, resulting in an increase in cases and deaths. The pandemic occurred at a time when global malaria development had reached a stalemate [25]. In 2020, there were an approximate 627,000 malaria fatalities and around 241 million malaria cases were recorded around the world. There were around 14 million more cases and 69,000 more deaths were recorded as compared to 2019. During the pandemic, malaria prevention, treatment and diagnosis were hamper this led to a two-third excess deaths (47,000). In 2020,

Sub-Saharan Africa continued having a malaria burden, roughly 95% cases and 96% deaths reported; 80% deaths in the region are the children.

Case incidence: The global malaria case incidence rate in 2020 was 59 per 1000 people at risk, compared to a target of 35 putting it, putting it 40% behind schedule.

The global mortality rate: 15.3 fatalities per 100,000 people at risk in 2020 compared to an 8.9 target, putting it 44% behind schedule.

5.2.2 Malaria in India: Statistics and Facts

Despite the fact that India has been independent for over 70 years and has made great progress in a variety of areas, however country's public health system continues to lag behind. Malaria has been a serious danger to India's public health since independence, with more than 70 million incident cases and millions of deaths annually owing to the disease National Vector Borne Disease Control Programme (NVBDCP). There is no well-developed malaria surveillance system in India. India is alone responsible for 4% of the global burden of this deadly disease. According to World malaria report (2018), India reported 9.5 million malaria cases in 2017, down to 24% from 2016. As a result of this achievement, India is no longer among the top three countries with the greatest malaria burden. Despite this, more than 90% of the population remains at high risk of developing the disease. According to a recent study published in the Lancet, India ranks fourth in the world in terms of the *total* global malaria cases. Malaria infections in India have reduced nearly half, according to current data from NVBDCP, with 0.42 million cases and 96 deaths in 2018, contrast to 0.84 million cases and 194 deaths in 2017. Nearly half of the patients tested positive for *P. falciparum*. Odisha, India's *worst* affected state is responsible for 40% of all the malaria cases (NVBDCP). The National Framework for Malaria Elimination (NFME) 2016-2030 was established in 2015 by India's Prime Minister, who joined an initiative by Asia Pacific countries to eradicate malaria by 2030. In order to attain the above-mentioned aim, the country devised a 5-year roadmap dubbed as the National Strategic Plan (NSP) for Malaria Elimination (2017-2022). The government is on course to achieve its goal of establishing a "Malaria-Free India", thanks to this substantial up scaling of efforts [26, 27].

RTS,S/AS01 (RTS,S), is the first vaccination against a human parasite to be recommended by WHO, has proven to be an innovative tool for accelerating malaria eradication progress. In the Sub-Saharan Africa WHO recommended RTS,S for children's in October 2021, with Malawi becoming the first to get the malaria vaccination [28]. The promising vaccine candidate suppresses hepatocyte invasion at the pre-erythrocytic stage. Although the vaccine is effective against all forms of clinical malaria, it falls short of the Malaria Vaccine Technology Roadmap's goal of 75% efficacy for two years. R21, a more immunogenic CSP-based particle vaccine, is also being developed as a next-generation RTS,S-like vaccine [28–30].

5.2.3 Providing Malaria Treatment Despite All Odds

Despite the challenge posed by COVID-19, by the end of 2020, approximately three quarter (72%) of insecticide treated mosquito nets were delivered. In 2020, during high-transmission rainy season, thirteen nations in Africa's sub-region get 11.8 million additional children with preventative anti-malarial drugs than they did in 2019. During the pandemic, some nations, particularly those with low malaria burdens and strong health systems, even saw gains against malaria. In 2021, WHO declared China and El Salvador as malaria-free, the Islamia Republic of Iran achieved three years of zero indigenous cases in 2020. In the Greater Mekong sub-region, there were around 82,000 cases reported by the end of 2020, which is lower than the cases reported in 20210 (65,000 cases) and 2019 (100,000 cases).

In spite of these gains, malaria mortality of the WHO African Region increased in 2020 by 12% compared to the 2019, showing the impact of even minor service delays in a malaria-prone area. In April–June 2020, 15 malaria-endemic countries reported a 20% decline in malaria testing in comparison to the data of same period in 2019. In 2020, compared to the previous year, hardly 48 million treatment courses were issued by national malaria programs. Only India, out of the world's 11 countries with the maximum malaria cases, made headway against the disease. The ten other African countries all reported an increase in cases and number of deaths.

5.2.4 Meeting the Global Goal

Based on the report, development towards the WHO global malaria strategy's 2020 goals is severely behind track. New tactics, technology, and better execution of existing ones will be required to meet the WHO malaria strategy's 2030 targets, which include a 90% decline in global malaria occurrence and death rates by 2030.

5.2.5 Elimination of Malaria

Between the years 2000 and 2020, twenty-three countries had no indigenous malaria cases for three years in a row. Turkmenistan (2010),

Kyrgyzstan (2016), United Arab Emirates (2007), Argentina (2019), Paraguay (2018), Morocco (2010), Sri Lanka (2016), Uzbekistan (2018), Algeria (2019), China (2021) and Armenia (2011) are among the countries that have been declared malaria-free by the WHO. El Salvador was the first Central American country to be certified malaria-free by the WHO. In more than 30 years, China became the first country that declared malaria-free in the WHO western pacific region. Azerbaijan and Tajikistan have formally requested that WHO certify them as malaria-free, and the procedure is underway. WHO started the E-2025 project in April 2021 to support 25 countries that have the capacity to eliminate malaria transmission by 2025 [25].

5.2.6 Status of Anti-Malarial Drugs

Since their discovery, antibiotics have saved the lives of countless people. Their influence on human health and well-being is immeasurable. Today, their function is so ingrained in the healthcare system that it's assumed to be an integral part of it. For many people, Alexander Fleming's (Nobel Prize winner) is a reminder of the humble beginnings of antibiotics and the groundbreaking work he did in his research laboratory.

At present, we have a vast arsenal of antibiotics with us, ready for use against the entire spectrum of microorganisms. The very necessity of generating newer and newer classes of antibiotics is the drug resistance. Antibiotics can be countered by rapidly proliferating organisms or bacteria with error-prone DNA replication, which can create a mutation in the population that defeats the drug's activity. All microbes share this characteristic, which was initially discovered in bacteria.

Peru, a South American country, was the first country to use anti-malarial drugs [31]. Therapeutic bark extract from Cinchona, a medicinal plant, has been reported to have been in use for medicinal purposes in the Americas prior to European colonization, which introduced malaria. The medicinal properties of this plant, later on, were discovered for malaria as well. Influencing South American socio-economic conditions, anti-malarial activity paved the way for European colonialism, which in turn brought the disease, which remains endemic to the continent today.

By 1820, an alkaloid, Quinine was identified as the active component of the bark extract that has anti-plasmodium activity [32]. To prevent the crystallization of hemozoin, quinine accumulates in the food vacuoles of parasites including *P. vivax*, *P. malariae* and *P. falciparum* [33]. Despite its modest gametocidal activity, quinine is mostly blood schizonticidal [34]. In the beginning, attempts to synthesize quinine were ineffective.

Chloroquine, as we know it now, was first synthesized by German chemists in the early twentieth century. Chloroquine was as effective as quinine, and a major success in the fight against malaria. Chloroquine is delivered *via* a membrane protein expressed by *P. falciparum* [35]. However, after the emergence of drug resistance in the population, Chloroquine could no longer be used. Soon after, Chloroquine derivatives were synthesized. One of them being Hydroxychloroquine, a chloroquine derivative was more tolerated than chloroquine and used as therapy for the early stages of COVID-19 infection. The quinoline nucleus, a heterocyclic aromatic ring, serves as a structural scaffold for the generation of novel anti-malarial classes. Other anti-malarials are detailed in the following paragraphs.

Proguanil is anti-malarial and also known as Dihydrofolate Reductase (DHFR) inhibitor [36]. DNA replication necessitates the enzyme DHFR, which recycles Tetrahydrate Folate (THF). Anti-DHFR inhibitors hinder the parasite from recycling THF, thereby killing it. Mefloquine has a longer half-life and is based on quinine. When chloroquine resistance was discovered, Mefloquine was utilized. However, resistant strains with increased copy number of the gene, *Pfmdr1* have been observed in the population.

The WHO has included Primaquine in its "list of essential medications", since its invention in 1946 [37]. Inhibition of DNA replication by Primaquine is one of its effects. This medicine is used to treat recurrent malaria by targeting parasite hypnozoites in the liver. Doxycycline disrupts the intra-erythrocytic stages of *Plasmodium* [38]. Prophylactically, it is useful to treat *Plasmodium* infections because of its delayed effect.

In contrast to the current synthetic anti-malarials, Artemisinin has its origins in China [39]. IC_{50} levels of this noteworthy medicine have contributed to the success in combating drug resistance. Artimisinin, a compound produced from the plant *Artemisia annua*, has been shown to be effective against *Plasmodium* trophozoites. Numerous Artimisinin derivatives are currently available, including Dihydroartemesinin, methyl ether derivative, ethyl ether derivative and hemisuccinate derivative. Artemisinin is a pro-drug. Dihydroartemesinin is the active molecule formed by rapid hydrolysis of artemisinin [40]. Plasmodicidal activity is attributed to a unique endoperoxide bridge present in artemisinin.

Anti-malarials can be broadly divided into four categories: Quinine and its derivatives, Antifolates, Artemisinin and Antimicrobials. The iron present in *Plasmodium* is a key factor in the effectiveness of Quinine-based and Artemisinin derivatives.

The major boost in the development of anti-malarials was the discovery of Artemisinin. However, since then, the parasite has developed a number of point mutations that allow it to evade the medication regime. The wise use of pharmacological combinations is the way to go, and it is already being done in clinical settings. Monotherapy against malaria has been discontinued in *Plasmodium* endemic countries for the last several decades. Artemisinin-based Combination Therapies (ACT) is the new paradigm in healthcare settings. Anti-malarial combination therapy confers drug strategy's long-term viability hinges on the discovery of newer drug classes beyond the present four.

5.3 Life Cycle of Malaria

Malaria is caused by *Plasmodium*, a parasitic eukaryotic protozoan. *P. falciparum* is the most harmful and deadliest of the malaria-causing *Plasmodium* species, followed by *P. vivax* [25]. On the other hand, *P. knowlesi* has been linked to a potentially fatal human illness. Malaria vaccines with high efficiency are needed for successful disease eradication, but the parasite's complicated biology and life cycle make the development difficult. The life cycle of a parasite is an excellent example of how the parasite and host relationships develop. The parasite infects two hosts in succession: Insect hosts, female Anopheles mosquitos, where the parasite goes through a sexual stage requiring gamete fusion and recombination; and vertebrate hosts, where the parasite develops in the liver before reproducing asexually in erythrocytes. Briefly, infected mosquitos inject sporozoites into the human dermis during a bite in order to feed on blood. The sporozoites enter the blood via the epidermis and prepare to infect and colonize liver cells (hepatocytes) in this transverse phase [41]. *Plasmodium* pore-forming proteins such as Sporozoite Protein Essential for Cell Traversal1 (SPECT1) , cell traversal protein for Ookinetes and Sporozoites (CelTOS) and Perforin-Like Protein 1 (PLP1) may assist the parasite avoid the host immune response during this most vulnerable period of transverse infection [42, 43]. After infecting hepatocytes, the parasite develops into hepatic schizonts, which contain thousands of merozoites. In the hepatocyte stage, some sporozoite surface proteins, such as Circumsporozoite Protein (CSP), Thrombospondin-Related Anonymous Protein (TRAP) and 6-cysteine protein, P36, assist parasite penetration [44–46]. RBCs are invaded when hepatic schizonts burst, releasing rapidly replicating merozoites into the bloodstream. Erythrocyte-Binding Like (EBL) and Reticulocyte-Binding Ligand (RBL) are the two protein families that function as parasite arsenals, assisting RBC invasion and immunological responses [47, 48]. The intra-erythrocytic stage, which lasts for 48 hours and is responsible for all sickness symptoms, involves RBC invasion, proliferation and re-infection [49]. During this intracellular stage, the parasite is protected from the host's immune reaction. The length of the blood stage and the number of merozoites produced by *Plasmodium* spp. varies. At this stage, certain parasites divide into gametocytes, which are then traversed to mosquitos during blood feeding on infected hosts [50, 51]. Male and female gametes develop from gametocytes, which fertilize in the mosquito midgut to produce the ookinete, a motile stage. The parasite then differentiates into sporozoites, which then transmit to vertebrates via blood. The life cycle of *P. falciparum* is schematically explained in Figure 5.1.

Developmental Stages Inside Mosquito

During their blood meal on an infected human carrier, young female Anopheles mosquitoes ingest the malaria parasite. In the mosquito's lumen, a mature gametocyte is converted into a gamete. The occurrence of gametogenesis is marked by the process of exflagellation, which involves the discharge of eight thread-like gametes from male gametocytes. The mobilization of Ca^{2+} and an increase in cGMP levels have been linked to the induction of exflagellation [52]. The motile male gamete combines with the female gamete after exflagellation, result is diploid zygotes. In the mosquito's gut, the zygote undergoes morphological changes and develops into a motile ookinete. By digesting chitin located in the peritrophic membrane

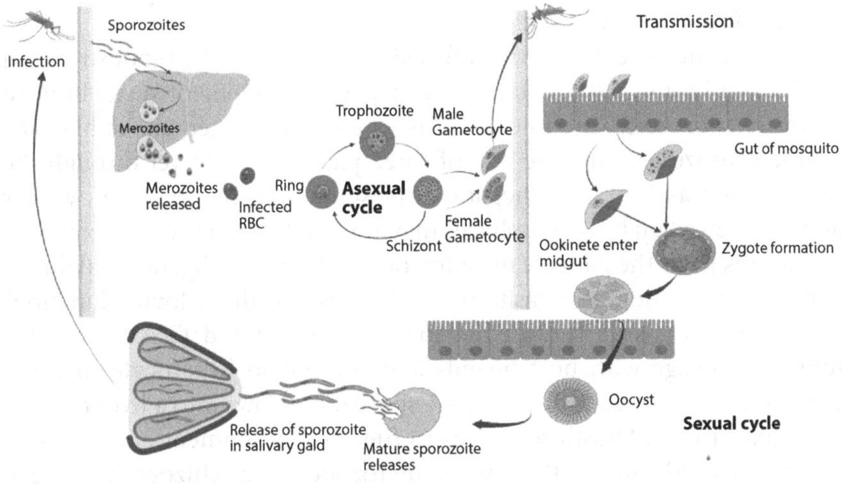

Figure 5.1 Malaria parasite's lifecycle in human and mosquito.

of the mosquito's midgut wall, the ookinete penetrates the midgut epithelium and develops an oocyst [53–55]. When an oocyst reaches maturity, it bursts, releasing sporozoites that go to salivary glands, where they can infect a new human host.

Sporozoite

Banana-shaped cell with three membranes: an exterior sporozoite membrane, two inner sporozoite membranes, and an outer sporozoite membrane [56]. The nature and appearance of sporozoites free from mature oocysts and those released from the salivary gland are strikingly different. The sporozoites produced from the oocyst are not infectious to mammals and do not have gliding motility, but the sporozoites released from the salivary gland are infected and have gliding motility [57, 58]. More than 30 genes are selectively elevated in salivary gland sporozoites, and these genes are referred to as UIS (Up-regulated in Infective Sporozoites) [59, 60]. The genes uis3 and uis4, which are only expressed in infective sporozoites, have been shown to be essential for sporozoite liver stage development [61].

Developmental Stages Inside the Human Host

P. falciparum uses humans as intermediate hosts, with two separate developmental phases, one in hepatocytes and the other in erythrocytes. While the liver stage is thought to be latent, the blood stage is the main reason for all the clinical symptoms.

Pre-Erythrocytic Cycle

The sporozoites injected by the infected mosquito into the bloodstream of the human host signal the start of the hepatic development stage. In most cases, 25 sporozoites are injected. After being discharged at the bite site, motile sporozoites can take one of three paths: some travel through the bloodstream and infiltrate hepatic cells, some go through lymph vessels and are chewed up by macrophages in lymph nodes, and the rest remain in the dermis [62]. The parasite must interact with the host ligands and finally construct a junction of parasite-host cell known as the "Moving junction" in order to successfully invade the host liver cell. CS and TRAP have been found to engage with host ligands and commit to hepatocyte invasion [59]. Following invasion, an infective sporozoite grows into a liver trophozoite that uses the host's resources to fuel its development and division, eventually producing 20,000-30,000 merozoites by a schizogony process. These merozoites are protected from immunological attack by phagocytic *Kupffer cells* by a host-derived membrane called a merosome.

Erythrocytic Cycle

Merozoites stick to and enter erythrocytes within 60 seconds of being discharged into the bloodstream, replicating and producing infectious merozoites [63]. The pathogenesis of this disease is caused by the parasite's asexual division in erythrocytes and the release of merozoites when the erythrocyte lyses.

Ring Stage

Merozoites grow into a tiny discoidal, cup-shaped ring with a thick cytoplasmic rim after entering an erythrocyte. The nucleus, mitochondrion, plastid, most ribosomes and ER are all found on this rim, while the smooth membranes are more abundant in the thinner center. The basophilic ring-like appearance in Giemsa-stained preparation reflects the thickened perimeter of the disc and the peripheral position of the RNA content. The nucleus appears in Giemsa preparations in a variety of shapes, ranging from an elongated body bent along the parasite's perimeter to a more spherical appearance, and occasionally two nuclei.

To obtain nutrients, the parasite feeds on hemoglobin. FV are small vacuoles with a diameter of 150–300 nm that catabolize hemoglobin to release a toxic substance called heme, which is then converted into a less toxic substance called hemozoin. The non-toxic hemozoin is kept in FVs all across the cytoplasm. All of the hemozoin content accumulates toward the conclusion of the ring stage, forming a single massive pigmented vacuole.

Gruring *et al.* showed that ring stage parasites move freely inside the host cell and undergo dynamic changes between amoeboid and disc-shaped forms [64]. This dynamic shift between the two forms is mediated by *F-actin*. The parasite's ring stage lasts for 24 to 28 hours after invading an erythrocyte and then transforms into the trophozoite stage.

Trophozoite Stage

the parasite evolves from the ring stage to the more complicated trophozoite stage, it takes on an irregular shape with bulges and deep tubular invaginations at its surface [65, 66]. It is the metabolically most active stage, as evident by extensive euchromatin in the nucleus, as well as rise in the amount of ribosomes and enlargement of the ER, both of which indicate higher cellular metabolic rate [67, 68]. At this stage, the parasite has a big pigmented vacuole with FVs and hemozoin in it. The parasite exports membranes and other structures to the surrounding erythrocytes as it grows and differentiates, such as the Maurer clefts, circular clefts, and dense protein-containing projections from the erythrocyte surface known as Knobs. These changes to the erythrocyte surface aid in trophozoite sequestration by endothelial cells lining blood vessel capillaries in the brain and placenta, which is the reason for the clinical demonstration of cerebral and placental malaria. 32-36 hours after invasion, the trophozoite enters the schizont stage.

Schizont Stage

The multi-nucleated schizont stage, which lasts for 8-10 hours, is the final stage of Plasmodium's asexual division. The nucleus, apicoplast, golgi complex and mitochondrion, multiply during this phase, the parasite prepared and assembles portion in preparation for the upcoming erythrocyte infection cycle. The nucleus divides four times during mitosis, yielding 16 nuclei per schizont; the number of nuclei can range from 8 to 32 within a single schizont. The erythrocyte membrane and cytoskeleton become deformed as the parasite prepares for the next cycle of infection, and the surface becomes rough and angular in shape. As the number of knobs on the surface grows, schizonts are more likely to congregate in blood capillaries. A mature, ready-to-burst schizont is made up of 2 membranes, one is an inner Parasitophorous Vacuole Membrane (PVM) and another is an outer erythrocyte membrane that encloses the residual body and merozoites with pigment vacuole.

Merozoite Stage

In terms of size and organelle content, merozoite is the simplest form of *asexual* developmental stage. Its length and width range from 1.3–1.6 μm and 0.9–1 μm, respectively. They're ellipsoidal, with one pole having a short flat-ended apical protrusion. The surface of the merozoite is coated with a 15 nm thick layer made up of protruding tufts of tiny filaments. A pellicle made up of three membranes, sits beneath this covering. It also has a single, elongated mitochondrion that lies longitudinally beneath the pellicle, and an apicoplast that is wider and shorter with an irregular form. Aside from these organelles, the nucleus of merozoites is irregular and variable in shape, with no signs of a nucleolus. Schematic representation of *P. falciparum* merozoite is shown in Figure 5.1.

A set of secretory vesicles: micronemes, rhoptries, and dense granules, are found near the merozoite's apical end, which is a distinctive trait of the apicomplexan family. Proteins released by these organelles are required for merozoite adhesion to host erythrocytes to be committed.

Merozoite Apical Secretory Organelles

P. falciparum belongs to the apicomplexan family of parasites. Members of this family are distinguished by a polarized apical end that houses a variety of unique organelles. These organelles are essential for invading and surviving within the host cells.

Rhoptries are membrane-bound, bulbous vesicles that are around 400 nm long and 200 nm broad [69]. These organelles, which are only two in number, house a slew of different proteins. In the basal bulb some proteins are found., while others are found in the narrow duct or neck region, example, Rhoptry Associated Protein (RAP1/2) complex is found in the bulb, whereas p225 protein and RHOP1-3 complex are found in the duct [70, 71]. It contributes to creation of the Parasitophorous Vacuole (PV) during invasion, in addition to its role in host cell invasion [72, 73].

Micronemes are the smallest apicomplexan secretory organelles, clustering towards the merozoite's apical end. They are significantly tiny and more in numbers than rhoptries (up to 40 per merozoite) and connect around the terminals of rhoptry ducts that is near to their junction with the parasite's apical membrane. They feature a bulbous base that is roughly 60 nm broad and 100 nm long in *P. falciparum*.

These are thick, spherical vesicles that lie freely inside the cytoplasm and are dispersed *throughout* the apical region of the merozoite. These constructions' content, which includes invasion related molecules through the pellicle into PV are exocytosed [74]. Ring Membrance Antigen (RIMA), and a Subtilisin-like protease (SUB2) are some of the proteins found in dense granules [75–77].

5.4 Drug Against Hepatic Malarial Stage

As liver stage *Plasmodium* is asymptomatic but essential for replication, differentiation and for further infection of sporozoite. Liver stage can be a good target against malarial parasite as a large population of *Plasmodium* parasite go for schizogony but some of them become dormant for days and years, they wait for the right favorable time to break their dormancy [78]. These dormant stages are known as Hypnozoite and they believe to turn as secondary schizogony to cause erythrocytic infection. Some research clarifies about some specific proteins in liver stage parasite. These proteins are specific to liver stage. Targeting liver stage parasite can also reduce resistivity issue in treating malaria. Due to low load of parasite in liver, drug will take less time to inhibit its action and due to short exposure it would be difficult for parasite to get resistance against these drugs. There are some whole sporozoite vaccines in developing process to target against surface antigen of sporozoite. These vaccine include RT,S and CSP(circum sporozoite protein). RT means repeat T cell epitope and "S" is B cell Surface protein [79]. These vaccine development are taking a bit more time to get accuracy point so some of the drug are allow to go for repurposing against *Plasmodium*. There are many drugs that are repurposed against malaria because of drug resistivity issue with malarial parasite. Also this repurposing of drug cut down the cytotoxicity problem. Ivermectin is one such drug ready to use as anti-malarial drug targeting liver stage of *Plasmodium* parasite.

5.5 About Ivermectin

Ivermectin is an antiparasitic drug also used as anti-insecticidal drug included in avermectin family. This family is macrocyclyic lactone based compound. It is first well established endectocide. It is a semisynthetic drug commonly used to treat some major tropic disease like loidiasis, lymphatic filariasis, and onchocerciasis (Figure 5.2). Being a lipophilic molecule, it is soluble throughout the body. It is mainly seen to have function within dermal region and adipose tissue while it shows only a small action in blood plasma. Due, to this reason Ivermectin is considered as a good vector control. It can inhibit sporozoite and block hypnozoite, so it is a good insecticide. The role of Ivermectin in different stages of the malaria parasite is given in Table 5.1. In parasite, Ivermectin target against microfilariae. Ivermectin is employed as an anti-malarial tool by Mass Drug

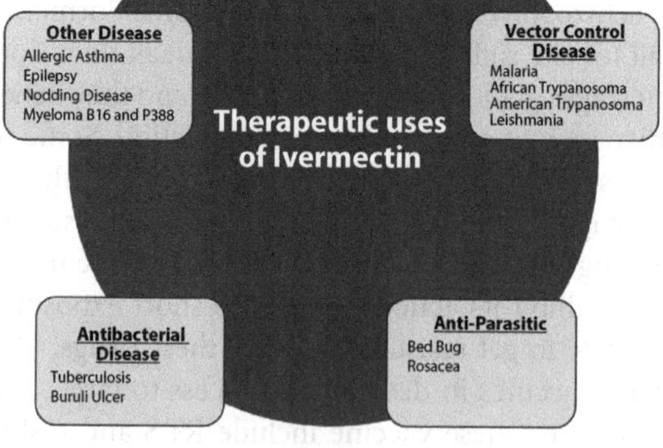

Figure 5.2 Therapeutic uses of ivermectin [5].

Table 5.1 Role of Ivermectin in different stages of Malaria.

Stages	Sexual stage of mosquito	Hepatic stage	Erythrocytic stage
Role of Ivermectin	Inhibit sporogony and lower down Ookinetes production [81].	Inhibit schizogony and hypnozoite by disrupting somatic integrity of Parasite [81].	Block nucleo-plasmid shuttling mechanism by inhibiting importin [82].

Administration (MDA) at malaria endemic region. Ivermectin was also recommended against SARS-Cov2. The infection of COVID seems to be drastically decrease by intake of Ivermectin.

5.5.1 Ivermectin Against Liver Cells of Malaria Parasite

Three drugs namely emamectin, eprinomectin, and Ivermectin of avermectin family were tested against *Plasmodium* out of which Ivermectin was found best effective drug. *In-vitro* and *in-vivo* experiments were performed to check Ivermectin effect on hepatoma cells. Through *in vitro* process where flow-cytometer was performed and using Green-Fluorescent protein intensity it was observed that Ivermectin does not disrupt parasitophorous vacuole membrane (pvm) neither affects it's localization but it harms the schizogony and somatic integrity of parasite. Through *in-vivo* assay where control and different concentration of Ivermectin treated mice were analyzed by immunofluorescence microscopy at 46hpi we can conclude that Ivermectin debilitate the development and differentiation of parasite in hepatic cell which lead to low erythrocytic infection [80]. Some recent studies are going on where Ivermectin is used as a combination drug to reduce malarial transmission.

5.6 Designing and Synthesis of Ivermectin Inhibitors

Soren *et al.* from Kitasato Institute, Japan have grown unique microorganisms. They isolate new compounds that give direction to the discovery of the antimicrobial pyrindicin, proteasome inhibitor lactacystin and the protein kinase-C inhibitor staurosporin [83]. In 1973s, Kitasato institute cultivated extraordinary microorganisms and extracted best compounds handover to MSD research laboratory *in vivo* test. The broth of soil bacteria

Streptomyces avermectinus, product was isolated and tested in mice having nematode *Nematospiroides dubius.* The product showed marvellous anathelmintic activity, it shows no toxic effect on mice. The active principle discovered to be eight avermectine (Figure 5.3) [84].

In all avermectin (1-8), B-series compounds having 5-hydroxy group were found to be most active among all. A mixture of 80% hydrogenated avermectin B_{1a} and 20% B_{1b} was given the name Ivermectin (9), actually it is a semi-synthetic derivative of Avermectin (Figure 5.3). Ivermectin is a potent endectocide and a fragmentation product of *Streptomyces avermitilis* [85]. Ivermectin shows broad-spectrum of activity such as lymohatic filariasis, myiasis, gnathostomiasis, pediculosis, and stongyloidiasis [4, 86, 87].

Because of the insecticidal effect of Ivermectin on the mosquito vector having *Plasmodium* parasites and at sublethal concentration *P. falciparum* in Anopheles gambiae inhibit sporogony, due to that Ivermectin have a significant impact as a malarial drug [17]. Ivermectin showed *in vitro* inhibition at blood stage of *P. falciparum* infection and *in vivo* inhibition of *Plasmodium berghei.* In *Plasmodium* hepatic infection it also shows both *in vitro* and *in vivo* inhibition [80, 82]. In comparison to Primaquine, In *in vitro*, infected human heptoma cell, Ivermectin reduces it's infection [80]. Ivermectin has emerged as a possible treatment for multistage malaria control due to its combined effects against *Plasmodium* sporogonic, erythrocytic and hepatic stages.

5.6.1 Hybrids of Ivermectin

Lovepreet *et al.* first time reported Ivermectin based anti-malarial hybrids, that shown dual action on the erythrocytic and heptic stages of

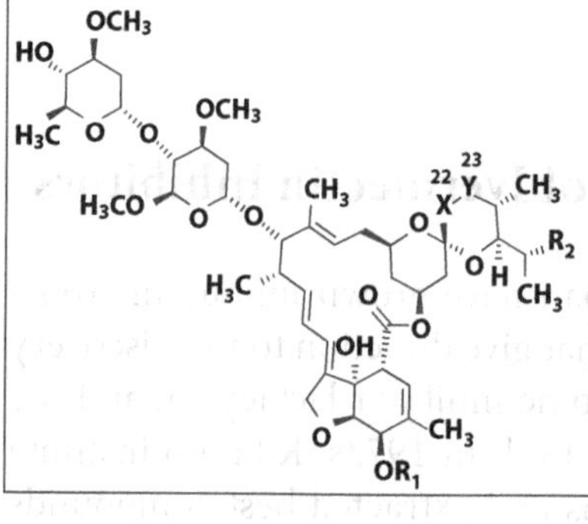

1. Avermectin A_{1a} $R_1 = CH_3$, $R_2 = C_2H_5$, X-Y = CH=CH
2. Avermectin A_{1b} $R_1 = CH_3$, $R_2 = CH_3$, X-Y = CH=CH
3. Avermectin A_{2a} $R_1 = CH_3$, $R_2 = C_2H_5$, X-Y = $CH_2CH(\alpha OH)$
4. Avermectin A_{2b} $R_1 = CH_3$, $R_2 = CH_3$, X-Y = $CH_2CH(\alpha OH)$
5. Avermectin B_{1a} $R_1 = H$, $R_2 = C_2H_5$, X-Y = CH=CH
6. Avermectin B_{1b} $R_1 = H$, $R_2 = CH_3$, X-Y = CH=CH
7. Avermectin B_{2a} $R_1 = H$, $R_2 = C_2H_5$, X-Y = $CH_2CH(\alpha OH)$
8. Avermectin B_{2b} $R_1 = H$, $R_2 = CH_3$, X-Y = $CH_2CH(\alpha OH)$
9. Ivermectin $R_1 = H$, $R_2 = C_2H_5$, X-Y = CH_2CH_2(80%)
$R_1 = H$, $R_2 = CH_3$, X-Y = CH_2CH_2(20%)

Figure 5.3 Avermectin (A_{1a}, A_{1b}, A_{2a}, A_{2b}, B_{1a}, B_{1b}, B_{2a}, B_{2b}) and Ivermectin.

P. falciparum and *P. berghei* infection. Initially, they converted Ivermectin into aglycon 10, then further modification on C-13 resulted into Ivermectin hybrid **11, 12, 13** and **14** (Figure 5.4). *In vitro* activity on all these hybrids and Ivermectin aglycon were performed on the *P. berghei* infection of hepatic stage and all of them showed better activity then the parental Ivermectin. To the hepatic cells, none of them was found toxic. Compound (**11**) bearing a ferrocene group on C-13 of triazole linker in the macrolide frame work of Ivermectin, shown (IC_{50} = 0.990 ± 0.068 μM) at the hepatic stage. Compounds (**12, 13** and **14**) having substituted pyrimidine unit, in this hybrid (**12**) having 3-NO_2 phenyl group at C-4 aryl group with *iso*-propyl ester at the C-5, shown (IC_{50} = 0.911 ± 0.076 μM) at liver stage. Compound (**14**), having phenyl group at the C-4 aryl group and ethyl ester at C-5, showed (IC_{50} = 0.990 ± 0.050 μM) on liver stage. Among them compound (**13**) showed best activity (IC_{50} = 0.503 ± 0.050 μM). This is about 3-fold more active than its parental Ivermectin

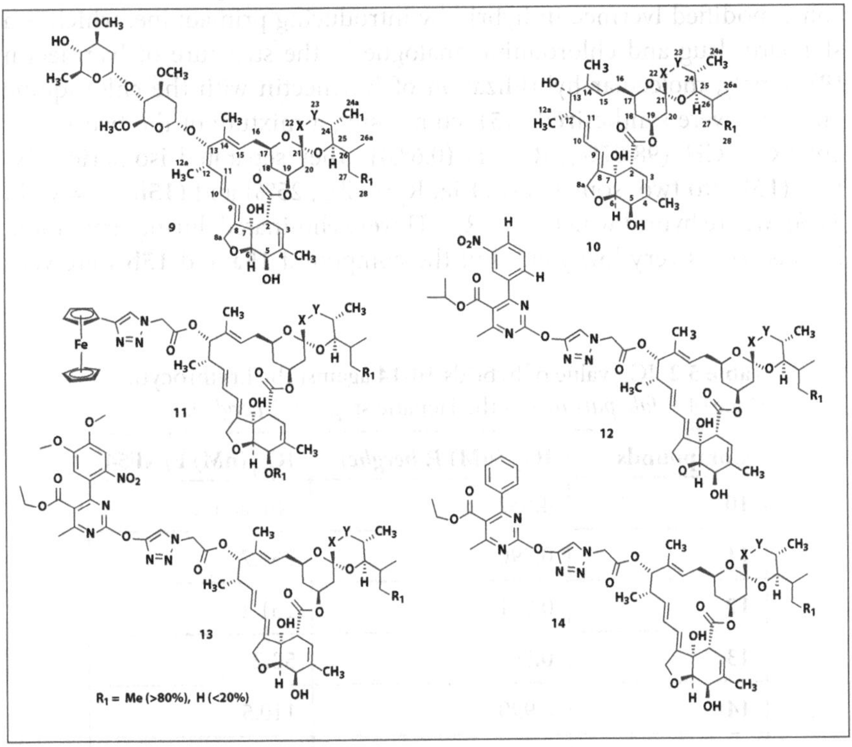

Figure 5.4 Ivermectin based anti-malarial hybrids.

(IC_{50} = 1.639 ± 0.1899 μM). Among all hybrids Ivermectin aglycon (**10**) presented a lowest activity.

Lovepreet *et al.* group also tested these compounds against the CQ^s *Pf* NF54 strain of *P. falciparium*, whereas they used chloroquine and Ivermectin as control (Table 5.2). All the compounds showed better activity than Ivermectin but were less active than chloroquine. Among all of them, compound (**12**) with 2-nitro-4,5-dimethoxy group on C-4 of pyrimide, in combination with ester which is on C-5 was displayed tremendous inhibition in both blood and liver stage of infection that is about 10-fold and 3-fold better than Ivermectin. This compound also displayed low level of cytotoxicity. Ivermectin aglycon (**10**) did not shown activity in antiplasmodial blood stage [88].

After getting good results of Ivermectin hybrids by appending the Ivermectin with orgamotallic and heterocyclic anti-malarial pharmacophores, even 3–10-fold better activity then parental Ivermectin against both liver-stage and blood stage infection by *Plasmodium* parasites was observed.

In another study Lovepreet *et al.* (2022) have design and synthesized some modified Ivermectin hybrid by introducing primaquine, which is a standard drug and chloroquine analogue in the structure of Ivermectin (**9**). Firstly, molecular hybridization of Ivermectin with the chloroquine analogue were synthesized (**15**), comprising a mixture of the two inherent [R_1 = CH_3 (98.87%), R_1 = H (0.6%)]. They separated isomeric mixture (**15**) into two isomers *i.e.* (**15a**, R_1 = CH_3, 25%) and (**15b**, R_1 = CH_3, 71%), where hybrid which have R_1 = H were eliminated during separation, because of its very low yield. But the compound **15a** and **15b** were very

Table 5.2 IC_{50} value of hybrids **10-14** against the Erythrocytic stage of *P. falciparum* and the Hepatic stage of *P. berghei.*

Compounds	**IC_{50} (μM) *P. berghei***	**IC_{50} (nM) *Pf*NF54**
10	4.557	Not active
11	0.990	256.7
12	0.911	161.2
13	**0.503**	**50.2**
14	0.990	110.5
Ivermectin	1.639	519.6

stable and pure. They also synthesized another hybrid by appending primaquine in Ivermectin, they get isomeric compound 16 and they successfully separated it in **16a** (R_1 = CH_3, 51.44%), **16b** (R_1 = CH_3, 48.56%) and isomer which have R_1 = H were eliminated during separation. **16a** and **16b** were stable and pure (Figure 5.5).

Ivermectin hybrid **15, 15a, 15b, 16, 16a** and **16b** were tested for in *vitro* anti-malarial activity against hepatic stage of *P. berghei* infection where Ivermectin (**9**), Ivermectin aglycon (**10**), primaquine and chloroquine used as a control. In all of them Ivermectin hybrid 15 (IC_{50} = 0.317 μM), **15a** (IC_{50} = 0.274 μM) and **15b** (IC_{50} = 0.186 μM) shows the best result, in comparison to other hybrids. Even **15b** shows about threefold more activity than Ivermectin hybrid (**13**). *In vitro* activity against *P. falciparium erythrocytic* infection were also performed, where again **15a** (IC_{50} = 48.2 nM) and 15b (IC_{50} = 64.8 nM) shows the best activity among all other hybrids, even better than the Ivermectin (7-fold higher). However, compound (10) displayed the lowest activity. These results showed that individual isomer shows a better result than its isomer against both the blood

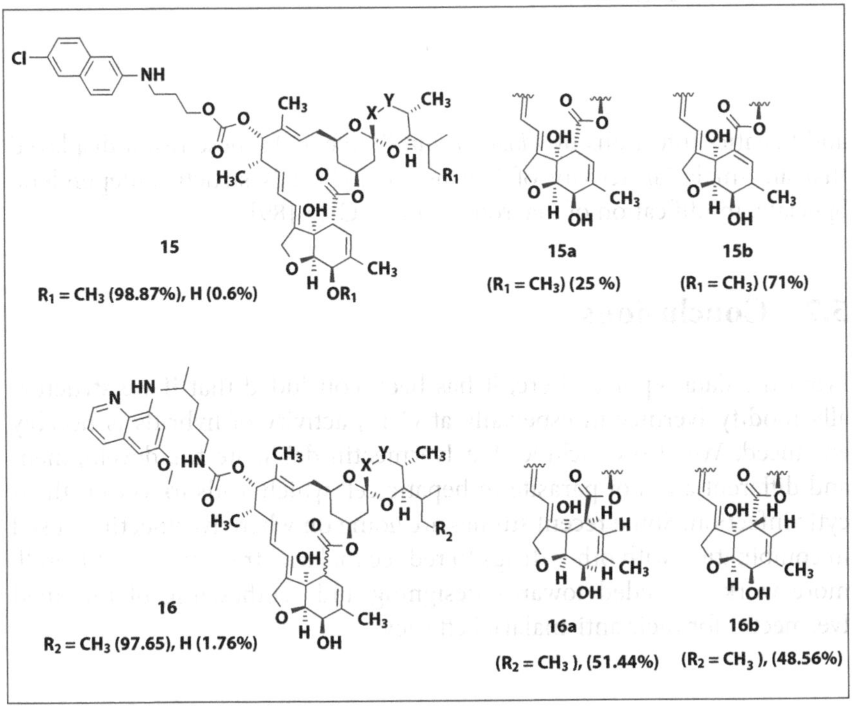

Figure 5.5 Ivermectin based anti-malarial hybrids.

Table 5.3 IC_{50} value of **10, 11, 12, 13** and **14** against the Hepatic stage of *P. berghei* and the Erythrocytic stage of *P. falciparum*.

Compounds	**IC_{50} (μM) *P. berghei***	**IC_{50} (nM) *Pf*NF54**
15	0.317	74.3
15a	0.274	48.2
15b	0.317	64.8
16	1.360	ND
16a	1.291	ND
16b	2.057	ND
Ivermectin	1.321	359.6
Aglycon	6.375	ND
Primaquine	8.428	ND
Chloroquine	ND	23.7

ND: not determined

and hepatic- infections by *Plasmodium* (Table 5.3). These result displayed that anti-malarial activity of Ivermectin hybrid is structure-dependent. Specially modification of macrolide unit at C-13 [89].

5.7 Conclusions

From the data reported here, it has been concluded that if we structurally modify Ivermectin especially at C-13, activity of hybrids is notably enhanced. We also conclude that Ivermectin debilitate the development and differentiation of parasite in hepatic cell which leads to low erythrocytic infection. Some recent studies are going on where Ivermectin is used in combination with other drugs to reduce malarial transmission. Overall, more work is needed towards designing and synthesizing of modified Ivermectin for their anti-malarial efficacy.

5.8 Acknowledgments

Mohammad Abid and Shailja Singh gratefully acknowledge the financial support in the form of Core Research Grant from the Science and Engineering Research Board (SERB), Government of India (Project No. CRG/2018/003967). The Drug and Pharmaceuticals Research Programme (DPRP) (Project No. P/569/2016-1/TDT) for Shailja Singh is acknowledged. Iram Irfan acknowledges DST for postdoctoral fellowship under women scientist scheme (WOS-A) (File no: SR/WOS-A/CS-116/2018).

References

1. Omura, S. and Crump, A., Ivermectin: Panacea for resource-poor communities? *Trends Parasitol.*, 30, 445, 2014.
2. Campbell, W.C., Fisher, M.H., Stapley, E.O., Albers-Schonberg, G., Jacob, T.A., Ivermectin: A potent new antiparasitic agent. *Science*, 221, 823, 1983.
3. Zouré, H.G.M., Noma, M., Tekle, A.H., Amazigo, U.V., Diggle, P.J., Giorgi, E., Remme, J.H.F., The geographic distribution of onchocerciasis in the 20 participating countries of the African Programme for Onchocerciasis Control: (2) pre-control endemicity levels and estimated number infected. *Parasites Vectors*, 7, 1, 2014.
4. Crump, A. and Omura, S., Ivermectin, "wonder drug" from Japan: The human use perspective. *Proc. Jpn. Acad. Ser. B Phys. Biol. Sci.*, 87, 2, 13, 2011.
5. Crump, A., Ivermectin: Enigmatic multifaceted "wonder" drug continues to surprise and exceed expectations. *J. Antibiot. (Tokyo)*, 70, 495, 2017.
6. Currie, B.J. and McCarthy, J.S., Permethrin and Ivermectin for Scabies. *N. Engl. J. Med.*, 362, 717, 2010.
7. Chosidow, O., Giraudeau, B., Cottrell, J., Izri, A., Hofmann, R., Mann, S.G., Burgess, I., Oral ivermectin versus malathion lotion for difficult-to-treat head lice. *N. Engl. J. Med.*, 362, 896, 2010.
8. Chaccour, C.J., Kobylinski, K.C., Bassat, Q., Bousema, T., Drakeley, C., Alonso, P., Foy, B.D., Ivermectin to reduce malaria transmission: A research agenda for a promising new tool for elimination. *Malar. J.*, 12, 1, 2013.
9. Lu, X.M., Batugedara, G., Lee, M., Prudhomme, J., Bunnik, E.M., Le Roch, K.G., Nascent RNA sequencing reveals mechanisms of gene regulation in the human malaria parasite Plasmodium falciparum. *Nucleic Acids Res.*, 45, 7825, 2017.
10. Blasco, B., Leroy, D., Fidock, D.A., Antimalarial drug resistance: Linking Plasmodium falciparum parasite biology to the clinic. *Nat. Med.*, 23, 917, 2017.

11. Kemirembe, K., Cabrera, M., Cui, L., Interactions between tafenoquine and artemisinin-combination therapy partner drug in asexual and sexual stage Plasmodium falciparum. *Int. J. Parasitol. Drugs Drug Resist.*, 7, 131, 2017.
12. Alout, H., Krajacich, B.J., Meyers, J.I., Grubaugh, N.D., Brackney, D.E., Kobylinski, K.C., Diclaro, J.W., Bolay, F.K., Fakoli, L.S., Diabaté, A., Dabiré, R.K., Bougma, R.W., Foy, B.D., Evaluation of ivermectin mass drug administration for malaria transmission control across different West African environments. *Malar. J.*, 13, 2014.
13. Kobylinski, K.C., Alout, H., Foy, B.D., Clements, A., Adisakwattana, P., Swierczewski, B.E., Richardson, J.H., Rationale for the coadministration of albendazole and ivermectin to humans for malaria parasite transmission control. *Am. J. Trop. Med. Hyg.*, 91, 655, 2014.
14. Kobylinski, K.C., Sylla, M., Chapman, P.L., Sarr, M.D., Foy, B.D., Ivermectin mass drug administration to humans disrupts malaria parasite transmission in Senegalese villages. *Am. J. Trop. Med. Hyg.*, 85, 3, 2011.
15. Pinilla, Y.T., Lopes, C.P., S., S., Andrade, F.S., Melo, G.C., Orfanó, A.S., Secundino, N.F.C., Guerra, M.G.V.B., Lacerda, M.V.G., Kobylinski, K.C., Escobedo-Vargas, K.S., López-Sifuentes, V.M., Stoops, C.A., Baldeviano, G.C., Tarning, J., Vasquez, G.M., Pimenta, P.F.P., Monteiro, W.M., Promising approach to reducing Malaria transmission by ivermectin: Sporontocidal effect against Plasmodium vivax in the South American vectors Anopheles aquasalis and Anopheles darlingi. *PLoS Negl. Trop. Dis.*, 12, 1, 2018.
16. Kobylinski, K.C., Ubalee, R., Ponlawat, A., Nitatsukprasert, C., Phasomkulsolsil, S., Wattanakul, T., Tarning, J., Na-Bangchang, K., McCardle, P.W., Davidson, S.A., Richardson, J.H., Ivermectin susceptibility and sporontocidal effect in Greater Mekong Subregion Anopheles. *Malar. J.*, 16, 1, 2017.
17. Kobylinski, K.C., Foy, B.D., Richardson, J.H., Ivermectin inhibits the sporogony of Plasmodium falciparum in Anopheles gambiae. *Malar. J.*, 11, 1, 2012.
18. Chaccour, C., Lines, J., Whitty, C.J.M., Effect of ivermectin on Anopheles gambiae mosquitoes fed on humans: The potential of oral insecticides in malaria control. *J. Infect. Dis.*, 202, 113, 2010.
19. Kobylinski, K.C., Deus, K.M., Butters, M.P., Hongyu, T., Gray, M., da Silva, I.M., Sylla, M., Foy, B.D., The effect of oral anthelmintics on the survivorship and re-feeding frequency of anthropophilic mosquito disease vectors. *Acta Trop.*, 116, 119, 2010.
20. Foley, D.H., Bryan, J.H., Lawrence, G.W., The potential of ivermectin to control the malaria vector Anopheles farauti. *Trans. R. Soc. Trop. Med. Hyg.*, 94, 625, 2000.
21. Sylla, M., Gray, M., Chapman, P.L., Sarr, M.D., Rasgon, J.L., Mass drug administration of ivermectin in south-eastern Senegal reduces the survivorship of wild-caught, blood fed malaria vectors. *Malar. J.*, 9, 1, 2010.
22. Killeen, G.F., Kiware, S.S., Okumu, F.O., Sinka, M.E., Moyes, C.L., Claire Massey, N., Gething, P.W., Marshall, J.M., Chaccour, C.J., Tusting, L.S., Going beyond personal protection against mosquito bites to eliminate malaria

transmission: Population suppression of malaria vectors that exploit both human and animal blood. *BMJ Glob. Health*, 2, 2, 2017.

23. Killeen, G.F., Tatarsky, A., Diabate, A., Chaccour, C.J., Marshall, J.M., Okumu, F.O., Brunner, S., Newby, G., Williams, Y.A., Malone, D., Tusting, L.S., Gosling, R.D., Developing an expanded vector control toolbox for malaria elimination. *BMJ Glob. Health*, 2, 1, 2017.
24. Smit, M.R., Ochomo, E.O., Aljayyoussi, G., Kwambai, T.K., Abong'o, B.O., Chen, T., Bousema, T., Slater, H.C., Waterhouse, D., Bayoh, N.M., Gimnig, J.E., Samuels, A.M., Desai, M.R., Phillips-Howard, P.A., Kariuki, S.K., Wang, D., Ward, S.A., ter Kuile, F.O., Safety and mosquitocidal efficacy of high-dose ivermectin when co-administered with dihydroartemisinin-piperaquine in Kenyan adults with uncomplicated malaria (IVERMAL): A randomised, double-blind, placebo-controlled trial. *Lancet Infect. Dis.*, 18, 615, 2018.
25. WHO World Malaria Report, 2021, https://www.mmv.org/newsroom/publications/world-malaria-report-2021.
26. The National Center for Vector Borne Diseases Control (NCVBDC), https://nvbdcp.gov. in/., 2021.
27. Feachem, R. G., Chen, I., Akbari, *et al.*, Malaria eradication within a generation: ambitious, achievable, and necessary. *The Lancet*, 394, 1056, 2019.
28. Van Den Berg, M., Ogutu, B., Sewankambo, N.K., Biller-Andorno, N., Tanner, M., RTS,S malaria vaccine pilot studies: Addressing the human realities in large-scale clinical trials. *Trials*, 20, 1, 2019.
29. Mahmoudi, S. and Keshavarz, H., Efficacy of phase 3 trial of RTS, S/AS01 malaria vaccine: The need for an alternative development plan. *Hum. Vaccines Immunother.*, 13, 2098, 2017.
30. Collins, K.A., Snaith, R., Cottingham, M.G., Gilbert, S.C., Hill, A.V.S., Enhancing protective immunity to malaria with a highly immunogenic virus-like particle vaccine. *Sci. Rep.*, 7, 1, 2017.
31. Griffing, S.M., Gamboa, D., Udhayakumar, V., The history of 20th century malaria control in Peru. *Malar. J.*, 12, 1, 1, 2013.
32. Svend Norn Edith Kruse, P.R.K.H.P., On the history of Cinchona bark in the treatment of Malaria. *Dan Med. Arb.*, 44, 9, 2016.
33. Slater, A.F.G., Chloroquine: Mechanism of drug action and resistance in plasmodium falciparum. *Pharmacol. Ther.*, 57, 203, 1993.
34. Kazim, M., Puri, S.K., Dutta, G.P., Comparative evaluation of blood schizontocidal activity of quinine and quinidine against drug resistant rodent malaria. *J. Commun. Dis.*, 23, 254, 1991.
35. Martin, R.E., Marchetti, R.V., Cowan, A.I., Howitt, S.M., Bröer, S., Kirk, K., Chloroquine transport via the malaria parasite's chloroquine resistance transporter. *Sci. (80-.)*, 325, 1680, 2009.
36. Plowe, C.V., Djimde, A., Bouare, M., Doumbo, O., Wellems, T.E., Pyrimethamine and proguanil resistance-conferring mutations in Plasmodium falciparum dihydrofolate reductase: Polymerase chain reaction methods for surveillance in Africa. *Am. J. Trop. Med. Hyg.*, 52, 565, 1995.

37. Vale, N., Moreira, R., Gomes, P., Primaquine revisited six decades after its discovery. *Eur. J. Med. Chem.*, 44, 3, 937, 2009.
38. Pang, L. W., Limsomwong, N., Boudreau, E. F., Singharaj, P., Doxycycline prophylaxis for falciparum malaria. *Lancet (London, England)*, 1, 8543, 1161, 1987.
39. Klayman, D.L., Qinghaosu (artemisinin): An antimalarial drug from China. *Sci. (80-.)*, 228, 1049, 1985.
40. White, N.J., Clinical pharmacokinetics and pharmacodynamics of artemisinin and derivatives. *Trans. R. Soc. Trop. Med. Hyg.*, 88, 41, 1994.
41. Ejigiri, I. and Sinnis, P., Plasmodium sporozoite-host interactions from the dermis to the hepatocyte. *Curr. Opin. Microbiol.*, 12, 4, 401, 2009.
42. Kumar, H. and Tolia, N.H., Getting in: The structural biology of malaria invasion. *PLoS Pathog.*, 15, 1, 2019.
43. Pirahmadi, S., Zakeri, S., Mehrizi, A., Djadid, N., Raz, A.A., Sani, J., Abbasi, R., Ghorbanzadeh, Z., Cell-traversal protein for ookinetes and sporozoites (CelTOS) formulated with potent TLR adjuvants induces high-affinity antibodies that inhibit Plasmodium falciparum infection in Anopheles stephensi. *Malar. J.*, 18, 146, 2019.
44. Dundas, K., Shears, M.J., Sun, Y., Hopp, C.S., Crosnier, C., Metcalf, T., Girling, G., Sinnis, P., Billker, O., Wright, G.J., Alpha-v–containing integrins are host receptors for the Plasmodium falciparum sporozoite surface protein, TRAP. *Proc. Natl. Acad. Sci. U. S. A.*, 115, 4477, 2018.
45. Frevert, U., Sinnis, P., Cerami, C., Shreffler, W., Takacs, B., Nussenzweig, V., Malaria circumsporozoite protein binds to heparan sulfate proteoglycans associated with the surface membrane of hepatocytes. *J. Exp. Med.*, 177, 1287, 1993.
46. Ishino, T., Chinzei, Y., Yuda, M., Two proteins with 6-cys motifs are required for malarial parasites to commit to infection of the hepatocyte. *Mol. Microbiol.*, 58, 5, 1264, 2005.
47. Gunalan, K., Gao, X., Yap, S.S.L., Huang, X., Preiser, P.R., The role of the reticulocyte-binding-like protein homologues of Plasmodium in erythrocyte sensing and invasion. *Cell. Microbiol.*, 15, 35, 2013.
48. Lopaticki, S., Maier, A.G., Thompson, J., Wilson, D.W., Tham, W.H., Triglia, T., Gout, A., Speed, T.P., Beeson, J.G., Healer, J., Cowman, A.F., Reticulocyte and erythrocyte binding-like proteins function cooperatively in invasion of human erythrocytes by malaria parasites. *Infect. Immun.*, 79, 1107, 2011.
49. Ashley, E.A. and White, N.J., The duration of Plasmodium falciparum infections. *Malar. J.*, 13, 500, 2014.
50. Baker, D.A., Malaria gametocytogenesis. *Mol. Biochem. Parasitol.*, 172, 2, 57, 2010.
51. Bennink, S., Kiesow, M.J., Pradel, G., The development of malaria parasites in the mosquito midgut. *Cell. Microbiol.*, 18, 905, 2016.
52. Kawamoto, F., Alejo-Blanco, R., Fleck, S. L., Kawamoto, Y., Sinden, R. E., Possible roles of Ca2+ and cGMP as mediators of the exflagellation of

Plasmodium berghei and Plasmodium falciparum. *Mol. Biochem. Parasitol.*, 42, 1, 101, 1990.
53. Huber, M., Cabib, E., Miller, L.H., Malaria parasite chitinase and penetration of the mosquito peritrophic membrane. *Proc. Natl. Acad. Sci. U. S. A.*, 88, 2807, 1991.
54. Shahabuddin, M. and Kaslow, D.C., Chitinase: A novel target for blocking parasite transmission? *Parasitol. Today*, 9, 252, 1993.
55. Shahabuddin, M., Toyoshima, T., Aikawa, M., Kaslow, D.C., Transmission-blocking activity of a chitinase inhibitor and activation of malarial parasite chitinase by mosquito protease. *Proc. Natl. Acad. Sci. U. S. A.*, 90, 4266, 1993.
56. Meszoely, C.A.M., Erbe, E.F., Beaudoin, L.M., Beaudoin, R.L., Freeze-fracture studies on the sporoblast and sporozoite development in the early oocyst. *Am. J. Trop. Med. Hyg.*, 41, 499, 1989.
57. Vanderberg, J.P., Development of infectivity by the Plasmodium berghei sporozoite. *J. Parasitol.*, 61, 43, 1975.
58. Vanderberg, J.P., Studies on the motility of Plasmodium sporozoites. *J. Protozool.*, 21, 1550, 1974.
59. Kaiser, K., Matuschewski, K., Camargo, N., Ross, J., Kappe, S.H.I., Differential transcriptome profiling identifies Plasmodium genes encoding pre-erythrocytic stage-specific proteins. *Mol. Microbiol.*, 51, 1221, 2004.
60. Matuschewski, K., Ross, J., Brown, S.M., Kaiser, K., Nussenzweig, V., Kappe, S.H.I., Infectivity-associated changes in the transcriptional repertoire of the malaria parasite sporozoite stage. *J. Biol. Chem.*, 277, 41948, 2002.
61. Mueller, A.K., Camargo, N., Kaiser, K., Andorfer, C., Frevert, U., Matuschewski, K., Kappe, S.H.I., Plasmodium liver stage developmental arrest by depletion of a protein at the parasite-host interface. *Proc. Natl. Acad. Sci. U. S. A.*, 102, 3022, 2005.
62. Amino, R., Thiberge, S., Martin, B., Celli, S., Shorte, S., Frischknecht, F., Ménard, R., Quantitative imaging of Plasmodium transmission from mosquito to mammal. *Nat. Med.*, 12, 220, 2006.
63. Cowman, A.F. and Crabb, B.S., Invasion of red blood cells by malaria parasites. *Cell*, 124, 755, 2006.
64. Grüring, C., Heiber, A., Kruse, F., Ungefehr, J., Gilberger, T.W., Spielmann, T., Development and host cell modifications of Plasmodium falciparum blood stages in four dimensions. *Nat. Commun.*, 2, 165, 2011.
65. Elford, B.C., Cowan, G.M., Ferguson, D.J.P., Parasite-regulated membrane transport processes and metabolic control in malaria-infected erythrocytes. *Biochem. J.*, 308, 361, 1995.
66. Slomianny, C., Three-dimensional reconstruction of the feeding process of the malaria parasite. *Blood Cells*, 16, 2, 1990.
67. Gull, D.E.A.K., The plasmodium cell cycle: Facts and questions. *Ann. Trop. Med. Parasitol.*, 92, 361, 1998.
68. White, J.H. and Kilbey, B.J., DNA replication in the malaria parasite. *Parasitol. Today*, 12, 151, 1996.

69. Bannister, L.H., Hopkins, J.M., Fowler, R.E., Krishna, S., Mitchell, G.H., Ultrastructure of rhoptry development in Plasmodium falciparum erythrocytic schizonts. *Parasitology*, 121, 3, 273, 2000.
70. Crewther, P.E., Culvenor, J.G., Silva, A., Cooper, J.A., Anders, R.F., Plasmodium falciparum: Two antigens of similar size are located in different compartments of the rhoptry. *Exp. Parasitol.*, 70, 193, 1990.
71. Preiser, P., Kaviratne, M., Khan, S., Bannister, L., Jarra, W., The apical organelles of malaria merozoites: Host cell selection, invasion, host immunity and immune evasion. *Microbes Infect.*, 2, 1461, 2000.
72. Bannister, L.H. and Mitchell, G.H., The fine structure of secretion by Plasmodium knowlesi merozoites during red cell invasion. *J. Protozool.*, 36, 362, 1989.
73. Soldati, D., Foth, B.J., Cowman, A.F., Molecular and functional aspects of parasite invasion. *Trends Parasitol.*, 20, 567, 2004.
74. Butcher, G.A., Dennis, E.D., Mitchell, G.H., Structure and invasive behaviour of Plasmodium knowlesi merozoites *in vitro*. *Parasitology*, 71, 483, 1975.
75. Aikawa, M., Torii, M., Sjölander, A., Berzins, K., Perlmann, P., Miller, L.H., Pf155/RESA antigen is localized in dense granules of Plasmodium falciparum merozoites. *Exp. Parasitol.*, 71, 326, 1990.
76. Trager, W., Rozario, C., Shio, H., Williams, J., Perkins, M.E., Transfer of a dense granule protein of Plasmodium falciparum to the membrane of ring stages and isolation of dense granules. *Infect. Immun.*, 60, 4656, 1992.
77. Blackman, M.J. and Holder, A.A., Secondary processing of the Plasmodium falciparum merozoite surface protein-1 (MSP1) by a calcium-dependent membrane-bound serine protease: Shedding of MSP133 as a noncovalently associated complex with other fragments of the MSP1. *Mol. Biochem. Parasitol.*, 50, 307, 1992.
78. Vaughan, A.M. and Kappe, S.H.I., Malaria parasite liver infection and exoerythrocytic biology. *Cold Spring Harb. Perspect. Med.*, 7, 6, 1, 2017.
79. Dame, J.B., Williams, J.L., McCutchan, T.F., Weber, J.L., Wirtz, R.A., Hockmeyer, W.T., Lee Maloy, W., David Haynes, J., Schneider, I., Roberts, D., Sanders, G.S., Premkumar Reddy, E., Diggs, C.L., Miller, L.H., Structure of the gene encoding the immunodominant surface antigen on the sporozoite of the human malaria parasite plasmodium falciparum. *Sci. (80-.)*, 225, 587, 1984.
80. Mendes, A.M., Albuquerque, I.S., Machado, M., Pissarra, J., Meireles, P., Prudêncio, M., Inhibition of Plasmodium liver infection by ivermectin. *Antimicrob. Agents Chemother.*, 61, 1, 2017.
81. Azevedo, R., Mendes, A.M., Prudêncio, M., Inhibition of Plasmodium sporogonic stages by ivermectin and other avermectins. *Parasites Vectors*, 12, 1, 1, 2019.
82. Panchal, M., Rawat, K., Kumar, G., Kibria, K.M., Singh, S., Kalamuddin, M., Mohmmed, A., Malhotra, P., Tuteja, R., Plasmodium falciparum signal recognition particle components and anti-parasitic effect of ivermectin

in blocking nucleo-cytoplasmic shuttling of SRP. *Cell Death Dis.*, 5, 1, 994, 2014.
83. Christensen, S.B., Natural products that changed society. *Biomedicines*, 9, 472, 2021.
84. Õmura, S., A splendid gift from the Earth: The origins and impact of the Avermectins (Nobel Lecture). *Angew. Chem. - Int. Ed.*, 55, 35, 10190, 2016.
85. Õmura, S., Ivermectin: 25 years and still going strong. *Int. J. Antimicrob. Agents*, 31, 91, 2008.
86. Shinohara, E.H., Martini, M.Z., De Oliveira Neto, H.G., Takahashi, A., Oral myiasis treated with ivermectin: Case report. *Braz. Dent. J.*, 15, 79, 2004.
87. Currie, M.J., Reynolds, G.J., Glasgow, N.J., Bowden, F.J., A pilot study of the use of oral ivermectin to treat head lice in primary school students in Australia. *Pediatr. Dermatol.*, 27, 595, 2010.
88. Singh, L., Fontinha, D., Francisco, D., Mendes, A.M., Prudêncio, M., Singh, K., Molecular design and synthesis of ivermectin hybrids targeting hepatic and erythrocytic stages of plasmodium parasites. *J. Med. Chem.*, 63, 1750, 2020.
89. Singh, L., Fontinha, D., Francisco, D., Prudêncio, M., Singh, K., Synthesis and antiplasmodial activity of regioisomers and epimers of second-generation dual acting ivermectin hybrids. *Sci. Rep.*, 12, 1, 2022.

6

Ivermectin: A Potential Repurposed Anti-Cancer Therapeutic

Mohd Kamil Hussain[1]*, Rashid Ali[2], Shakir Ahamad[3], Mohammad Faheem Khan[4] and Mohammad Saquib[5]†

[1]*Department of Chemistry Govt. Raza P.G. College (MJP Rohilkhand University Breilly), Rampur, India*

[2]*Department of Chemistry Jamia Millia Islamia (A Central University) Jamia Nagar, Okhla, New Delhi, India*

[3]*Department of Chemistry, Aligarh Muslim University, Aligarh, India*

[4]*Department of Biotechnology, Era's Lucknow Medical College, Era University, Lucknow, UP, India*

[5]*Department of Chemistry, University of Allahabad, Prayagraj (Allahabad), UP, India*

Abstract

Ivermectin (IVM), a lactone-based macrolide broad-spectrum antiparasitic drug, is frequently used to treat various parasitic disorders, including scabies, elephantiasis, and river blindness. The excellent effectiveness of IVM against parasite diseases was discovered by S. Ōmura and W. C. Campbell, who shared the 2015 Nobel Prize in Physiology and Medicine. IVM has recently been found to regulate several signaling pathways, preventing the development of various cancer types, indicating enormous potential as a therapy for treating cancers. Here, we discussed the mechanism of anti-carcinogenesis through which IVM induces programmed cell death to prevent the growth of various tumors. IVM exerts its therapeutic effect against various cancer types *in vitro* and *in vivo* by inhibiting mitochondrial respiration, angiogenesis, cancer stem cells (CSCs), PAH2-SID, P-Glycoprotein, Yes-Associated Protein 1 (YAP1), Wnt-TCF pathway responses, RNA helicase, KPNB1 protein or Akt/mTOR pathway. IVM also exerts its anti-carcinogenic effect by activating the SIN3 domain, chloride ion channels, caspase-dependent apoptosis,

**Corresponding author*: mkhcdri@gmail.com

†*Corresponding author*: saquibem@gmail.com

Rashid Ali and Shahid ul-Islam (eds.) Chemistry and Biological Activities of Ivermectin, (121–140)

transcription factor E3, P2X7 receptors, and induction of Immunogenic Cell Death (ICD), PAK1-mediated cytostatic autophagy and mitochondrial dysfunction.

Keywords: Ivermectin, cancer, wonder drug, drug repurposing, apoptosis, autophagy, cell death, doses

Abbreviations

CSCs	Cancer stem cells
DAMP	Damage-associated molecular pattern
EGFR	Epidermal growth factor receptor
GABA	Gamma-aminobutyric acid
HMGB1	High mobility group box-1 protein
IVM	Ivermectin
MDR	Multidrug resistance
OCT-4	Octamer-binding protein 4
PAK1	P-21-activated kinases 1
P-gp	P-glycoprotein
ROS	Reactive oxygen species
SID	SIN3-interaction domain
TNBC	Triple-negative breast cancer
YAP1	Yes-associated protein 1

6.1 Introduction

Repurposing existing drugs, originally developed for other diseases, has proven successful in medical sciences. In contrast to the traditional drug discovery process, drug repurposing or therapeutic switching offers several benefits, including speedier medicine access to patients while lowering costs, reducing drug development and minimizing the risk of failure, and a highly efficient process. However, due to a vast number of current therapies and diseases, as well as the diversity of patients and diseases, particularly cancers, drug repurposing could be inefficient and time-consuming. The main problem is how to repurpose an existing pharmaceutical to treat a particular condition efficiently. Historically, drug repurposing has been coincidental and opportunistic; as soon as a drug was found to demonstrate an off-target effect or a new on-target effect, it was encouraged for commercialization [1–4].

Cancer is a group of diseases characterized by uncontrolled cell growth and proliferation that, if untreated, can result in patient death. Cancer is the leading cause of death worldwide, accounting for more than ten million mortalities in 2020, or roughly one in every six. Breast, lung, colon, rectum, and prostate cancers are the most prevalent. Nearly one-third of cancer-related deaths are caused by the use of alcohol, smoking, having a high body mass index, eating little fruits and vegetables, and not exercising enough. In low- & lower-middle-income countries, infections that cause cancer, like the human papillomavirus (HPV) and hepatitis, account for about 30% of cancer cases. Many can be cured if tumors are diagnosed early and appropriately managed [5, 6]. The COVID-19 pandemic had a negative impact on cancer diagnosis and treatment in 2020. Due to the shutdown of healthcare facilities and concerns about COVID-19 exposure, access to care was restricted, which led to delays in diagnosis and treatment. This might have led to a temporary drop in the reported incidence of cancer, followed by a rise in the number of people who developed the advanced disease and, ultimately, a higher death rate [5, 7]. The advent of drug resistance in cancer chemotherapy, combined with the issue of toxicity and low tumor selectivity, has fuelled the hunt for more potent and less toxic anti-breast cancer drugs that target the interaction mechanisms involved in cancer cell division and metastasis. The discovery and development of novel therapeutic compounds are critical in the fight against mortality and illness, and repurposing currently available drugs is a critical step in that direction [8–11].

Few drugs can legitimately claim the moniker of "Wonder drug," IVM, however, can be listed alongside those worthy rivals because of its adaptability, safety, and significant global impact, particularly on hundreds of millions of the world's poorest people [5]. IVM is a polycyclic lactone insecticide produced by the bacterium *Streptomyces avermitilis* and is used as a general antiparasitic medication that led to the 2015 Nobel Prize in Physiology and Medicine.

IVM was introduced in 1981 to manage nematodes and arthropods in farmed animals. After a long period with its remarkable therapeutic impact against numerous animal parasitic diseases, IVM was FDA-approved for human use in 1987 [12–15]. IVM consists of a mixture of two homologs containing > 80%, 22, 23-dihydro-AVM B1a and < 20%, 22, 23-dihydro-AVM B1b, and synthesized via selective catalytic reduction of the *Z*-22,23-C=C bond present in the spiropyranose ring residue of the AVM B1a and B1b (Figure 6.1) [16, 17]. Initial research into IVM's biological function revealed that it activates the glutamate-gated chloride channel, which blocks communication in nerve cells and ultimately kills the parasites.

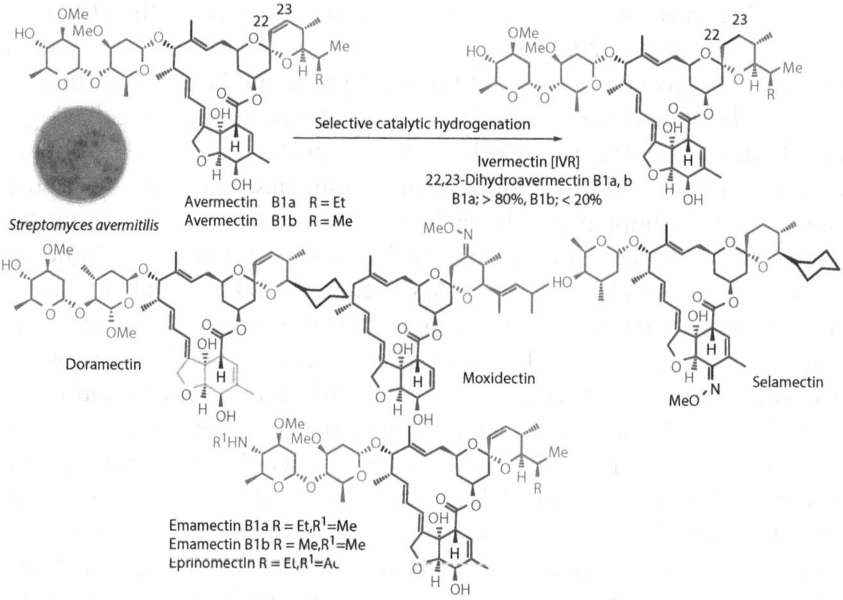

Figure 6.1 Structure of avermectins (AVM), IVM & other derivatives such as doramectin, moxidectin, selamectin, emamectin, and eprinomectin. The two components of IVM are closely related. At the C-26 position, IVM B1a has $-CH_2CH_3$, whereas the IVM B1b has CH_3 substituent. IVM consists of >80% of B1a and <20% of B1b.

These research findings provided the sufficient evidences for the selectivity of drugs targeting GABA receptors in mammals [18]. Interestingly, various *in vitro* and *in vivo* investigations have demonstrated the potential of IVM in treating numerous pathologic conditions besides parasitic infections. For example, IVM increases insulin sensitivity in diabetic rats, decreases serum cholesterol and blood glucose, blocks the replication of several flaviviruses, and reduces the life expectancy of the important vectors of malaria and trypanosomiasis. Additionally, IVM also demonstrates anti-inflammatory activities in T cell-induced skin diseases [13].

6.2 Mechanism of Anti-Carcinogenesis

IVM has long been recognized as an effective antiparasitic agent. In the previous years, a growing number of researchers have shown that IVM may have broad applications as an anti-carcinogenic agent against various cancers, including breast cancer, glioblastoma, [19] ovarian cancer, [20]

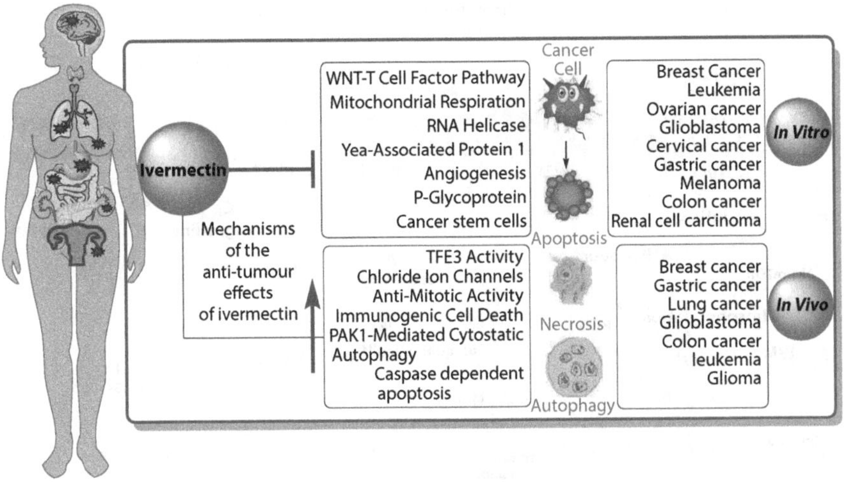

Figure 6.2 Anticancer potential of IVM: IVM exerts its effect (*in vitro* or *in vivo*) against multiple types of malignancies through inhibition or activation of various biological processes, enzymes, and ion channels, exerts anti-mitotic activity, and induces caspase-dependent apoptosis.

neurofibromatosis type 2 tumors, [21] and leukemia, [22] gastric cancer, colon cancer, glioma, melanoma, and renal cell carcinoma. IVM exerts its effect against various types of cancer by inhibiting mitochondrial respiration, angiogenesis, P-glycoprotein, RNA-helicase, Yes-Associated Protein 1, and cancer stem cells. IVM activates mammalian chloride ion channels, exerts anti-mitotic activity, induces autophagy, immunogenic cell death, caspase-dependent apoptosis, increases TFE3 activity, and serves as a small-molecule mimetic of the SID peptide (Figures 6.2 and 6.3) [23, 24]. This finding suggests that IVM is a potent anti-cancer agent.

6.3 Activation of Chloride Ion Channels

Targeting potassium or chloride ion channels is a crucial anti-cancer therapeutic strategy as several studies have indicated that they are involved in the biochemical processes to promote the growth of cancer cells. A study reported that treatment of leukemia cells (OCI-AML2) with IVM raised the concentration of intracellular Cl^- resulting in hyperpolarization of the plasma membrane, mitochondrial membranes and the generation of reactive oxygen species (ROS) [22]. IVM affected Ca^{2+} influx in leukemia cells

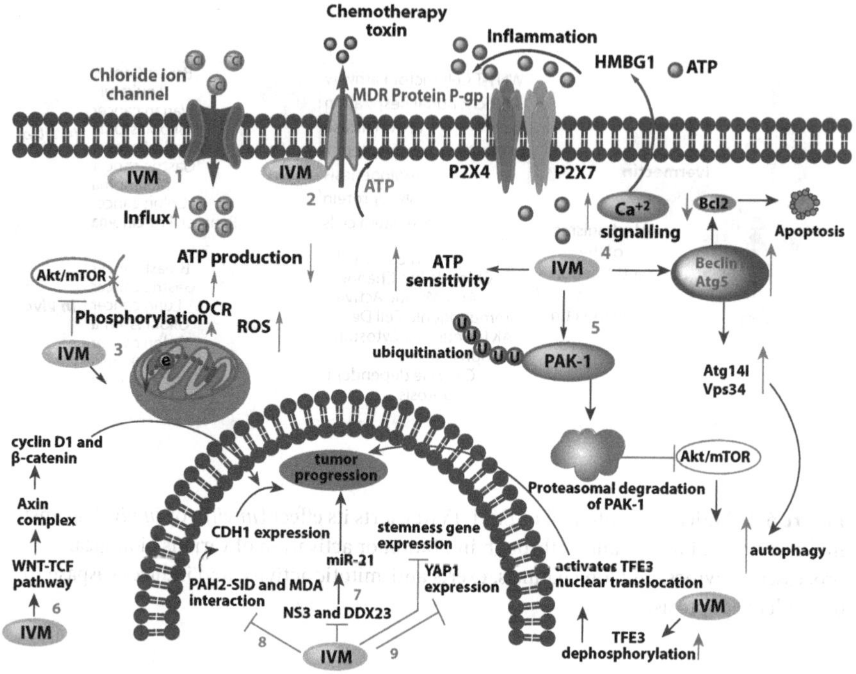

Figure 6.3 Mechanism of anti-carcinogenesis: (1) IVM serves as an ionophore, activating chloride channels and causing apoptosis and osmotic cell death. (2) IVM inhibits the P-glycoprotein pump's substrate and prevents multidrug resistance. (3) IVM restricts electron transfer in oxidative phosphorylation, which raises oxygen consumption rate (OCR) to produce ATP for the cell by reducing the functioning of the mitochondrial complex I. Concurrently, there is a decrease in Akt phosphorylation, affecting mitochondria biogenesis. Furthermore, changes in the mitochondrial processes are linked to higher quantities of ROS, which damage DNA. (4) IVM induces immunogenic cell death (ICD) by activating an ATP- and HMGB1-rich milieu, which causes inflammation. IVM also enhances ATP sensitivities and calcium signaling in P2X membrane receptors, including P2X4 and P2X7, to promote ATP-dependent immunological responses. (5) IVM induces poly-ubiquitination of the kinase PAK1, directing proteasomal degradation. In turn, faulty PAK1 inhibits the Akt/mTOR pathway. Simultaneously IVM increases the expression of Beclin1 and Atg5, which are associated with autophagy activation while decreasing the activity of the apoptosis negative regulator Bcl-2. This results in autophagy and apoptosis. (6) IVM inhibits cancer development by inhibiting positive regulators (AXIN2, LGR5, and ASCL2) of the WNT-TCF signaling pathway. (7) IVM inhibits the action of the RNA helicases (NS3 and DDX23), which are involved in ribosome synthesis, post-transcriptional modifications, and mRNA degradation. DDX23 functions as a promoter of miR-21, a well-known tumor growth stimulant. (8) IVM acts as a mimetic of the SID peptide. (9) IVM inhibits proliferation by inhibiting Yes-Associated Protein-1 (YAP-1), which is associated with various human malignancies [20–80].

because hyper membrane polarization might produce Ca^{2+} influx. IVM raised intracellular calcium, but this cytotoxic effect was not associated with cytotoxicity since chelating extracellular Ca^{2+} did not prevent IVM-induced cell death (Figure 6.3) [22].

6.4 Anti-Mitotic Effect and Inhibition of Angiogenesis

Microtubule is a crucial target for developing novel drugs to treat cancers because they are critical in maintaining signal transduction, cell morphology, mitosis, and material transport [25]. IVM has been found in a tubulin polymerization study to interact with nematode (Haemonchus contortus) tubulin, promoting tubulin polymerization in mammals [26]. Additionally, IVM stopped the HeLa cell line's *in vitro* development and protected mammalian tubulin from the depolymerizing effects of low temperatures. The antiproliferative effect of IVM against HeLa cells was found to be reversible [26]. IVM has been approved for human use, thus employing it as an anti-mitotic drug which could be an attractive surrogate for the cancer treatment (Figure 6.3).

6.5 Inhibition of Mitochondrial Respiration

Glioblastoma is a highly vascular brain tumor that is resistant to existing treatment. Targeting angiogenesis and glioblastoma cells may be an efficient and successful option for treating glioblastoma. IVM efficiently inhibits the proliferation of glioblastoma cells *in vitro/in vivo* and inhibits angiogenesis. In glioblastoma cells (U87 and T98G), IVM strongly suppresses cell proliferation and induces apoptosis in both cells via a caspase-dependent mechanism. IVM significantly inhibits the formation of capillary network, proliferation, and survival of human brain microvascular endothelial cells (HBMEC) [27, 28]. In glioblastoma and HBMEC cells, IVM inhibits mitochondrial respiration and membrane potential and suppresses ATP formation while increasing mitochondrial superoxide. IVM's inhibitory effects were considerably reversed in cells treated with antioxidants or cells lacking in mitochondria. In conclusion, IVM exhibited its effect by suppressing phosphorylation of Akt, mTOR, and ribosomal S6 in U87, T98G, and HBMEC cells (Figure 6.3) [27].

6.6 Inhibitor of Cancer Stem Cells (CSCs)

Embryonic stem (ES) cells have a remarkable capacity for self-regeneration, while simultaneously producing embryonic germ layers and germ lines during the development of the embryo. Various ES cell-specific transcription factors, including homeobox protein Nanog (NANOG) and octamer-binding protein 4 (Oct4), are required for self-renewal ES cells and pluripotency. These transcription factors are coupled with distinct repressor complexes on their targets and interact with distinct repression complexes to regulate expressions critical for ES cell fate determination [29]. IVM specifically and dramatically reduced the expression of "stemness" genes and cancer stem cell markers in breast tumors, reducing the proliferation of cancer stem-like cells (CSCs) in breast carcinoma. IVM significantly reduced the proliferation of CSCs-enriched populations compared to the total cell population [14].

6.7 Induction of Immunogenic Cell Death (ICD)

The immune system places the tumor cells under intense selection pressure at the beginning of the tumor as part of tumor immunosurveillance. The immune system is expected to eradicate the majority of early neoplastic lesions before they may become clinically significant tumors. Despite the immune system's constant selection pressure, which destroys all other immunosusceptible tumor cells, immunoevasive tumor cells can develop, proliferate, and escape [30–33]. ICD is a novel concept that has previously surfaced. The release of high mobility group protein B1 (HMGB1), surface-exposed calreticulin (CRT), and secreted ATP are only a few instances where the DAMPs significantly influence the immunogenic properties of ICD. Pattern recognition receptors identify the bulk of DAMPs [30]. Through the IVM-sensitive P2X 4/P2X 7-gated Pannexin-1 channels, IVM induces ICD by stimulating autophagy and releasing ATP from tumor cells undergoing ICD [34, 35]. When ATP and Ca2+ signaling are needed to control the cell volume, the release of ATP may be momentarily protective, but only in cells particularly susceptible to IVM-induced cell swelling. High concentrations of ATP instantaneously increase the cytotoxic effect of IVM, where the size of cancer cells remains constant. The release of ATP and the downstream signaling of P2X7 receptor death signals were revealed to be necessary for the cancer cell death. IVM triggers autophagy and the release of ATP and HMGB1, which are important inflammatory mediators (Figure 6.3) [34].

6.8 Epigenetic Modulator

A crucial additional mode of action for IVM is its function in cancer treatments based on epigenetic regulation. In a study of MDA-MB-231, it was found that IVM mimics the SIN3-interaction domain (SID), which binds to the paired amphipathic helix-2 (PAH2) motif of the breast cancer-related epigenetic regulators SIN3A and SIN3B [36–38]. The NMR study identified that IVM inhibits the PAH2-SID interaction [19]. SIN3A is a complex component that is a positive regulator of the stem cell and self-regeneration prognostic markers *NANOG* and *SOX2*. It was found that IVM inhibits the expression of *NANOG* and *SOX2* in D3H2LN cells by 80% at 0.5 μM concentration with a 90-100% reduction in clonogenic tumorsphere growth. IVM also enhanced the expression of ESR1, the ERα, and PGR receptors and restored the sensitivity of tamoxifen in MDA-MB-231 and MMTV-Myc mouse cells *in vitro* (Figure 6.3) [19].

6.9 Induction PAK1-Mediated Cytostatic Autophagy

Autophagy is a self-degradative intracellular process essential for controlling cell growth and metabolism. In this process, cytoplasmic constituents macromolecules like damaged organelles, pathogens, and aggregated proteins are catabolized in a conserved manner by lysosomal enzymes to create their components, such as amino acids (AAs), nucleotides, sugars, ATP, and fatty acids, which are then recycled back into the cytoplasm [39, 40]. Tumor cells have several characteristics in common, including high metabolic demands, cell proliferation, resistance to cell death, allowing replicative immortality to evade growth suppressors, inducing angiogenesis, metastasis, and high cellular stress, all of which activate autophagy in order to balance cell energy sources and regulate tissue homeostasis [39–41]. The p21-activated kinase 1 (PAK1) is a crucial modulator of cytoskeletal reorganization and nuclear signal transduction. It impacts a wide array of signals associated with numerous biological processes [42]. PAK1 play a crucial role in the growth and development of >70% of all cancers in human beings, such as thyroid, cervical, lung, breast, prostate, colon, gastric, pancreatic as well as melanoma, glioma, hepatoma, neurofibromatosis, and multiple myeloma. Blocking PAK1 signals results in tumor cell death [43]. IVM exerts its effect by inhibiting the Akt/mTOR pathway by enhancing the degradation of PAK1 through ubiquitination process, which ultimately raises the autophagic flux [44]. Another investigation also found that IVM inhibited the proliferation of breast cancer cells via activation of cytostatic autophagy, disrupting the signaling process, and possibly inhibiting PAK1 expression. Considering that IVM-induced cytostatic autophagy inhibits breast cancer tumor growth, the drug has the potential to lower the proliferation of breast cancer xenograftsux (Figure 6.3) [45].

6.10 Inhibition of P-glycoprotein (P-gp)

The failure of various types of chemotherapy is primarily attributed to multidrug resistance (MDR), the primary mechanism through which various cancers become resistant to the drugs used in chemotherapy. Patients with a range of solid tumors such as ovarian, breast, lung, and gastrointestinal tract cancers and blood cancers are affected by MDR. The majority of cancerous cells in tumors are heterogeneous populations, some of which are drug and some of which are drug-resistant. Only drug-sensitive cancer cells are killed by chemotherapy, but a more considerable percentage of drug-resistant cancer cells are left behind. As a result of these drug-resistant cells continuing to multiply, it ultimately led to the failure of cancer chemotherapy [46–49]. MDR development is still a key impediment to long-term cancer therapeutic success. A well-known membrane transporter called P-gp (MDR1 or ABCB1) can efflux drugs out of cancer cells, decreasing the effectiveness of cancer chemotherapy. The P-gp expression is increased in cancer cells as a protective measure to prevent chemotherapy-induced cell death [50–54]. IVM significantly reduced tumor cells resistance to chemotherapeutics at low doses *in vitro* and *in vivo*. IVM lowered P-gp expression by blocking the EGFR by binding to its extracellular domain, which prevented EGFR activation and ERK/Akt/NF-B downstream signaling rather than directly inhibiting P-gp activity. This inhibition led to the suppression of P-gp transcription [55]. Additionally, it was discovered that IVM reduced P-ATPase gp's activity and reversed *in vitro* the P-gp-related MDR. The specific mechanisms by which IVM increases a cell's sensitivity to chemotherapy agents and counteracts a tumor cell's resistance are still widely undefined (Figure 6.3) [56–59].

6.11 Inhibition of Yes-Associated Protein 1 (YAP1)

YAP1 is the primary transcriptional regulator in the Hippo-signaling pathway, and it play a significant role in controlling organ growth by limiting cell proliferation and promoting apoptosis [60, 61]. *YAP1* delocalized between the cytoplasm and the nucleus of the cell. YAP and TAZ are located inside the nucleus, where they interact with the transcription factors and regulate the target expression of the gene to modulate cell growth, proliferation, and survival. Large tumor suppressor 1/2 (LATS1/2) kinases phosphorylate YAP1 in the cytoplasm and prevent it from migrating to the nucleus to boost the activity of YAP1's transcription [62]. The Hippo pathway was discovered to be dysregulated in many solid malignant tumors, and YAP1 was found to be a crucial oncogenic role in colon, lung, liver, ovary, and prostate cancers [63, 64]. The anti-cancer benefits of IVM in gastric cancer (GC) were assessed in a study by Nambara *et al.* and its mechanism of action via suppression of YAP1 expression was established that IVM significantly suppressed the growth of GC *in vitro* (MKN1 GC cells IC_{50} = 10.2 μM) and *in vivo* (*10 mg kg*$^{-1}$*, i.p./day up to 22 days*) through the inhibition of YAP1 expression. These findings provided insight into the anti-proliferative activity of IVM as a YAP1 inhibitor and offered a conceptual framework for preclinical evaluations of IVM for the treatment of GC [65].

6.12 Inhibition of RNA Helicase

The highly conserved non-coding single-stranded RNA molecules known as microRNAs (miRNAs) are involved in post-transcriptional gene expression. An elevated level of miR-21 in glioblastoma tumors has been reported compared to non-neoplastic and brain tissues. A study suggested that the knockdown of miR-21 in glioblastoma cells activates caspases, raising apoptotic cell death [66]. The onset of glioma cancer is closely associated with the upregulation of miR-21. Moreover, Yin *et al.* verified that DDX23 (DEADbox RNA helicase) encourages miR-21 synthesis in malignant glioma cells at the post-transcriptional level. The expression of DEADbox RNA helicase was closely related to the low survival rates of glioma malignancy, and DDX23 expression was increased in glioma tissues compared to normal brain tissues [67]. A study shows that IVM works by inhibiting RNA helicases, particularly the NS3 DEAD-box RNA helicase. IVM functions as an RNA helicase inhibitor, significantly reducing the invasion and growth of tumor cells in U87MG glioma cells by targeting miR-21 without affecting the expression of the DEADbox RNA helicase [67]. Additionally, IVM dramatically reduced the proliferation of the cell lines DU145, HeLa, A549, and PANC-1, as well as X01 and CSC2 cells (Figure 6.3) [67].

6.13 Caspase-Dependent Apoptosis

Cancer cells exhibit uncontrolled proliferation, apoptosis evasion, and angiogenesis, regardless of their source or type [68, 69]. One of the primary purposes of apoptosis is cancer prevention. In cancer, the intrinsic pathways are generally inhibited; however, there are various ways to inhibit apoptosis. Loss of apoptotic control in tumor cells allows them to live for longer periods, giving mutations more opportunity to accumulate. This can increase tumor invasiveness, promote angiogenesis, obstruct differentiation, and deregulate cell proliferation [70, 71]. In addition to regulating several mitotic events, the nuclear import receptor karyopherin 1 (KPNB1) is associated with the nuclear import of several proteins. Up-regulation of KPNB1 has been reported in malignancies such as glioblastoma. Inhibition or suppression of KPNB1 induces mitotic arrest and apoptosis in the malignant cell, but the underlying mechanism is fully confirmed [72]. IVM significantly inhibited the multi-stage cell cycle and induced apoptosis in epithelial ovarian cancer cells by suppressing KPNB1 [72]. Another study also showed that IVM efficiently inhibited the proliferation of HeLa and glioblastoma cells by inducing apoptosis, boosting the release of cytochrome c, increasing Bax and p53 expression, decreasing Bcl-2 expression, and lowering the levels of cyclin E, cyclin D1, CDK2, CDK4, and CDK6 (Figure 6.3) [73, 74].

6.14 Activation of Transcription Factor E3 (TFE3)

TFE3, a transcription factor for the basic helix-loop-helix leucine zipper that belongs to the MiT/TFE family, is regulated by mTOR-induced phosphorylation and contributes to oxidative stress response and oxidative metabolism [75]. According to a study, IVM-induced dephosphorylation of $TFE3^{(Ser321)}$, activation of TFE3 nuclear translocation, and enhanced TFE3 reporter activity resulted in biosynthesis of lysosomes and the activation of autophagy-related genes, which then induce autophagy in human melanoma (SK-MEL-28) cells (Figure 6.3) [76].

6.15 Inhibition of Wnt-TCF Pathway Responses

The Wnt pathway is crucial for development, tissue homeostasis, and cancer development [77–79]. Several essential processes, including differentiation, metastasis, cell migration, cell proliferation, aging, cell death, tumor development, and tumor growth, are linked to Wnts and their downstream effectors. IVM increases apoptosis and inhibits proliferation in many human cancer types. It inhibits the levels of CYCLIN D1 and C-terminal -CATENIN phosphoforms in an okadaic acid-sensitive fashion, demonstrating that protein phosphatases are involved in its action. IVM specifically inhibits TCF-dependent but not TCF-independent xenograft development *in vivo* without manifesting any adverse effects (Figure 6.3) [80].

6.16 Conclusions

Repurposing already approved drugs for different indications, such as switching non-oncology pharmaceuticals to oncology therapies for malignancies, is one effective strategy used to find therapeutics in the short run. IVM is a 'wonder drug' that shows remarkable anti-cancer properties either by inhibiting or activating various signaling pathways. For instance, inhibition of mitochondrial respiration, angiogenesis, cancer stem cells (CSCs), PAH2-SID, P-Glycoprotein, Yes-Associated Protein 1 (YAP1), Wnt-TCF pathway responses, RNA helicase, KPNB1 protein, or Akt/mTOR pathway. It also activates the SIN3 domain, chloride ion channels, caspase-dependent apoptosis, transcription factor E3, P2X7 receptors, and induces immunogenic cell death (ICD), PAK1-mediated cytostatic autophagy, and mitochondrial failure, indicating its enormous potential as an anti-cancer drug. However, more studies and clinical investigations are required to establish its potency and efficacy against various malignancies (Figure 6.3). In this chapter, we have thoroughly explored the potential clinical applications of IVM in cancer therapy, and the mechanism by which it encourages programmed cell death and prevents tumor growth. We are in the opinion that this particular chapter would benefit the young researchers globally to get insight into the anti-cancer activity of IVM – a wonder drug, and would surely stimulate further scientific research in this wonderful arena.

References

1. Sleire, L., Førde, H.E., Netland, I.A., Leiss, L., Skeie, B.S., Enger, P.Ø., Drug repurposing in cancer. *Pharmacol. Res.*, 124, 74–91, 2017, https://doi.org/10.1016/j.phrs.2017.07.013.
2. Bouche, G., Gedye, C., Meheus, L., Pantziarka, P., Drug repurposing in oncology. *Lancet Oncol.*, 21, e542, 2020, https://doi.org/10.1016/S1470-2045(20)30561-1.

3. Falvo, P., Orecchioni, S., Roma, S., Raveane, A., Bertolini, F., Drug repurposing in oncology, an attractive opportunity for novel combinatorial regimens. *Curr. Med. Chem.*, 28, 2114–36, 2021, https://doi.org/10.2174/0929867327999200817104912.
4. Pushpakom, S., Iorio, F., Eyers, P.A., Escott, K.J., Hopper, S., Wells, A. *et al.*, Drug repurposing: Progress, challenges and recommendations. *Nat. Rev. Drug Discovery*, 18, 41–58, 2019, https://doi.org/10.1038/nrd.2018.168.
5. Yabroff, K.R., Wu, X.-C., Negoita, S., Stevens, J., Coyle, L., Zhao, J. *et al.*, Association of the COVID-19 pandemic with patterns of statewide cancer services. *JNCI J. Natl. Cancer Inst.*, 114, 907–9, 2022, https://doi.org/10.1093/jnci/djab122.
6. Siegel, R.L., Miller, K.D., Fuchs, H.E., Jemal, A., Cancer statistics, 2022. *CA Cancer J. Clin.*, 72, 7–33, 2022, https://doi.org/10.3322/caac.21708.
7. Ahamad, S., Branch, S., Harrelson, S., Hussain, M.K., Saquib, M., Khan, S., Primed for global coronavirus pandemic: Emerging research and clinical outcome. *Eur. J. Med. Chem.*, 209, 112862, 2021, https://doi.org/10.1016/j.ejmech.2020.112862.
8. Saquib, M., Baig, M.H., Khan, M.F., Azmi, S., Khatoon, S., Rawat, A.K. *et al.*, Design and synthesis of bioinspired benzocoumarin-chalcones chimeras as potential anti-breast cancer agents. *ChemistrySelect*, 6, 8754–65, 2021, https://doi.org/10.1002/slct.202101853.
9. Ansari, M.I., Khan, M.M., Saquib, M., Khatoon, S., Hussain, M.K., Dithiolethiones: A privileged pharmacophore for anticancer therapy and chemoprevention. *Future Med. Chem.*, 10, 1241–60, 2018, https://doi.org/10.4155/fmc-2017-0281.
10. Hussain, M.K., Ansari, M.I., Yadav, N., Gupta, P.K., Gupta, A.K., Saxena, R. *et al.*, Design and synthesis of ERα/ERβ selective coumarin and chromene derivatives as potential anti-breast cancer and anti-osteoporotic agents. *RSC Adv.*, 4, 8828–45, 2014, https://doi.org/10.1039/C3RA45749D.
11. Saquib, M., Ansari, M.I., Johnson, C.R., Khatoon, S., Kamil Hussain, M., Coop, A., Recent advances in the targeting of human DNA ligase I as a potential new strategy for cancer treatment. *Eur. J. Med. Chem.*, 182, 111657, 2019, https://doi.org/10.1016/j.ejmech.2019.111657.
12. Ōmura, S. and Crump, A., Ivermectin and malaria control. *Malar. J.*, 16, 172, 2017, https://doi.org/10.1186/s12936-017-1825-9.
13. Laing, R., Gillan, V., Devaney, E., Ivermectin – old drug, new tricks? *Trends Parasitol.*, 33, 463–72, 2017, https://doi.org/10.1016/j.pt.2017.02.004.
14. Dominguez-Gomez, G., Chavez-Blanco, A., Medina-Franco, J., Saldivar-Gonzalez, F., Flores-Torrontegui, Y., Juarez, M. *et al.*, Ivermectin as an inhibitor of cancer stem-like cells. *Mol. Med. Rep.*, 17, 2017, https://doi.org/10.3892/mmr.2017.8231.
15. Crump, A. and Omura, S., Ivermectin, "Wonder drug" from Japan: The human use perspective. *Proc. Jpn. Acad. Ser. B*, 87, 13–28, 2011, https://doi.org/10.2183/pjab.87.13.

16. Springer, J.P., Arison, B.H., Hirshfield, J.M., Hoogsteen, K., The absolute stereochemistry and conformation of avermectin B2a aglycone and avermectin B1a. *J. Am. Chem. Soc.*, 103, 4221–4, 1981, https://doi.org/10.1021/ja00404a041.
17. Campbell, W.C., History of avermectin and ivermectin, with notes on the history of other macrocyclic lactone antiparasitic agents. *Curr. Pharm. Biotechnol.*, 13, 853–65, 2012, https://doi.org/10.2174/138920112800399095.
18. Geary, T., Ivermectin 20 years on: Maturation of a wonder drug. *Trends Parasitol.*, 21, 530–2, 2005, https://doi.org/10.1016/j.pt.2005.08.014.
19. Kwon, Y.-J., Petrie, K., Leibovitch, B.A., Zeng, L., Mezei, M., Howell, L. *et al.*, Selective inhibition of SIN3 corepressor with avermectins as a novel therapeutic strategy in triple-negative breast cancer. *Mol. Cancer Ther.*, 14, 1824–36, 2015, https://doi.org/10.1158/1535-7163.MCT-14-0980-T.
20. Kodama, M., Kodama, T., Newberg, J.Y., Katayama, H., Kobayashi, M., Hanash, S.M. *et al.*, *In vivo* loss-of-function screens identify KPNB1 as a new druggable oncogene in epithelial ovarian cancer. *Proc. Natl. Acad. Sci.*, 114, E7301–E7310, 2017, https://doi.org/10.1073/pnas.1705441114.
21. Hashimoto, H., Messerli, S.M., Sudo, T., Maruta, H., Ivermectin inactivates the kinase PAK1 and blocks the PAK1-dependent growth of human ovarian cancer and NF2 tumor cell lines. *Drug Discovery Ther.*, 3, 243–6, 2009, https://doi.org/22495656.
22. Sharmeen, S., Skrtic, M., Sukhai, M.A., Hurren, R., Gronda, M., Wang, X. *et al.*, The antiparasitic agent ivermectin induces chloride-dependent membrane hyperpolarization and cell death in leukemia cells. *Blood*, 116, 3593–603, 2010, https://doi.org/10.1182/blood-2010-01-262675.
23. Juarez, M., Schcolnik-Cabrera, A., Dueñas-Gonzalez, A., The multitargeted drug ivermectin: From an antiparasitic agent to a repositioned cancer drug. *Am. J. Cancer Res.*, 8, 317–31, 2018, https://doi.org/29511601.
24. Liu, J., Zhang, K., Cheng, L., Zhu, H., Xu, T., Progress in understanding the molecular mechanisms underlying the antitumour effects of ivermectin. *Drug Des. Devel. Ther.*, 14, 285–96, 2020, https://doi.org/10.2147/DDDT.S237393.
25. Jordan, M.A. and Wilson, L., Microtubules as a target for anticancer drugs. *Nat. Rev. Cancer*, 4, 253–65, 2004, https://doi.org/10.1038/nrc1317.
26. Ashraf, S. and Prichard, R., Ivermectin exhibits potent anti-mitotic activity. *Vet. Parasitol.*, 226, 1–4, 2016, https://doi.org/10.1016/j.vetpar.2016.06.015.
27. Liu, Y., Fang, S., Sun, Q., Liu, B., Anthelmintic drug ivermectin inhibits angiogenesis, growth and survival of glioblastoma through inducing mitochondrial dysfunction and oxidative stress. *Biochem. Biophys. Res. Commun.*, 480, 415–21, 2016, https://doi.org/10.1016/j.bbrc.2016.10.064.
28. Goradel, N.H., Mohammadi, N., Haghi-Aminjan, H., Farhood, B., Negahdari, B., Sahebkar, A., Regulation of tumor angiogenesis by microRNAs: State of the art. *J. Cell. Physiol.*, 234, 1099–110, 2019, https://doi.org/10.1002/jcp.27051.

29. Liang, J., Wan, M., Zhang, Y., Gu, P., Xin, H., Jung, S.Y. *et al.*, Nanog and Oct4 associate with unique transcriptional repression complexes in embryonic stem cells. *Nat. Cell Biol.*, 10, 731–9, 2008, https://doi.org/10.1038/ncb1736.
30. Krysko, D.V., Garg, A.D., Kaczmarek, A., Krysko, O., Agostinis, P., Vandenabeele, P., Immunogenic cell death and DAMPs in cancer therapy. *Nat. Rev. Cancer*, 12, 860–75, 2012, https://doi.org/10.1038/nrc3380.
31. Gillies, R.J., Verduzco, D., Gatenby, R.A., Evolutionary dynamics of carcinogenesis and why targeted therapy does not work. *Nat. Rev. Cancer*, 12, 487–93, 2012, https://doi.org/10.1038/nrc3298.
32. Zitvogel, L., Tesniere, A., Kroemer, G., Cancer despite immunosurveillance: Immunoselection and immunosubversion. *Nat. Rev. Immunol.*, 6, 715–27, 2006, https://doi.org/10.1038/nri1936.
33. Garg, A.D., Nowis, D., Golab, J., Vandenabeele, P., Krysko, D.V., Agostinis, P., Immunogenic cell death, DAMPs and anticancer therapeutics: An emerging amalgamation. *Biochim. Biophys. Acta*, 1805, 53–71, 2010, https://doi.org/10.1016/j.bbcan.2009.08.003.
34. Draganov, D., Gopalakrishna-Pillai, S., Chen, Y.-R., Zuckerman, N., Moeller, S., Wang, C. *et al.*, Modulation of P2X4/P2X7/Pannexin-1 sensitivity to extracellular ATP via Ivermectin induces a non-apoptotic and inflammatory form of cancer cell death. *Sci. Rep.*, 5, 16222, 2015, https://doi.org/10.1038/srep16222.
35. Martins, I., Wang, Y., Michaud, M., Ma, Y., Sukkurwala, A.Q., Shen, S. *et al.*, Molecular mechanisms of ATP secretion during immunogenic cell death. *Cell Death Differ.*, 21, 79–91, 2014, https://doi.org/10.1038/cdd.2013.75.
36. Lee, J.-Y. and Kong, G., Roles and epigenetic regulation of epithelial-mesenchymal transition and its transcription factors in cancer initiation and progression. *Cell. Mol. Life Sci.*, 73, 4643–60, 2016, https://doi.org/10.1007/s00018-016-2313-z.
37. Tam, W.L. and Weinberg, R.A., The epigenetics of epithelial-mesenchymal plasticity in cancer. *Nat. Med.*, 19, 1438–49, 2013, https://doi.org/10.1038/nm.3336.
38. Silverstein, R.A. and Ekwall, K., Sin3: A flexible regulator of global gene expression and genome stability. *Curr. Genet.*, 47, 1–17, 2005, https://doi.org/10.1007/s00294-004-0541-5.
39. Li, X., He, S., Ma, B., Autophagy and autophagy-related proteins in cancer. *Mol. Cancer*, 19, 12, 2020, https://doi.org/10.1186/s12943-020-1138-4.
40. Desantis, V., Saltarella, I., Lamanuzzi, A., Mariggiò, M.A., Racanelli, V., Vacca, A. *et al.*, Autophagy: A New mechanism of prosurvival and drug resistance in multiple myeloma. *Transl. Oncol.*, 11, 1350–7, 2018, https://doi.org/10.1016/j.tranon.2018.08.014.
41. Boya, P., Reggiori, F., Codogno, P., Emerging regulation and functions of autophagy. *Nat. Cell Biol.*, 15, 713–20, 2013, https://doi.org/10.1038/ncb2788.

42. Crump, A., Ivermectin: Enigmatic multifaceted 'wonder' drug continues to surprise and exceed expectations. *J. Antibiot. (Tokyo)*, 70, 495–505, 2017, https://doi.org/10.1038/ja.2017.11.
43. Hashimoto, H., Sudo, T., Maruta, H., Nishimura, R., The direct PAK1 inhibitor, TAT-PAK18, blocks preferentially the growth of human ovarian cancer cell lines in which PAK1 is abnormally activated by autophosphorylation at Thr 423. *Drug Discovery Ther.*, 4, 1–4, 2010, https://doi.org/22491145.
44. Wang, K., Gao, W., Dou, Q., Chen, H., Li, Q., Nice, E.C. *et al.*, Ivermectin induces PAK1-mediated cytostatic autophagy in breast cancer. *Autophagy*, 12, 2498–9, 2016, https://doi.org/10.1080/15548627.2016.1231494.
45. Dou, Q., Chen, H.-N., Wang, K., Yuan, K., Lei, Y., Li, K. *et al.*, Ivermectin induces cytostatic autophagy by blocking the PAK1/Akt axis in breast cancer. *Cancer Res.*, 76, 4457–69, 2016, https://doi.org/10.1158/0008-5472.CAN-15-2887.
46. Bukowski, K., Kciuk, M., Kontek, R., Mechanisms of multidrug resistance in cancer chemotherapy. *Int. J. Mol. Sci.*, 21, 3233, 2020, https://doi.org/10.3390/ijms21093233.
47. Emran, T.B., Shahriar, A., Mahmud, A.R., Rahman, T., Abir, M.H., Siddiquee, M.F.-R. *et al.*, Multidrug resistance in cancer: Understanding molecular mechanisms, immunoprevention and therapeutic approaches. *Front. Oncol.*, 12, 891652, 2022, https://doi.org/10.3389/fonc.2022.891652.
48. Vaidya, F.U., Sufiyan Chhipa, A., Mishra, V., Gupta, V.K., Rawat, S.G., Kumar, A. *et al.*, Molecular and cellular paradigms of multidrug resistance in cancer. *Cancer Rep. (Hoboken, NJ)*, e1291, 2020, https://doi.org/10.1002/cnr2.1291.
49. Mansoori, B., Mohammadi, A., Davudian, S., Shirjang, S., Baradaran, B., The different mechanisms of cancer drug resistance: A brief review. *Adv. Pharm. Bull.*, 7, 339–48, 2017, https://doi.org/10.15171/apb.2017.041.
50. Amawi, H., Sim, H.-M., Tiwari, A.K., Ambudkar, S.V., Shukla, S., ABC transporter-mediated multidrug-resistant cancer. *Adv. Exp. Med. Biol.*, 1141, 549–80, 2019, https://doi.org/10.1007/978-981-13-7647-4_12.
51. Vasan, N., Baselga, J., Hyman, D.M., A view on drug resistance in cancer. *Nature*, 575, 299–309, 2019, https://doi.org/10.1038/s41586-019-1730-1.
52. Robinson, K. and Tiriveedhi, V., Perplexing role of P-glycoprotein in tumor microenvironment. *Front. Oncol.*, 10, 265, 2020, https://doi.org/10.3389/fonc.2020.00265.
53. Maia, R.C., Vasconcelos, F.C., Souza, P.S., Rumjanek, V.M., Towards comprehension of the ABCB1/P-glycoprotein role in chronic myeloid leukemia. *Molecules*, 23, 119, 2018, https://doi.org/10.3390/molecules23010119.
54. Gottesman, M.M., Fojo, T., Bates, S.E., Multidrug resistance in cancer: Role of ATP-dependent transporters. *Nat. Rev. Cancer*, 2, 48–58, 2002, https://doi.org/10.1038/nrc706.
55. Jiang, L., Wang, P., Sun, Y.-J., Wu, Y.-J., Ivermectin reverses the drug resistance in cancer cells through EGFR/ERK/Akt/NF-κB pathway. *J. Exp. Clin. Cancer Res.*, 38, 265, 2019, https://doi.org/10.1186/s13046-019-1251-7.

56. Korystov, Y.N., Ermakova, N.V., Kublik, L.N., Levitman, M.K., Shaposhnikova, V.V., Mosin, V.A. *et al.*, Avermectins inhibit multidrug resistance of tumor cells. *Eur. J. Pharmacol.*, 493, 57–64, 2004, https://doi.org/10.1016/j.ejphar.2004.03.067.
57. Pouliot, J.F., L'Heureux, F., Liu, Z., Prichard, R.K., Georges, E., Reversal of P-glycoprotein-associated multidrug resistance by ivermectin. *Biochem. Pharmacol.*, 53, 17–25, 1997, https://doi.org/10.1016/s0006-2952(96)00656-9.
58. Lespine, A., Dupuy, J., Orlowski, S., Nagy, T., Glavinas, H., Krajcsi, P. *et al.*, Interaction of ivermectin with multidrug resistance proteins (MRP1, 2 and 3). *Chem. Biol. Interact.*, 159, 169–79, 2006, https://doi.org/10.1016/j.cbi.2005.11.002.
59. Lespine, A., Martin, S., Dupuy, J., Roulet, A., Pineau, T., Orlowski, S. *et al.*, Interaction of macrocyclic lactones with P-glycoprotein: Structure-affinity relationship. *Eur. J. Pharm. Sci.*, 30, 84–94, 2007, https://doi.org/10.1016/j.ejps.2006.10.004.
60. Porazinski, S., Wang, H., Asaoka, Y., Behrndt, M., Miyamoto, T., Morita, H. *et al.*, YAP is essential for tissue tension to ensure vertebrate 3D body shape. *Nature*, 521, 217–21, 2015, https://doi.org/10.1038/nature14215.
61. Guimei, M., Alrouh, S., Saber-Ayad, M., Hafezi, S.A., Vinod, A., Rawat, S. *et al.*, Inhibition of Yes-associated protein-1 (YAP1) enhances the response of invasive breast cancer cells to the standard therapy. *Breast Cancer (Dove Med. Press*, 12, 189–99, 2020, https://doi.org/10.2147/BCTT.S268926.
62. Hagenbeek, T.J., Webster, J.D., Kljavin, N.M., Chang, M.T., Pham, T., Lee, H.-J. *et al.*, The Hippo pathway effector TAZ induces TEAD-dependent liver inflammation and tumors. *Sci. Signal.*, 11, eaaj1757, 2018, https://doi.org/10.1126/scisignal.aaj1757.
63. Steinhardt, A.A., Gayyed, M.F., Klein, A.P., Dong, J., Maitra, A., Pan, D. *et al.*, Expression of Yes-associated protein in common solid tumors. *Hum. Pathol.*, 39, 1582–9, 2008, https://doi.org/10.1016/j.humpath.2008.04.012.
64. Zanconato, F., Cordenonsi, M., Piccolo, S., YAP/TAZ at the roots of cancer. *Cancer Cell*, 29, 783–803, 2016, https://doi.org/10.1016/j.ccell.2016.05.005.
65. Nambara, S., Masuda, T., Nishio, M., Kuramitsu, S., Tobo, T., Ogawa, Y. *et al.*, Antitumor effects of the antiparasitic agent ivermectin via inhibition of Yes-associated protein 1 expression in gastric cancer. *Oncotarget*, 8, 107666–77, 2017, https://doi.org/10.18632/oncotarget.22587.
66. Chan, J.A., Krichevsky, A.M., Kosik, K.S., MicroRNA-21 is an antiapoptotic factor in human glioblastoma cells. *Cancer Res.*, 65, 6029–33, 2005, https://doi.org/10.1158/0008-5472.CAN-05-0137.
67. Yin, J., Park, G., Lee, J.E., Choi, E.Y., Park, J.Y., Kim, T.-H. *et al.*, DEAD-box RNA helicase DDX23 modulates glioma malignancy via elevating miR-21 biogenesis. *Brain*, 138, 2553–70, 2015, https://doi.org/10.1093/brain/awv167.
68. Arbiser, J.L., Bonner, M.Y., Gilbert, L.C., Targeting the duality of cancer. *NPJ Precis. Oncol.*, 1, 2017, https://doi.org/10.1038/s41698-017-0026-x.

69. Xu, W., Jing, L., Wang, Q., Lin, C.-C., Chen, X., Diao, J. *et al.*, Bax-PGAM5L-Drp1 complex is required for intrinsic apoptosis execution. *Oncotarget*, 6, 30017–34, 2015, https://doi.org/10.18632/oncotarget.5013.
70. Hassan, M., Watari, H., AbuAlmaaty, A., Ohba, Y., Sakuragi, N., Apoptosis and molecular targeting therapy in cancer. *Biomed. Res. Int.*, 2014, 150845, 2014, https://doi.org/10.1155/2014/150845.
71. Lopez, J. and Tait, S.W.G., Mitochondrial apoptosis: Killing cancer using the enemy within. *Br. J. Cancer*, 112, 957–62, 2015, https://doi.org/10.1038/bjc.2015.85.
72. Zhu, Z.-C., Liu, J.-W., Li, K., Zheng, J., Xiong, Z.-Q., KPNB1 inhibition disrupts proteostasis and triggers unfolded protein response-mediated apoptosis in glioblastoma cells. *Oncogene*, 37, 2936–52, 2018, https://doi.org/10.1038/s41388-018-0180-9.
73. Song, D., Liang, H., Qu, B., Li, Y., Liu, J., Zhang, Y. *et al.*, Ivermectin inhibits the growth of glioma cells by inducing cell cycle arrest and apoptosis *in vitro* and *in vivo*. *J. Cell. Biochem.*, 120, 622–33, 2019, https://doi.org/10.1002/jcb.27420.
74. Zhang, P., Zhang, Y., Liu, K., Liu, B., Xu, W., Gao, J. *et al.*, Ivermectin induces cell cycle arrest and apoptosis of HeLa cells via mitochondrial pathway. *Cell Prolif.*, 52, e12543, 2019, https://doi.org/10.1111/cpr.12543.
75. Slade, L. and Pulinilkunnil, T., The MiTF/TFE family of transcription factors: Master regulators of organelle signaling, metabolism, and stress adaptation. *Mol. Cancer Res.*, 15, 1637–43, 2017, https://doi.org/10.1158/1541-7786.MCR-17-0320.
76. Deng, F., Xu, Q., Long, J., Xie, H., Suppressing ROS-TFE3-dependent autophagy enhances ivermectin-induced apoptosis in human melanoma cells. *J. Cell. Biochem.*, 120, 1702–1715, 2018, https://doi.org/10.1002/jcb.27490.
77. Polakis, P., Wnt signaling and cancer. *Genes Dev.*, 14, 1837–51, 2000.
78. Arias, A.M., Brown, A.M., Brennan, K., Wnt signalling: Pathway or network? *Curr. Opin. Genet. Dev.*, 9, 447–54, 1999, https://doi.org/10.1016/s0959-437x(99)80068-9.
79. Funayama, N., Fagotto, F., McCrea, P., Gumbiner, B.M., Embryonic axis induction by the armadillo repeat domain of beta-catenin: Evidence for intracellular signaling. *J. Cell Biol.*, 128, 959–68, 1995, https://doi.org/10.1083/jcb.128.5.959.
80. Melotti, A., Mas, C., Kuciak, M., Lorente-Trigos, A., Borges, I., Ruiz i Altaba, A., The river blindness drug Ivermectin and related macrocyclic lactones inhibit WNT-TCF pathway responses in human cancer. *EMBO Mol. Med.*, 6, 1263–78, 2014, https://doi.org/10.15252/emmm.201404084.

7

Ivermectin as an Anti-Inflammatory Agent

Mudasir Nabi Peerzada and Saurabh Verma*

Tumor Biology Department, National Institute of Pathology, Indian Council of Medical Research, New Delhi, India

Abstract

Inflammation is the protective response occurring as a result of microbial infection, tissue injury and any noxious condition. Unchecked inflammation may result in autoimmune or auto-inflammatory disorders, neurodegenerative disease, or cancer. The process of inflammation is complex, involving the innate and adaptive immune systems in response to any infection. Several mediators are involved in inflammatory response such as chemokines, inflammatory cytokines like tumor necrosis factor (TNF), interleukin-1 (IL-1), interleukin-6 (IL-6), and transcription factor NK-κB. Anti-inflammatorily agents currently in use or under development include histone deacetylase inhibitors, PPAR agonists, and small RNAs. A variety of safe and effective anti-inflammatory agents are available, including NSIDs, with many more drugs under development. Anti-parasitic drug ivermectin has shown anti-inflammatory activity and inhibit the function of several targets involved in inflammation progression. In this chapter, we discuss the preclinical and clinical investigations related anti-inflammatory action of ivermectin.

Keywords: Inflammation, cytokine, prostaglandins, ivermectin, drug, doses

7.1 Introduction

Inflammation is the protective support that takes place in the higher organisms as a result of microbial infection, tissue injury and other noxious conditions. It is one of the important responses from the immune system of the organism which initiates the several mechanisms for the removal of any harmful stimuli and ultimately leads to the healing of damaged cells.

**Corresponding author*: saurabhverma.nip@gov.in; svarmasv1@rediffmail.com

Rashid Ali and Shahid ul-Islam (eds.) Chemistry and Biological Activities of Ivermectin, (141–162)

The first line of defense against any foreign material, antigen, or harmful molecules translates into acute inflammation. The process of inflammation is a complex as it involves initiation, promotion, and resolution. Broad-spectrum of inflammation has been studied that occur through the various stimuli or external factors and involve various mechanisms. In response to any harmful condition, the response through means of inflammation leads to restoration of homeostasis. Inflammation is therefore recognized as the adaptive response by the body to any harmful act that can cause damage to the cellular homeostasis.

As the adaptive response concurs at the expense of normal cell functioning, the host body therefore suffers from large damages due to response persistence [1]. Inflammation response takes place swiftly on receiving stimuli that can be any harmful agent such as microbial infection, or any foreign irritant. The components of innate immune system such as macrophages, dendritic cells, lymphocytes, neutrophils and mast cells play significant role in inflammatory response process [2]. In addition, the non-immune cells such as endothelial cells, epithelial cells, and fibroblasts also contribute to the process of inflammatory response. The cell signaling pathways that are involved in the inflammation vary according to the initiating stimuli. In case of stimulus form bacterial infection, the activation of pathogen-specific receptors occurs that leads to the production of inflammatory mediators. These mediators can be chemokines and inflammatory cytokines like tumor necrosis factor (TNF), interleukin-6 (IL-6) and interleukin-1 (IL-1). Secretion of these mediators causes the modification of vascular endothelial permeability and recruitment of neutrophils, plasma containing antibodies and therefore promotes the infection rate [3]. The time taken by the inflammatory response to come into play deepens on the extent of harm caused by infection. Inflammatory response mostly involves the secretion of acute phase proteins such as C-reactive protein and coagulation factors by the liver cells due to production of inflammatory cytokines. Productions of these proteins assist in the generation of prostaglandins and the symptoms of pain, fever appear as it involves central nervous system [4].

Viral infections represent the similar effect but involve the signaling pathways that lead to the production of cytokines such as cytotoxic lymphocytes and type-1 interferons (IFNs). Cytokine type-1 IFNs play pivotal role in the progression of viral infections. However, the parasitic infections or allergens stimulate the production of IL-4, IL-5, IL-13 and histamine while inflammatory response process [5]. Inflammatory stimulus is recognized by the pattern recognition receptors (PRRs) that are expressed by both innate and adaptive immune systems. These receptors sense the presence of any foreign irritant that can cause the cellular damage with the aid of danger-associated molecular patterns and pathogen associated molecular patterns (PAMPs) which can be Toll-like receptors (TLRs), RIG-1-like receptors (RLRs), C-type lectin receptors (CLRs) and NOD-like receptors (NLRs). Recruitment of these receptors leads to the signal transduction to cell nucleus to initiate the transcriptional and posttranscriptional mechanisms that produce pro-inflammatory cytokines such as TNF, IL-1β and IL-6. However, the signaling from the PRRs cause the activation of transcription factors to produce the pro-inflammatory chemokines and cytokines.

The transcription factor NK-κB is one of the most explored factors involved in the production of pro-inflammatory products. Mammalian NK-κB is composed of five proteins including p50, p52, RelA, c-Rel, and RelB and all these have the structural homology in the N-terminal side with that of the retroviral oncoprotein v-Rel [6]. One of the important aspects of NK-κB family proteins is that they are retained in the cytoplasm of un-stimulated cells because of ankyrin repeat-containing IκB proteins. In addition to NK-κB, the other factors involved in the inflammatory response include activator protein-1 (AP-1), E2F, serum responses factor (SRF), cyclic-AMP (cAMP) response element binding protein (CREB), and the ternary complex factors (TCFs) [7]. These factors get activated on the receiving the inflammatory stimuli, and are associated with the phosphorylation or dephosphorylation of these factors.

The mechanisms involved in selective transcription process of inflammatory genes widen to the differential binding of transcription factors to the particular DNA base-pairs. Investigation has suggested that the process of transcription in eukaryotic genes depends on the structure of chromatin. Chromatin has repeating units made of histones H2A, H2B, H3 and H4, linker histones and other nonhistone proteins [8]. Regulation of inflammation in humans is controlled by a series of epigenetic mechanisms. During inflammation, the acetylation of histone H3 at promotor site causes the increase in recruitment of NF-κB proteins. This histone acetylation causes the activation of several inflammation related genes, whereas the histone deacetylase (HDAC) promotes the deactivation of these genes.

Reversal of inflammatory response occurs once the stimulating factors are eradicated and this process occurs quickly through an inflammation resolution process. Reversal process is mediated by anti-inflammatory factors such as IL-10, TGF-β, glucocorticoids, monocytes [9]. The acute inflammation advances towards the chronic sage if the reversal process fails to eradicate the triggering stimulus at any phase. The damage caused to the host in chronic stage of inflammation is majorly due to the response of host but not due to the foreign stimulus [10]. Chronic inflammatory conditions are associated with several diseases of concern such as obesity, atherosclerosis, type 2 diabetes, inflammatory bowel diseases, asthma, neurodegenerative diseases, cancer, and rheumatoid arthritis. Inflammation is also associated with Alzheimer's disease, Parkinson's disease and multiple sclerosis [11]. In addition, inflammatory bowel disease, glomerulonephritis caused by the inflammation of the glomeruli, small blood vessels in the kidneys are the ailments having close link with inflammation. It has been established that inflammation is associated with progression of cancer, an inflammatory microenvironment is frequently observed in tumors [12, 13]. Inflammation caused by the infections is demonstrated to have link with the development to some cancers. For instance, lung cancer caused by tobacco is triggered by the JNK1 and IKK-β mediated chronic inflammation which indicates that pathways responsible for cancer development and inflammation are linked with each other [14, 15].

Chronic inflammation has the significant linkage with the pathogenesis of inflammatory diseases. The mechanism of response to inflammation is different for the inflammatory diseases and more complex some times. Therefore, the asthma and allergic inflammations occur due to deregulated interactions between the mucosal epithelia and innate immune cells. Adaptive immunity cells like CD4 T helper 2 (Th2) and Th2-associated cytokines are strongly linked to inflammation due to asthma [16, 17].

Current methods of treatment for the inflammation include use of selective glucocorticoid receptor agonist, immune selective anti-inflammatory derivatives, protectins, resolvins, and TNF inhibitors [18]. Anti-inflammatory drugs inhibit the activity of cyclooxygenases (COX) that are associated with biosynthesis of prostaglandins which in turn accelerate the inflammation progression. The prostaglandin biosynthesis or COX pathway is an essential component of inflammatory responses [19]. COX pathway gets activated by means of lipopolysaccharides (LPS), cytokines, xenobiotics, and fluorides. The COX-2 isoform is associated with inflammation, and is inhibited by celecoxib and rofecoxib [20]. However, the conventional non-steroidal anti-inflammatory drugs (NSAIDs) suffer from side effects related to gastric and renal mechanisms, as they inhibit both the COX-1 and COX-2 isoenzymes. Maximum of the selective COX-2 inhibitors suffer from no gastric side-effects but do have linkage with cardiac issues on long-term use [21, 22]. The research is ongoing for the quest for anti-inflammatory drugs without side-effects (Figure 7.1) [23].

The development of novel anti-inflammatory agents is generally designed in a way to inhibit the prostaglandin E2 biosynthesis through the cyclooxygenases. Several inhibitors crossed the clinical investigations

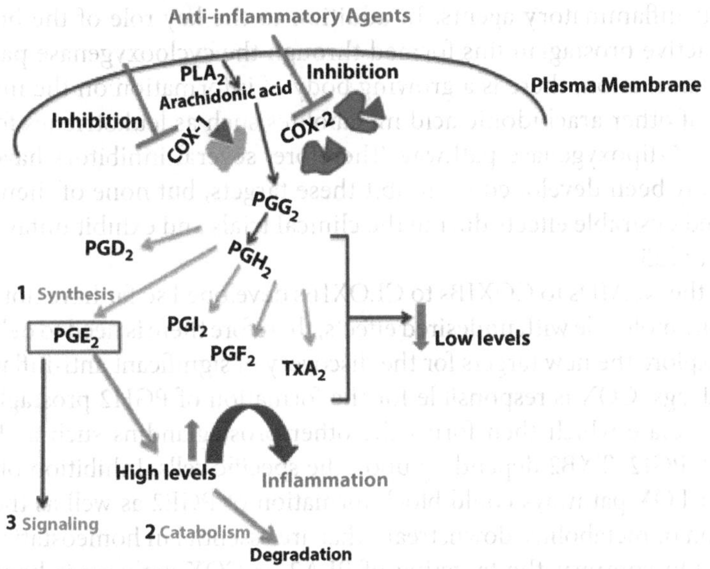

Figure 7.1 Pathway for the formation of PGE2 and other metabolites responsible for inflammation.

successfully but show unwanted side-effects. Therefore, research is going on - to identify the new targets that can limit the PGE2 production anyway. Scientists are trying to target the 5-LOX pathway as an alternative to develop the inhibitors of desired anti-inflammatory effects. 5-LOX pathway is involves in the generation of inflammatory leukotrienes which makes this pathway the potential target to develop the new anti-inflammatory agents. In addition, the development of phospholipase inhibitors that are involved in the production of arachidonic acid is one of the important areas for the development of anti-inflammatory agents. Further, COX-2 and 5-LOX dual inhibitors (CLOXIBs) are also developed as anti-inflammatory agents [24].

However, several drugs developed for these targets are available in the market, but these have undesirable side-effects, as they affect the formation of other key metabolites mediating physiological functions. Therefore, the other targets that are now focused include PGE synthases, 15-PGDH and PGE receptors. The mPGES-1 is considered as one of the important PGE synthases in the development of new anti-inflammatory agents. The inflammation mediator, EP4 is presumed significant target among PGE receptors in anti-inflammatory drug discovery. Arachidonic acid metabolites such as leukotrienes that are produced through 5-lipoxygenase pathway are indicated as significant targets for the development of anti-inflammatory agents. In addition to the key role of the biologically active prostaglandins formed through the cyclooxygenase pathway in inflammation, there is a growing body of information on the involvement of other arachidonic acid metabolites such as leukotrienes formed via the 5-lipoxygenase pathway. Therefore, several inhibitors have successfully been developed to inhibit these targets, but none of them have showed desirable effects during the clinical trials and exhibit unfavorable toxicity [25].

As the NSAIDs to COXIBs to CLOXIBs developed so far have not delivered the molecule with undesired effects, therefore there is need to delineate and explore the new targets for the discovery of significant anti-inflammatory drugs. COX is responsible for the formation of PGH2 prostaglandin intermediate which then forms the other prostaglandins such as PGE2, PGD2, PGI2, TXB2 depending upon the specific cells. Inhibition of COX and or LOX pathways could block formation of PGE2 as well as the production of metabolites downstream that are essential in homeostatic functions. On contrary, the targeting of PLA2 or COX pathway is known to affect the histamine production, therefore the target selection and its validation for other functions is extremely necessary to develop the safe and effective drugs (Figure 7.2) [26, 27].

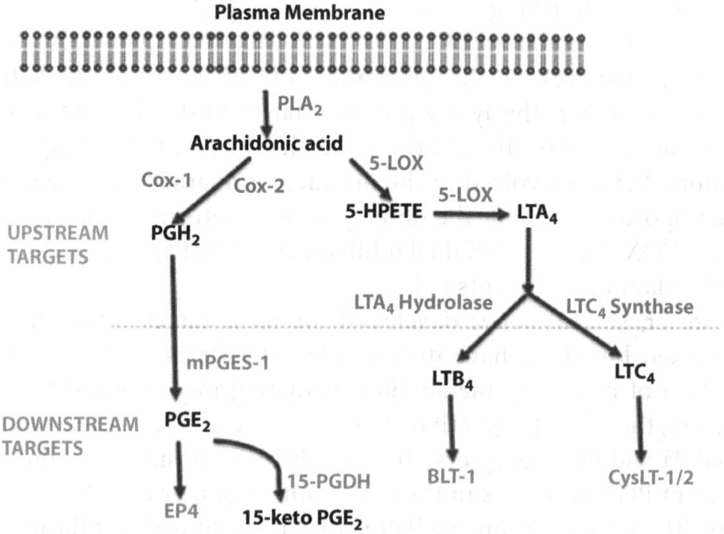

Figure 7.2 Representation of the metabolism of arachidonic acid through cyclooxygenase and lipoxygenase pathways indicating the probable up-stream down-stream targets for the discovery and development of anti-inflammatory drugs.

7.2 Ant-Inflammatory Action of Ivermectin

Several anti-inflammatory agents are used to prevent the inflammation depending on the stimuli and type of disease. Emergence of the resistance and target selectivity prompted the researchers to find out the novel anti-inflammatory agents. Macrolide antibiotics are widely used as the anti-microbial agents because they can bind reversibly to 50s ribosomal subunit in preventing the process of protein synthesis. Macrolides have showed anti-inflammatory action both *in vivo* and in clinical investigations. Ivermactin (Figure 7.3), the semi-synthetic molecule of macrolides is effective in the treatment of strongyloides, scabies, and some types of filariasis. Ivermectin belongs to a group of avermectins, the compounds produced by *Streptomyces avermitilis*. The avermectins produced naturally comprises of avermectin A_1, A_2, B_1, and B_2. Each of these avermectins exists in variants (a) and (b). Designations A and B represent the presence of either methoxy or hydroxy group at the C5 position. Whereas, labels 1 and 2 represents the presence of a double bond between C22 and C23 or a hydrogen atom at C22 and hydroxyl group at C23 positions, respectively. The 'a' variants contains secondary butyl group at C25 position whereby -

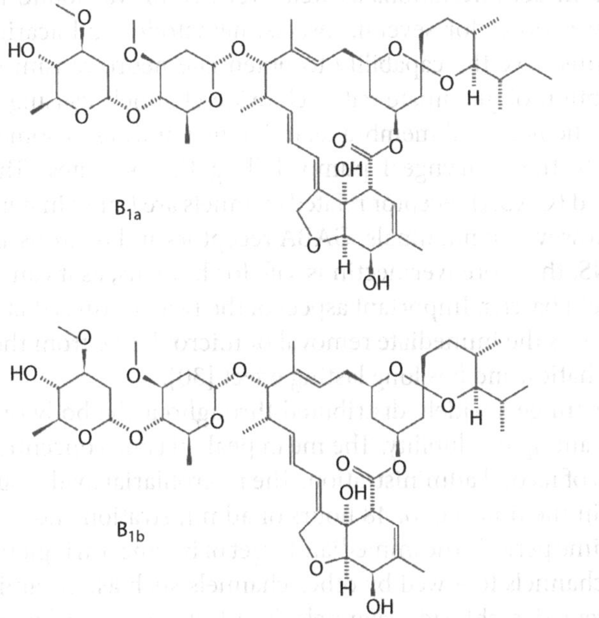

Figure 7.3 Structures of ivermectin having the mixture of avermectin B_{1a} and B_{1b} in 80:20 ratio.

the 'b' variants contains isopropyl group at this position. These differences in chemical structure were found to have significant effect on the activities of the avermectins [28].

It has been found that B series of avermectins have showed more activity against the gastrointestinal nematodes of sheep. However, avermectin B1 is more active than B2 when administered orally, however avermectin B2 is more active than B1 in parenteral administration. Therefore, development of anthelmintic drug candidates was focused on the 'B' series and the structure at the C22 and C23 positions of the avermectins. Ivermectin is the modified derivative of avermectin B1 which comprises of 80% avermectin B1a and 20% avermectin B1b (Figure 7.2). This derivative has showed potent activity against a broad spectrum of parasitic nematodes orally as well as in parenteral administration. Ivermectin was found safe and effective and was therefore commercialized in veterinary, agriculture and aquaculture fields in 1981. Later it was found that this drug has the potential to be effective in human health and was therefore used for the treatment of onchocerciasis throughout the tropics among poverty stricken population [29].

Ivermectin was found as broad-spectrum antiparasitic agent, to combat parasitic worms in veterinary and human medicine. This compound has been used in the treatment of lymphatic filariasis, worm-related infections, and insect infestations as well. Ivermectin was found highly effective at lower doses for several, insects, nematodes and acarine parasites. Avermectins have the capability to potentiate neurotransmission though the disruption of glutamate-gated chloride channels causing hyperpolarization of the neuronal membrane, inducing paralysis of somatic muscles, particularly the pharyngeal pump, killing the parasites. The γ-aminobutyric acid (GABA) receptor related channels are found in nematodes and insects. However in mammals, GABA receptors and neurons are restricted to the CNS, therefore ivermectin is safe for humans, as it cannot cross the blood–brain barrier. Important aspect of the Ivermectin is that its administration causes the immediate removal of microfilariae from the peripheral skin lymphatics, and has long lasting effect [30].

Ivermectin gets widely distributed throughout the body organs due to its significant lipid solubility. The mean peak plasma concentration results in 4 hours of its oral administration. The microfilarial load is suppressed to 78% within the time gap of 48 hours of administration and to 98% with 2 weeks of time period. The immediate target of ivermectin is glutamate-gated chloride channels followed by other channels such as, histamine- GABA- and pH-sensitive chloride channels. The fast removal of microfilariae is because of the suppression of its ability to target the immune defense mechanism of the host. Ivermectin is used for the treatment of onchocerciasis, strongyloidiasis and lymphatic filariasis. Ivermectin is widely used as an antiparasitic agent in mammals since decades as it has showed minimal adverse effects. Ivermectin is being evaluated for its inhibitory profile against several parasites causing different diseases. It has been examined for myiasis, trichinosis, malaria, leishmaniasis, African trypanosomiasis, American trypanosomiasis, and schistosomiasis. Further it has been evaluated against epilepsy, amyotrophic lateral sclerosis, HIV, dengue, encephalitis, tuberculosis, Buruli ulcer, and cancer [29].

Since, the macrolide based compounds possess anti-inflammatory properties, therefore it was speculated that ivermectin could be the drug candidate for treatment of various inflammations. In this context, as a part of ongoing discovery of the new anti-inflammatory agents, ivermectin was established to have anti-inflammatory properties. Ivermectin is known as wonder drug and is in clinical use since 1981 for treatment of parasitic infections in humans as described above [31]. The impact of ivermectin in controlling onchocerciasis and lymphatic filariasis lead its discoverers to get the Nobel Prize in Medicine in 2015 [32]. Ivermectin is included on the world health organization's list of essential medicines. Ivermectin has been found effective in several human diseases and is explored for the treatment of other diseases a well [33].

In one of the research, the impact of ivermectin in mice challenged with intraperitoneal lipopolysaccharide (LPS) was investigated. Administration of ivermectin orally for two hours in mice before giving LPS causes the reduction in mortality up to half than untreated mice. This is because ivermectin blocks the production of cytokine by LPS challenged macrophage and demonstrates the anti-inflammatory properties of this drug [34]. In addition, the impact of ivermectin on LPS signaling in the RAW 264.7 murine macrophage cells was investigated. Ivermectin inhibits LPS-induced production of inflammatory cytokines by blocking NF-κB pathway. It was found that at 0.625 to 5mg/L concentration of ivermectin, activation of NF-κB and the stress-activated MAP kinases such as JNK and p38.6 was suppressed which indicated that ivermectin has anti-inflammatory effects. It is also reported that ivermectin significantly reduced the production of IL-1, IL-6 and TNF-alpha both *in vivo* and *in vitro* conditions and this action support its anti-inflammatory activity [35].

Some reports are there which suggest that ivermectin shows anti-inflammatory activity in animal model of allergic inflammation such as asthma and atopic dermatitis when administered systemically or topically [36, 37]. The anti-inflammatory effect is observed at dosage of 2mg/kg when given systemically *in vivo* sitting. Dosage of 2 mg/kg of ivermectin reduces the hyper-secretion of mucus in the respiratory tract, decrease in the recruitment of immune cells, suppression in the production of cytokines and IgE/IgG1 in bronchoalveolar lavage (BAL). This effect demonstrates that ivermectin has an anti-inflammatory effect at a systemic level as well as on lung tissue [36]. Such anti-inflammatory effects of ivermectin are limited to LPS or toll-like receptor 4 (TLR4) signaling; however its inflammatory effects involve other proinflammatory signaling pathways downstream as well [38]. The anti-inflammatory effects of ivermectin could be because of the inhibition of IL-4 production in case of allergic asthma. Investigations have showed that ivermectin suppresses allergic inflammation effectively because its administration causes the movement of inflammatory cells into the region around airways. Its administration leads to rise in the level of goblet cells within the airways and elevation of Th2 cytokines in the BALF *in vivo.* It also inhibits the development of airway hyper-responsiveness in mouse model studies [39].

Ivermectin causes decrease in ovalbumin specific immunoglobulins IgE and IgG1, in addition to having effect on marker bronchoalveolar lavage fluid (BALF) cytokines. Cells communicate with each other by means of the cytokines, therefore they are considered as the important factors in treatment of diseases including inflammation. In case of allergic asthma, overproduction of Th2 cytokines occurs and is characterized as the T cell dependent condition. This over production of cytokines cause inflammatory responses by means of eosinophil activation, airway infiltration, mucus production and immunoglobulin E production. Ivermectin causes decrease in the OVA-specific IgE levels, IL-4 and IL-13 levels in the BALF *in vivo.* It is important to mention that IL-13 and IL-13 have significant role in causing airway inflammation, goblet cell hyperplasia and airway luminal narrowing. Further, IL-4 is considered responsible for the overproduction of IgE leading to the allergic asthma. Therefore, inhibition of these factors by use of ivermectin indicates that it is one of the significant drug candidates with potent anti-inflammatory effects [36, 40].

Ivermectin reduces the IL-4, IL-13 levels and count of PAS-positive mucus-producing goblet cells in the airways for the prevention of inflammation. IgG1 is regulated with IgE by means of IL-4/Th2 pathway in mice and ivermectin reduces the production of OVA-specific IgG1 which demonstrates that ivermectin show anti-allergic effect through by reducing the IgG1 production. In addition, IL-5 plays important role in the growth, survival and function of eosinophils. Th2 cell-derived IL-5 initiates maturation of CD34+/IL-5Ra+ precursor cells to eosinophils followed by mobilization from bone marrow into the blood stream. Ivermectin inhibits the IL-5 production having effect on eosinophil maturation and inhibition of inflammation effect [41]. MAPKs regulate the cell growth and differentiation, and have the link with the cytokines. Additionally, the NF-κB is involved in the expression of cytokines responsible for inflammatory response. Ivermectin cause the inhibition of cytokine production through suppression of MAPK and NF-κB signaling, therefore it is speculated that it cause the inhibition of allergic asthma inflammation trough the same mechanism [35, 42].

Furthermore, it has been found that topical ivermectin leads to the improvement in skin inflammation associated with allergy through the suppression of priming and activation of allergen-specific T cells and the production of cytokines [37]. Rosacea occurs predominantly in adults and is one of the common chronic skin inflammatory diseases. The effected region is usually the face and it has sensational burning accompanied with inflammatory lesions. Papulopustular rosacea type is most common type of rosacea, and is associated with persistent erythemia and transient pustules. For the treatment of rosacea, agents such as topical metronidazole or azelaic acid, oral tetracycline class of anti-bacterials are used. Antiparasitic agents are also used because of growth of Demodex mite in rosacea inflammation condition. Ivermectin has the anti-inflammatory effect for rosacea condition, and it is speculated that its effect is because of the anti-parasitic activity. It is also considered that ivermectin could be recovering the inflammation through some alternative mechanism in case of rosacea skin lesions. Ivermectin does not inhibit the activity of CYP isoenzymes and may not interact with drugs that are metabolized by such isozymes. But, CYP3A4 has the role in ivermectin metabolism, therefore caution is to be taken while administration of ivermectin with CYP3A4 inhibitors to maintain the concentration of ivermectin in plasma [43, 44].

Ivermectin 1%, a novel topical agent is indicated for the treatment of inflammatory lesions caused by rosacea. This drug has shown significant effectiveness and tolerance in the topical treatment of inflammatory lesions of rosacea [45]. Rosacea is a common chronic inflammatory skin disease that occurs mainly in adults. Rosacea is mainly an inflammatory facial dermatoses associated with centrofacial erythema [46].

Available treatment options for the rosacea turned out to be infective often due to multi-factorial nature of this dermal condition. At present, majority of available treatment options include the topical drug administration. Rosacea patients can experience symptoms of concern and are inherently sensitive to topical treatment. However, the ivermectin seems to be one of the potent treatment options for this dermal inflammatory condition. Ivermectin 1%, is approved for the treatment of inflammatory rosacea by US-FDA [47]. Ivermectin reduces the inflammation associated demodex folliculorum, and also kills this microscopic mite. Investigation has revealed that ivermectin 1% cream reduces the inflammatory lesions in papulopustular rosacea, and improve the patient satisfaction and quality of the life [48]. In another event, it was found that ivermectin 1% reduces the inflammation more satisfactorily than metronidazole 0.75% cream [49]. Ivermectin is well tolerated and advantageous for the treatment of perioral dermatitis. Topical ivermectin has shown to down regulate the pro-inflammatory genes cathelicidin, IL-8, LL-37, TLR-4, HBD 3 and TNF-α in papulopustular rosacea along with its anti-parasitic action against mite Demodex [50].

Ivermectin has showed properties against inflammation in several studies. *In vitro* and *in vivo* analysis has demonstrated that ivermectin inhibit the lipopolysaccharide (LPS)-induced production of pro-inflammatory cytokines, including tumor necrosis factor α (TNF-α), interleukin (IL)-1b and IL-6 [35]. Ivermectin also suppressed LPS-induced secretion of inflammatory mediator's nitric oxide and prostaglandin E2 in macrophages *in vitro* [51]. *In vitro* investigation had shown that the ivermectin bind efficiently to the serum albumin plasma protein and has no significant affin ity for erythrocytes [52]. Ivermectin is metabolized by cytochrome P450 (CYP) 3A4 significantly and give rise to dimethyl-derivative of ivermectin and hydroxy ivermectin as two metabolites. These metabolic products reach steady state after 2 weeks of parent drug administration and are not accumulated over the 12 weeks of treatment time [53].

Investigations have revealed that ivermectin function by means of GABA receptors and ligand-gated channels including glutamate-gated chloride channels. Moreover, ivermectin has exhibited effectiveness in regulating the glucose, cholesterol, inhibition of malignant cells and inhibition of viral proliferation. Studies have also exclusively outlined that ivermectin causes the inhibition of the inflammatory cytokine generation through the suppression of the NF-κB pathway. Additionally, it has been revealed that controlling the inflammation in acetic acid colitis condition *in vivo* is possible through the activation of GABAB signaling pathway with the help of ivermectin or baclofen or by using both agents in combination therapy. However, the combination therapy is useful in that - the desired anti-inflammatory effect can be observed at lower doses of these agents, thereby reducing threat of any likely side effects of higher doses of either ivermectin or baclofen in mono-therapy. This strategy could be applied in the clinical setting of these agents safely [54].

Further, macrophages, neutrophils, and T-cells play important role in inflammation. They express pattern-recognition receptors (PRRs) that can perceive microbial components pathogen-associated molecular patterns (PAMPs). PRRs can detect the damage-associated molecular patterns

(DAMPs), released by damaged tissues and necrotic cells. There are five types of PRR families, namely, toll-like receptors (TLRs), NOD-like receptors (NLRs), RIG-I-like receptors, cytosolic DNA sensors and C-type lectin-like receptors. Different PRRs respond to the different PAMPs and DAMPs - though sharing some common signatures downstream of the signaling pathway [55, 56]. A common signaling event in the PRRs includes activation of the canonical NF-κB pathway. This active action leads to the transcriptional induction of pro-inflammatory cytokines, chemokines and additional inflammatory mediators. These inflammatory mediators can directly take on the induction of inflammation and can also act indirectly by promoting the inflammatory T cell differentiation mechanism. A signaling event that integrates the various PRR pathways for the activation of NF-κB is the transformation of growth factor-β-activated kinase 1 (TAK1). On activation, TAK1 activates the downstream IKK kinase and cause the IκBα phosphorylation and ultimately NF-κB activation. The pro-inflammatory function of NF-κB is well emphasized in macrophages that function in an immune response against infections. Macrophages secrete large array of cytokines and chemokines in response to various PAMPs and DAMPs. Macrophage M1 secrete pro-inflammatory cytokines including IL-1, IL-6, IL-12, TNF-α and chemokines that are involved in various inflammatory processes (Figure 7.4) [57, 58].

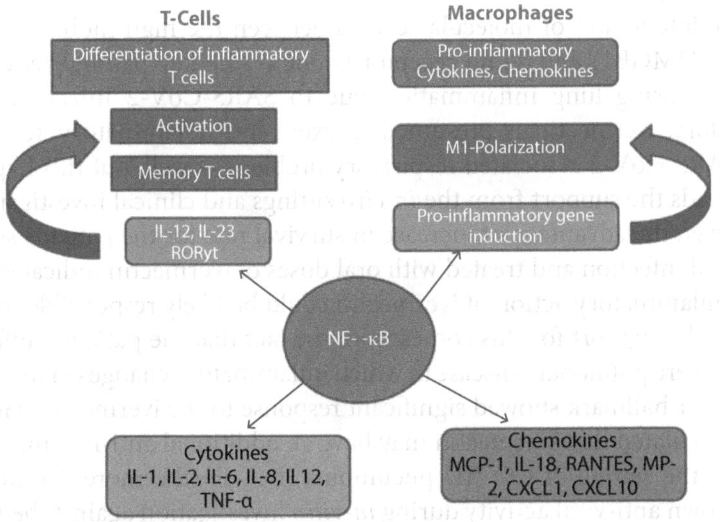

Figure 7.4 Involvement of NF-κB in the development and progression of inflammation. NF-κB stimulates the transcription of several genes for inflammation regulation on activation. NF-κB increases the production of inflammatory cytokines and chemokines.

Quite recently, ivermectin has also been investigated against the inflammatory conditions associated with the COVID-19 by acting as a positive allosteric modulator of the alpha-7 nicotinic acetylcholine receptor (α7 nAChR) [59]. Recently, *in silico* docking studies have indicated a potential direct interaction between the SARS-CoV-2 spike glycoprotein and α7 nAChR, indicating a potential candidate for the treatment of inflammatory phase of COVID-19 [60]. It is to be pertinent to mention that core receptor of the cholinergic anti-inflammatory pathway is α7 nAChR, which is under the control of the vagus nerve, and plays a crucial role in balancing of the body's response to inflammation and sepsis [55]. This anti-inflammatory pathway connects the involuntary parasympathetic nervous system innervating all major organs to cytokine-producing cells such as IL1, IL6 and TNF-secreting macrophages, mast cells and lymphocytes, which are reported to play a major role during the inflammatory phase of COVID-19 infection such as cytokine production [61]. It is speculated that administration of the ivermectin could interfere with nAChR pathway and could rapidly lower the pro-inflammatory cytokine levels and decrease expressions of chemokines as well as adhesion molecules at the inflammatory sites. Importantly, the marked increase in Ca^{2+} due to the acetylcholine (ACh) in the presence of micromolar concentrations of ivermectin may be one of the reasons for the significant clinical activity of the ivermectin during the late inflammatory phase of severe COVID-19 [62].

The interaction of molecular events between the high mobility group box 1 (HMGB1) and TLR4 receptor is one of the key pathogenetic processes causing lung inflammation due to SARS-CoV-2 infection [63]. Therefore, ivermectin is presumed to exert anti-inflammatory action in the SARS-CoV-2 associated respiratory problem as well, but this hypothesis needs the support from the *in vivo* sittings and clinical investigations. However, the advantage of increase in survival rate for the patients suffering viral infection and treated with oral doses of ivermectin indicated that anti-inflammatory action of ivermectin could be likely responsible for the effect. The support for this comes from the fact that the patients suffering from severe pulmonary disease in which inflammatory changes in the lungs are major hallmark showed significant response to the ivermectin. Hence, it is speculated that ivermectin may have an additional anti-inflammatory role in the setting of COVID pneumonia [64]. Furthermore, ivermectin has shown anti-viral activity during *in-vitro* investigation against the DNA and RNA viruses such as HIV-1, Venezuelan equine encephalitis virus, West Nile virus, influenza virus and dengue [65].

7.3 Conclusions

Ivermectin has shown a significant inhibition profile for the inhibition of a variety of targets involved in various pathways during inflammation responses. Ivermectin can be used at low doses and it causes the mild side effects, thus it may be effective as a therapy for patients with allergic airway inflammation. Ivermectin is one of the effective drugs for the topical treatment of inflammatory lesions occurring in rosacea. Ivermectin shows the anti-inflammatory effect through inhibition of NF-κB signaling and can reduce the production of IL-1, IL-6 and TNF-alpha. It has the capability to act against the LPS and is therefore significant candidate for the treatment of inflammatory conditions associated with late stage COVID-19 phase. Ivermectin has emerged as a wonder drug because of its applicability in the possible treatment for several diseases including inflammation. We hope that this chapter on the anti-inflammatory action of ivermectin will be very useful for the researchers working in this area.

Acknowledgements

M. N. Peerzada is thankful to the Indian Council of Medical Research (ICMR) New Delhi, Department of Health Research, Government of India for a postdoctoral fellowship (No. 3/1/3PDF(24)/2021-HRD-6).

References

1. Antonelli, M. and Kushner, I., It's time to redefine inflammation. *FASEB J.*, 31, 1787–1791, 2017, https://doi.org/10.1096/fj.201601326R.
2. Akira, S., Uematsu, S., Takeuchi, O., Pathogen recognition and innate immunity. *Cell*, 124, 783–801, 2006, https://doi.org/10.1016/j.cell.2006.02.015.
3. Turner, M.D., Nedjai, B., Hurst, T., Pennington, D.J., Cytokines and chemokines: At the crossroads of cell signalling and inflammatory disease. *Biochim. Biophys. Acta - Mol. Cell Res.*, 1843, 2563–2582, 2014, https://doi.org/10.1016/j.bbamcr.2014.05.014.
4. Sproston, N.R. and Ashworth, J.J., Role of C-reactive protein at sites of inflammation and infection. *Front. Immunol.*, 9, 754, 2018, https://doi.org/10.3389/fimmu.2018.00754.
5. Arrouchi, H., Lakhlili, W., Ibrahimi, A., Re-positioning of known drugs for Pim-1 kinase target using molecular docking analysis. *Bioinformation*, 15, 116–120, 2019, https://doi.org/10.6026/97320630015116.

6. Page, A., Navarro, M., Garín, M., Pérez, P., Llanos Casanova, M., Moreno, R., Jorcano, J.L., Cascallana, J.L., Bravo, A., Ramírez, A., IKKβ leads to an inflammatory skin disease resembling interface dermatitis. *J. Invest. Dermatol.*, 130, 1598–1610, 2010, https://doi.org/10.1038/jid.2010.28.
7. Dalton, S. and Treisman, R., Characterization of SAP-1, a protein recruited by serum response factor to the c-fos serum response element. *Cell*, 68, 597–612, 1992, https://doi.org/10.1016/0092-8674(92)90194-H.
8. Dillon, N. and Grosveld, F., Chromatin domains as potential units of eukaryotic gene function. *Curr. Opin. Genet. Dev.*, 4, 260–264, 1994, https://doi.org/10.1016/S0959-437X(05)80053-X.
9. Zizzo, G. and Cohen, P.L., IL-17 stimulates differentiation of human anti-inflammatory macrophages and phagocytosis of apoptotic neutrophils in response to IL-10 and glucocorticoids. *J. Immunol.*, 190, 5237–5246, 2013, https://doi.org/10.4049/jimmunol.1203017.
10. Chen, L., Deng, H., Cui, H., Fang, J., Zuo, Z., Deng, J., Li, Y., Wang, X., Zhao, L., Inflammatory responses and inflammation-associated diseases in organs. *Oncotarget*, 9, 7204–7218, 2018, https://doi.org/10.18632/oncotarget.23208.
11. Furman, D., Campisi, J., Verdin, E., Carrera-Bastos, P., Targ, S., Franceschi, C., Ferrucci, L., Gilroy, D.W., Fasano, A., Miller, G.W., Miller, A.H., Mantovani, A., Weyand, C.M., Barzilai, N., Goronzy, J.J., Rando, T.A., Effros, R.B., Lucia, A., Kleinstreuer, N., Slavich, G.M., Chronic inflammation in the etiology of disease across the life span. *Nat. Med.*, 25, 1822–1832, 2019, https://doi.org/10.1038/s41591-019-0675-0.
12. Singh, N., Baby, D., Rajguru, J.P., Patil, P.B., Thakkannavar, S.S., Pujari, V.B., Inflammation and cancer. *Ann. Afr. Med.*, 18, 121–126, 2012, n.d., https://doi.org/10.4103/aam.aam_56_18.
13. Greten, F.R. and Grivennikov, S.I., Inflammation and cancer: Triggers, mechanisms, and consequences. *Immunity*, 51, 27–41, 2019, https://doi.org/10.1016/j.immuni.2019.06.025.
14. Takahashi, H., Ogata, H., Nishigaki, R., Broide, D.H., Karin, M., Tobacco smoke promotes lung tumorigenesis by triggering IKKβ- and JNK1-dependent inflammation. *Cancer Cell*, 17, 89–97, 2010, https://doi.org/10.1016/j.ccr.2009.12.008.
15. Zhao, H., Wu, L., Yan, G., Chen, Y., Zhou, M., Wu, Y., Li, Y., Inflammation and tumor progression: Signaling pathways and targeted intervention. *Signal Transduction Targeting Ther.*, 6, 263, 2021, https://doi.org/10.1038/s41392-021-00658-5.
16. Locksley, R.M., Asthma and Allergic Inflammation. *Cell*, 140, 777–783, 2010, https://doi.org/10.1016/j.cell.2010.03.004.
17. Lloyd, C.M. and Hessel, E.M., Functions of T cells in asthma: More than just TH2 cells. *Nat. Rev. Immunol.*, 10, 838–848, 2010, https://doi.org/10.1038/nri2870.

18. Dhingra, A., Chopra, B., Dass, R., Mittal, S., An update on anti-inflammatory compounds: A review. *Antiinflamm. Antiallergy Agents Med. Chem.*, 14, 81–97, 2015, https://doi.org/10.2174/1871523014666150514102027.
19. Zarghi, A. and Arfaei, S., Selective COX-2 inhibitors: A review of their structure-activity relationships. *Iran. J. Pharm. Res. IJPR*, 10, 655–83, 2011, https://doi.org/24250402.
20. Baek, S.-H., Park, T., Kang, M.-G., Park, D., Anti-inflammatory activity and ROS regulation effect of sinapaldehyde in LPS-stimulated RAW 264.7 macrophages. *Molecules*, 25, 4089, 2020, https://doi.org/10.3390/molecules25184089.
21. Hörl, W.H., Nonsteroidal anti-inflammatory drugs and the kidney. *Pharmaceuticals (Basel)*, 3, 2291–2321, 2010, https://doi.org/10.3390/ph3072291.
22. Patrono, C., Cardiovascular effects of cyclooxygenase-2 inhibitors: A mechanistic and clinical perspective. *Br. J. Clin. Pharmacol.*, 82, 957–64, 2016, https://doi.org/10.1111/bcp.13048.
23. Wongrakpanich, S., Wongrakpanich, A., Melhado, K., Rangaswami, J., A comprehensive review of non-steroidal anti-inflammatory drug use in the elderly. *Aging Dis.*, 9, 143–150, 2018, https://doi.org/10.14336/AD.2017.0306.
24. Ahmed, A.U., An overview of inflammation: Mechanism and consequences. *Front. Biol. (Beijing)*, 6, 274, 2011, https://doi.org/10.1007/s11515-011-1123-9.
25. Wang, B., Wu, L., Chen, J., Dong, L., Chen, C., Wen, Z., Hu, J., Fleming, I., Wang, D.W., Metabolism pathways of arachidonic acids: Mechanisms and potential therapeutic targets. *Signal Transduction Targeting Ther.*, 6, 94, 2021, https://doi.org/10.1038/s41392-020-00443-w.
26. Zarghi, A. and Arfaei, S., Selective COX-2 inhibitors: A review of their structure-activity relationships. *Iran. J. Pharm. Res. IJPR*, 10, 655–83, 2011, http://www.ncbi.nlm.nih.gov/pubmed/24250402.
27. Mahesh, G., Anil Kumar, K., Reddanna, P., Overview on the discovery and development of anti-inflammatory drugs: Should the focus be on synthesis or degradation of PGE2? *J. Inflamm. Res.*, 14, 253–263, 2021, https://doi.org/10.2147/JIR.S278514.
28. González Canga, A., Sahagún Prieto, A.M., Diez Liébana, M.J., Fernández Martínez, N., Sierra Vega, M., García Vieitez, J.J., The pharmacokinetics and interactions of ivermectin in humans—A mini-review. *AAPS J.*, 10, 42–46, 2008, https://doi.org/10.1208/s12248-007-9000-9.
29. Laing, R., Gillan, V., Devaney, E., Ivermectin – old drug, new tricks? *Trends Parasitol.*, 33, 463–472, 2017, https://doi.org/10.1016/j.pt.2017.02.004.
30. Crump, A., Ivermectin: Enigmatic multifaceted 'wonder' drug continues to surprise and exceed expectations. *J. Antibiot. (Tokyo)*, 70, 495–505, 2017, https://doi.org/10.1038/ja.2017.11.
31. Crump, A. and Ōmura, S., Ivermectin, "wonder drug" from Japan: The human use perspective. *Proc. Jpn. Acad. Ser. B. Phys. Biol. Sci.*, 87, 13–28, 2011, https://doi.org/10.2183/pjab.87.13.

32. Tambo, E., Khater, E.I.M., Chen, J.-H., Bergquist, R., Zhou, X.-N., Nobel prize for the artemisinin and ivermectin discoveries: A great boost towards elimination of the global infectious diseases of poverty. *Infect. Dis. Poverty*, 4, 58, 2015, https://doi.org/10.1186/s40249-015-0091-8.
33. DiNicolantonio, J.J., Barroso, J., McCarty, M., Ivermectin may be a clinically useful anti-inflammatory agent for late-stage COVID-19. *Open Heart*, 7, e001350, 2020, https://doi.org/10.1136/openhrt-2020-001350.
34. Tucureanu, M.M., Rebleanu, D., Constantinescu, C.A., Deleanu, M., Voicu, G., Butoi, E., Calin, M., Manduteanu, I., Lipopolysaccharide-induced inflammation in monocytes/macrophages is blocked by liposomal delivery of Gi-protein inhibitor. *Int. J. Nanomed.*, 13, 63–76, 2017, https://doi.org/10.2147/IJN.S150918.
35. Zhang, X., Song, Y., Ci, X., An, N., Ju, Y., Li, H., Wang, X., Han, C., Cui, J., Deng, X., Ivermectin inhibits LPS-induced production of inflammatory cytokines and improves LPS-induced survival in mice. *Inflamm. Res.*, 57, 524–529, 2008, https://doi.org/10.1007/s00011-008-8007-8.
36. Yan, S., Ci, X., Chen, N., Chen, C., Li, X., Chu, X., Li, J., Deng, X., Anti-inflammatory effects of ivermectin in mouse model of allergic asthma. *Inflamm. Res.*, 60, 589–596, 2011, https://doi.org/10.1007/s00011-011-0307-8.
37. Ventre, E., Rozières, A., Lenief, V., Albert, F., Rossio, P., Laoubi, L., Dombrowicz, D., Staels, B., Ulmann, L., Julia, V., Vial, E., Jomard, A., Hacini-Rachinel, F., Nicolas, J.-F., Vocanson, M., Topical ivermectin improves allergic skin inflammation. *Allergy*, 72, 1212–1221, 2017, https://doi.org/10.1111/all.13118.
38. Zhang, W., Xie, X., Wang, J., Song, N., Lv, T., Wu, D., Zhang, N., Cao, Y., Increased inflammation with crude E. coli LPS protects against acute leptospirosis in hamsters. *Emerg. Microbes Infect.*, 9, 140–147, 2020, https://doi.org/10.1080/22221751.2019.1710435.
39. Tanabe, T. and Rubin, B.K., Airway goblet cells secrete pro-inflammatory cytokines, chemokines, and growth factors. *Chest*, 149, 714–720, 2016, https://doi.org/10.1378/chest.15-0947.
40. Pelaia, C., Heffler, E., Crimi, C., Maglio, A., Vatrella, A., Pelaia, G., Canonica, G.W., Interleukins 4 and 13 in asthma: Key pathophysiologic cytokines and druggable molecular targets. *Front. Pharmacol.*, 13, 851940, 2022, https://doi.org/10.3389/fphar.2022.851940.
41. Fulkerson, P.C., Schollaert, K.L., Bouffi, C., Rothenberg, M.E., IL-5 triggers a cooperative cytokine network that promotes eosinophil precursor maturation. *J. Immunol.*, 193, 4043–4052, 2014, https://doi.org/10.4049/jimmunol.1400732.
42. Xiao, K., Liu, C., Tu, Z., Xu, Q., Chen, S., Zhang, Y., Wang, X., Zhang, J., Hu, C.-A.A., Liu, Y., Activation of the NF- κ B and MAPK signaling pathways contributes to the inflammatory responses, but not cell injury, in IPEC-1 cells challenged with hydrogen peroxide. *Oxid. Med. Cell. Longev.*, 2020, 1–14, 2020, https://doi.org/10.1155/2020/5803639.

43. Cardwell, L., Alinia, H., Moradi Tuchayi, S., Feldman, S., New developments in the treatment of rosacea – Role of once-daily ivermectin cream. *Clin. Cosmet. Investig. Dermatol.*, 9, 71, 2016, https://doi.org/10.2147/CCID.S98091.
44. Rivero, A.L. and Whitfeld, M., An update on the treatment of rosacea. *Aust. Prescr.*, 41, 20–24, 2018, https://doi.org/10.18773/austprescr.2018.004.
45. Raedler, L.A., Soolantra (Ivermectin) 1% cream: A novel, antibiotic-free agent approved for the treatment of patients with Rosacea. *Am. Health Drug Benefits*, 8, 122–5, 2015, https://doi.org/26629276.
46. Rainer, B.M., Kang, S., Chien, A.L., Rosacea: Epidemiology, pathogenesis, and treatment. *Dermatoendocrinol.*, 9, e1361574, 2017, https://doi.org/10.1080/19381980.2017.1361574.
47. Raedler, L.A., Soolantra (ivermectin) 1% cream: A novel, antibiotic-free agent approved for the treatment of patients with rosacea. *Am. Health Drug Benefits*, 8, 122–5, 2015, http://www.ncbi.nlm.nih.gov/pubmed/26629276.
48. Ebbelaar, C.C.F., Venema, A.W., Van Dijk, M.R., Topical ivermectin in the treatment of papulopustular rosacea: A systematic review of evidence and clinical guideline recommendations. *Dermatol. Ther. (Heidelb)*, 8, 379–387, 2018, https://doi.org/10.1007/s13555-018-0249-y.
49. Schaller, M., Dirschka, T., Kemény, L., Briantais, P., Jacovella, J., Superior efficacy with ivermectin 1% cream compared to metronidazole 0.75% cream contributes to a better quality of life in patients with severe papulopustular rosacea: A subanalysis of the randomized, investigator-blinded ATTRACT study. *Dermatol. Ther. (Heidelb)*, 6, 427–436, 2016, https://doi.org/10.1007/s13555-016-0133-6.
50. Siddiqui, K., Gold, L.S., Gill, J., The efficacy, safety, and tolerability of ivermectin compared with current topical treatments for the inflammatory lesions of rosacea: A network meta-analysis. *Springerplus*, 5, 1151, 2016, https://doi.org/10.1186/s40064-016-2819-8.
51. Zhang, X., Song, Y., Xiong, H., Ci, X., Li, H., Yu, L., Zhang, L., Deng, X., Inhibitory effects of ivermectin on nitric oxide and prostaglandin E2 production in LPS-stimulated RAW 264.7 macrophages. *Int. Immunopharmacol.*, 9, 354–359, 2009, https://doi.org/10.1016/j.intimp.2008.12.016.
52. Chambers, E., Hoey, E.M., Trudgett, A., Fairweather, I., Timson, D.J., Binding of serum albumin to the anthelmintic drugs albendazole, triclabendazole and their sulphoxides. *Vet. Parasitol.*, 171, 172–175, 2010, https://doi.org/10.1016/j.vetpar.2010.02.043.
53. Rendic, S.P., Metabolism and interactions of Ivermectin with human cytochrome P450 enzymes and drug transporters, possible adverse and toxic effects. *Arch. Toxicol.*, 95, 1535–1546, 2021, https://doi.org/10.1007/s00204-021-03025-z.
54. Aryannejad, A., Tabary, M., Noroozi, N., Mashinchi, B., Iranshahi, S., Tavangar, S.M., Mohammad Jafari, R., Rashidian, A., Dehpour, A.R., Anti-inflammatory effects of ivermectin in the treatment of acetic acid-induced

colitis in rats: Involvement of GABAB receptors. *Dig. Dis. Sci.*, 67, 3672–3682, 2022, https://doi.org/10.1007/s10620-021-07258-x.

55. Hoover, D.B., Cholinergic modulation of the immune system presents new approaches for treating inflammation. *Pharmacol. Ther.*, 179, 1–16, 179, 2017, https://doi.org/10.1016/j.pharmthera.2017.05.002.
56. Amarante-Mendes, G.P., Adjemian, S., Branco, L.M., Zanetti, L.C., Weinlich, R., Bortoluci, K.R., Pattern recognition receptors and the host cell death molecular machinery. *Front. Immunol.*, 9, 9, 2379, 2018, https://doi.org/10.3389/fimmu.2018.02379.
57. Yu, H., Lin, L., Zhang, Z., Zhang, H., Hu, H., Targeting NF-κB pathway for the therapy of diseases: Mechanism and clinical study. *Signal Transduction Targeting Ther.*, 5, 209, 2020, https://doi.org/10.1038/s41392-020-00312-6.
58. Mussbacher, M., Salzmann, M., Brostjan, C., Hoesel, B., Schoergenhofer, C., Datler, H., Hohensinner, P., Basílio, J., Petzelbauer, P., Assinger, A., Schmid, J.A., Cell type-specific roles of NF-κB linking inflammation and thrombosis. *Front. Immunol.*, 10, 85, 2019, https://doi.org/10.3389/fimmu.2019.00085.
59. Chaccour, C., Abizanda, G., Irigoyen-Barrio, Á., Casellas, A., Aldaz, A., Martínez-Galán, F., Hammann, F., Gil, A.G., Nebulized ivermectin for COVID-19 and other respiratory diseases, a proof of concept, dose-ranging study in rats. *Sci. Rep.*, 10, 17073, 2020, https://doi.org/10.1038/s41598-020-74084-y.
60. Azam, F., Taban, I.M., Eid, E.E.M., Iqbal, M., Alam, O., Khan, S., Mahmood, D., Anwar, M.J., Khalilullah, H., Khan, M.U., An in-silico analysis of ivermectin interaction with potential SARS-CoV-2 targets and host nuclear importin α. *J. Biomol. Struct. Dyn.*, 40, 2851–2864, 2022, https://doi.org/10.1080/07391102.2020.1841028.
61. Courties, A., Boussier, J., Hadjadj, J., Yatim, N., Barnabei, L., Péré, H., Veyer, D., Kernéis, S., Carlier, N., Pène, F., Rieux-Laucat, F., Charbit, B., Bondet, V., Duffy, D., Berenbaum, F., Terrier, B., Sellam, J., Regulation of the acetylcholine/α7nAChR anti-inflammatory pathway in COVID-19 patients. *Sci. Rep.*, 11, 11886, 2021, https://doi.org/10.1038/s41598-021-91417-7.
62. Shirazi, F.M., Mirzaei, R., Nakhaee, S., Nejatian, A., Ghafari, S., Mehrpour, O., Repurposing the drug, ivermectin, in COVID-19: Toxicological points of view. *Eur. J. Med. Res.*, 27, 21, 2022, https://doi.org/10.1186/s40001-022-00645-8.
63. Aboudounya, M.M. and Heads, R.J., COVID-19 and toll-like receptor 4 (TLR4): SARS-CoV-2 may bind and activate TLR4 to increase ACE2 expression, facilitating entry and causing hyperinflammation. *Mediators Inflamm.*, 2021, 1–18, 2021, https://doi.org/10.1155/2021/8874339.
64. Hariyanto, T.I., Halim, D.A., Rosalind, J., Gunawan, C., Kurniawan, A., Ivermectin and outcomes from COVID-19 pneumonia: A systematic review and meta-analysis of randomized clinical trial studies. *Rev. Med. Virol.*, 32, 1–15, 2022, https://doi.org/10.1002/rmv.2265.

65. Heidary, F. and Gharebaghi, R., Ivermectin: A systematic review from antiviral effects to COVID-19 complementary regimen. *J. Antibiot. (Tokyo)*, 73, 593–602, 2020, https://doi.org/10.1038/s41429-020-0336-z.

8

Ivermectin: An Anthelminthic and Insecticide

Manisha Arora Pandit* **and Tarkeshwar†**

Kalindi College, University of Delhi, Delhi, India

Abstract

Ivermectin was isolated from a soil sample (from a bacterium, actinomycete, *Streptomyces avermitilis*, or *S. avermectinius*) in the Shizuoka region of Japan as a result of joint project between Kitasato Institute, Japan, and Merck, USA. This drug exhibited antiparasitic properties even at very low doses around 80s and has been used in several mass eradication programs for saving lives of millions of humans and preventing significant loss of livestock due to helminths, mosquitoes, bacteria, fungi, flatworms and protozoans. In addition to antiparasitic activities, ivermectin has been proved to be very effective against arthropods and nematodes vector of human and livestock. Presently this drug is being used in 35 countries not only as an antiparasitic agent but also as an antimalarial, antiviral, anticancer, anthelminthic, and antimetabolic at varied concentrations. Ivermectin blocks neurotransmission at neuromuscular junction by inhibiting closure of glutamate-gated chloride channels found in arthropods and nematodes.

Anthelminthic activities of ivermectin have been showed against *onchocerciasis, lymphatic filariasis, trichuriasis, strongyloidiasis, mansonelliasis,* and *loiasis.* Further, various Mass Drug Administration program in association with several NGOs, governmental agencies and local administration were conducted to eradicate these diseases completely from almost all the parts of the world.

As an insecticide, ivermectin has shown its promising effects and all the developing stages of insects belonging to several orders and also against tick and mites. Ivermectin has been successfully applied against arthropods using various modes methodologies to protect humans and their livelihood as well as economic resources such as agriculture and livestock.

However, it is obvious to estimates it deleterious effects on environment biodiversity, any environmental contamination arising due to extensive use as drug

**Corresponding author*: manishaarorapandit@kalindi.du.ac.in
†*Corresponding author*: tarkeshwar@kalindi.du.ac.in

Rashid Ali and Shahid ul-Islam (eds.) Chemistry and Biological Activities of Ivermectin, (163–198)

in human, agriculture practices and in cattle industries in terms of its appearance in food chain, farmland system which may affect non-target organisms and other biological agents.

Moreover, a detailed research is required to access the combination therapies with stringent guidelines for the use as therapies and knowing its potential risks, for the implementation to improve the efficacy of the treatment in human, livestock, and utilization for crop management. Further, a cost effective formulation should be developed so as to increase its acceptance particularly in underdeveloped or developing nations of the world which helps the nodal agencies to ensure adequate availability and amount of ivermectin to meet its growing demand. In this chapter, we have detailed - uses of ivermectin as an anthelminthic and insecticide.

***Keywords*:** Ivermectin, antiparasitic, insecticide, neurotransmission, helminth, insect, resistance, Mass Drug Administration (MDA)

8.1 Introduction

The establishment of a research collaboration between Kitasato Institute in Japan and the Merck, Sharpe and Dohme (MSD) research laboratories in the United States (US) in 1973 paved way for the discovery of Ivermectin a year later in 1974. Ivermectin was isolated from a soil sample in the Shizuoka region of Japan [1]. The drug exhibited exceptional antiparasitic properties at very low doses and since then it has been used in several mass eradication programs improving the lives of millions of humans globally, and preventing a significant loss of livestock.

Ivermectin was first introduced for therapeutic use as an antiparasitic in 1981 [2] and was certified for use in humans in 1987 post its donation by Merck & Co. Inc. [3]. William Campbell of Drew University New Jersey and Satoshi Ōmura of Kitasato University in Tokyo were instrumental not only in its discovery but also in its development as the most preferred drug for the treatment of river blindness or onchocerciasis. Both were awarded with a Nobel Prize in 2015 for their efforts [4].

The main source of the avermectins that form Ivermectin is the actinomycete, *Streptomyces avermitilis*, renamed as *S. avermectinius* [5]. Today it is widely used in more than 35 countries not only as an antiparasitic agent but also as an antimalarial, antiviral, anticancer, anthelminthic and antimetabolic at varied concentrations [6]. It is derived from a naturally occurring fermentation product, avermectin B1 [2] and is composed of 80% 22,23-dihydroavermectin B1a and B1b [6]. It lacks activity against bacteria, fungi, flatworms and protozoans while being extremely effective against arthropods and nematodes [1, 2].

8.2 Ivermectin as an Anthelmintic

8.2.1 Mode of Action

Ivermectin is a part of a category of drugs called as endectocides that are highly effective against both ecto- and endoparasites. It can be administered through different routes and works well even at very low doses. It blocks neurotransmission at the neuromuscular junction by preventing the closure of glutamate-gated chloride channels generally found in arthropods and nematodes. The enhanced permeability of chloride channels causes hyperpolarization and results in reduced or absent neuronal transmission [1]. The parasite is eventually killed due to subsequent paralysis of the somatic muscles. It`s mode of action in vertebrates involves the release of gamma-amino butyric acid (GABA) in neurons of the brain, which renders it safe for use in mammals due to the existence of the blood brain barrier [5]. In nematodes, ticks and insects the GABA-related chloride channels are present in the peripheral nervous system but are not influenced by Ivermectin at low doses [1].

8.2.2 Ivermectin and Public Health

i. Onchocerciasis Control by Ivermectin

Onchocerciasis or river blindness occurs due to infection by the parasitic worm *Onchocerca volvulus,* passed on to humans through the bite of black flies of the genus *Similium* that breeds commonly in water bodies such as rivers and streams (Figure 8.1). It is predominantly seen in sub-Saharan Africa with foci in South America and Yemen [3]. The infection results in itching, lesions on the skin, and eye disease that can ultimately result in blindness [3]. Earlier treatment involved the use of the Diethylcarbamazine (DEC) which caused severe side effects [7].

Ivermectin on the other hand is highly effective even at very low doses in eliminating microfilariae or the immature worms that cause the deadly infection. Since it does not affect the adult worms, it has to be administered once a year for the life span of adult female worms which is around 15 years to completely clear the infection. The implementation of ivermectin in the mass eradication programs against onchocerciasis has helped millions of people in the African and South American continents escape the fatal effects of the disease [5].

Use of ivermectin as the drug of choice post its donation in 1987 by MSD for the eradication of river blindness has proved to be highly beneficial and made it the dominant medication for onchocerciasis treatment.

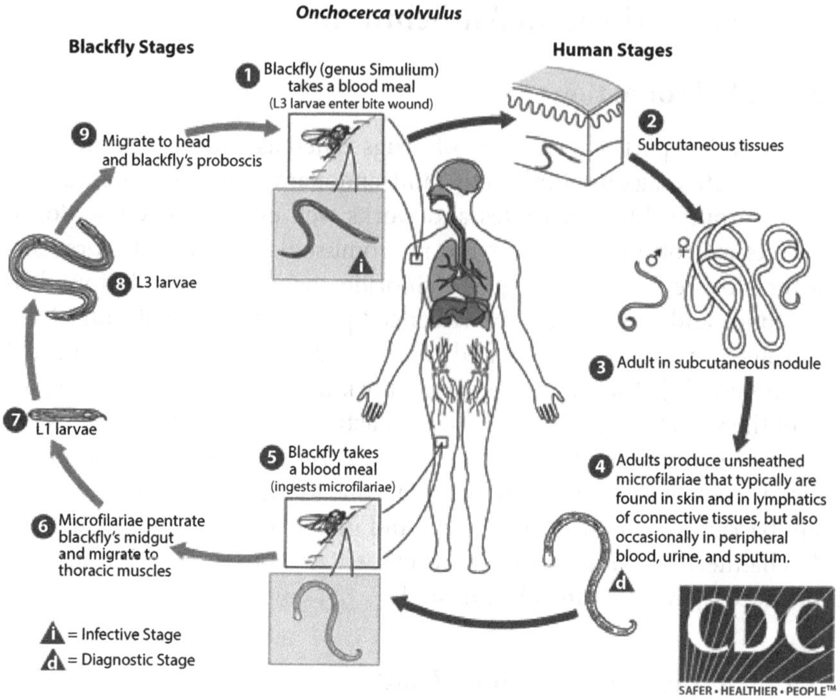

Figure 8.1 Life cycle of the filarial worm, *Onchocerca volvulus.* Source: https://www.cdc.gov/dpdx/onchocerciasis/index.html.

A number of distribution programs were put into play in countries such as Guatemala, Ecuador and Mexico [8]. Later, the Onchocerciasis Elimination Program for the Americas (OEPA) was established in 1993, while control of onchocerciasis in Africa was initiated in 1974 with the creation of the Onchocerciasis Control Programme in West Africa (OCP) that focused mainly on vector control [9], followed by the African Programme for Onchocerciasis Control (APOC) that implemented Community-Directed Treatment with ivermectin (CDTi), as its main strategy for mass delivery of ivermectin [10]. APOC completed in 2015 and was highly successful in significantly reducing the health burden of onchocerciasis in the 19 participating nations of Africa [11].

In Yemen, onchocerciasis is predominantly endemic in the western region of the country and even though *S. damnosum s.l.* has been identified as the prominent vector for the disease in this region, adequate information is still lacking about its feeding habit, biting rates, sites for breeding and disease transmission [3]. In 2010, a national action plan was launched in the country to eradicate the disease but its proper application could not be carried out due to the prevalent political unrest in the country. In 2018, a full Mass Drug Administration (MDA) was launched in the country [12].

APOC was considered a success in Africa as studies done in Kaduna (Nigeria), Mali and Senegal showed that large scale treatment with ivermectin could lead to complete eradication of infection in these areas [13]. Following its success, WHO launched the Expanded Special Project for Elimination of Neglected Tropical Diseases (ESPEN) in 2016. The project focuses on the elimination of five tropical diseases including onchocerciasis. The main aim is to completely eliminate the disease from most of the endemic African countries by 2025 [3].

The 2016 WHO guidelines lay emphasis on MDA interruption and verifying eradication of disease, these revised guidelines will allow endemic countries take evidence based decisions for onchocerciasis elimination. The current eradication strategy, therefore lays emphasis on the abatement of *O. volvulus* infection and its transference in order to stop drug interventions but maintain post-intervention vigilance [14].

Current Eradication Status in Africa

The WHO guidelines encompass three phases for elimination of onchocerciasis. These are:

1. **First phase:** The first phase involves active treatment with Ivermectin of at least 80% of the affected population till the time transference of the parasite stops. The treatment period could last for 12-15 years or longer. Vector eradication can also be carried out simultaneously. Testing is an important aspect of this phase and usually involves testing children younger than 10 years. Black flies are also tested for the presence of *O. volvulus* DNA by using O-150 PCR (polymerase chain reaction) and ELISA (enzyme-linked immunosorbent assay) [15].
2. **Second phase:** is mainly the vigilance phase and involves surveillance for emergence of new infections. It usually lasts for 3-5 years and involves O-150 PCR testing for presence of *O. volvulus*. In case the PCR results are in the range of < one infected black fly in a total of 2000 flies (threshold) then disruption of transmission is confirmed by Ov-16 testing in children. Absence of parasite transference due to vector elimination can also be confirmed by using the Ov-16 serology test to ensure movement into the third phase [15].

3. **Third phase:** involves keeping surveillance for resurgence of the parasite [14]. It establishes enduring eradication of the infection from the given area or country [15].

Three-phase elimination guidelines postulated by WHO are shown in graph (Figure 8.2):

The main aim of ESPEN has been to eradicate onchocerciasis from the African subcontinent by 2025 and to achieve this goal, African countries altered their policies to switch from control to elimination [16]. As per the latest reports and studies it is now clear that the inaugural goals of the program to eradicate onchocerciasis have been achieved in a number of endemic foci in Africa [17–20] but certain challenges remain that prevent its complete eradication.

Some of the challenges faced in the elimination program include [3]:

1. Cross-border transmission and incomplete transmission zone mapping
2. Co-endemicity shown with *Loa loa*
3. Political conflict
4. Lack of skilled workforce

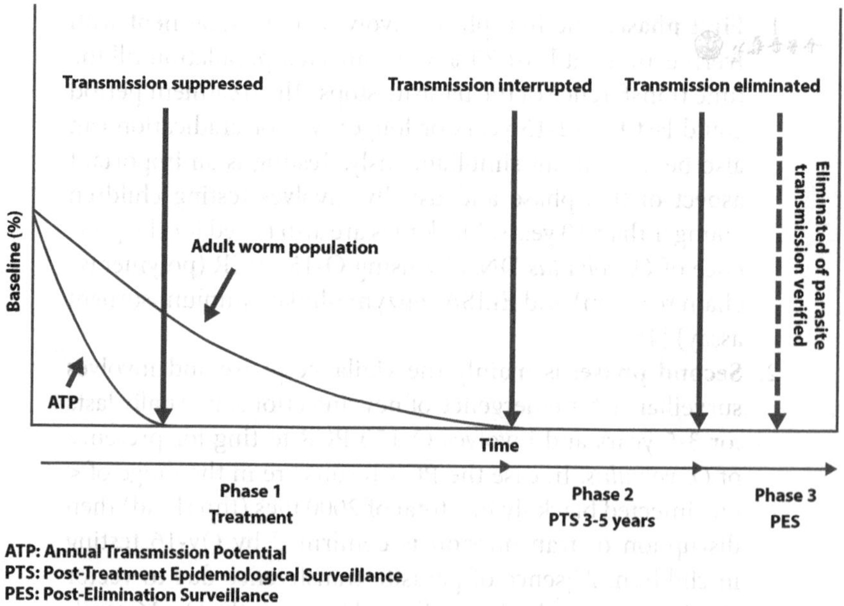

Figure 8.2 Phases in the elimination of human onchocerciasis [14].

5. Maintaining a robust drug distribution mechanism at the ground level
6. Possibility of resistance to Ivermectin
7. Lack of proper tools for diagnosis

Issues such as aberrant transmission zone mapping and cross border transmission have greatly affected the elimination strategy currently adopted by the African nations for eradication of onchocerciasis. Hypo and nonendemic zones, with less than 20% onchocercal nodule prevalence where treatment was not given due to low infection rates have slowed the progress made towards elimination as they now contribute towards infection spread and continual transmission [21, 22]). Cross border transmission is caused not only by human movement but also due to vector (black fly) movement as these flies can journey up to 600 km in wind bolstered flights [9].

Another factor that challenges the complete dismissal of the disease from Africa is its co-endemicity with loiasis, a disease caused by the filarial nematode *Loa loa* [23]. The east and west foci of the African continent are the two main regions where this co-endemicity exists [15]. In these areas, treatment with ivermectin results in severely detrimental neurological side effects that may lead to death. Therefore, most of the individuals infected with onchocerciasis remain untreated due to a high refusal rate leading to continuous transmission and a possibility of cross transmission into other disease free zones [15].

Political unrest also contributes to the failure of eradication goals and has been responsible for delays and reoccurrence of infections in several countries like Angola, DRC, South Sudan, Côte d'Ivoire and CAR [24]. Other issues that have hampered the goal of eradication of onchocerciasis include inadequate drug supply, logistical issues that prevent the delivery of the drugs to the required areas on time and poor drug coverage [25]. Individuals refusing to take the drug due to age or possible risk of side effects create a population of systemic non-compliers that act as reservoirs of infection that sustain transmission [26].

Moreover, most countries where river blindness is endemic are unable to provide the financial support required to sustain the eradication program. APOC was implemented directly by WHO and supported by a number of Non-Governmental Development Organizations (NGDOs) [27]. Under ESPEN, the international donor organizations which had earlier lent support to APOC are not keen on extending further financial aid to support ESPEN [15, 25]. The lack of funding could prove to be detrimental to the thorough removal of onchocerciasis from Africa. Low worker motivation and lack of incentive also adds to the problem as it reduces the interest of volunteers in continuing with the distribution of the drugs in the community [28].

In the last stages of the eradication program, skilled entomologists, proper diagnostic kits and laboratory facilities are essential to confirm the lack of transmission and assess the success rate of the program. These give rise to further technical difficulties in the already financially strapped African nations [25, 29].

To achieve the goal of complete onchocerciasis elimination by 2025 under ESPEN, continuous monitoring and evaluation needs to be carried out by the endemic countries along with identifying and correcting the individual challenges faced by them. Some of the corrective measures that need focus are: taking into account the political unrest in certain areas and modifying the drug distribution strategy accordingly, carrying out a more comprehensive and collaborative mapping of the disease and its elimination and continued financial support.

Current Scenario: Measures to Achieve Elimination Goals by 2025

Some foci in Africa need the application of ATSs (alternative treatment strategies) that differ from CDTI in order to achieve total eradication by 2025 even though its application requires higher financial and human input and commitment over the annual CDTI. ATSs involve giving ivermectin treatment in a biannual or pluriannual manner and emphasize on vector control. It also involves treatment with new drugs or different drug combinations and the use of 'test-and-treat' (TNT) strategies that can follow either the Loa-first TNT or the Oncho-first TNT depending upon which of the two diseases is endemic in the area. Successful ATSs can be employed post adequate modeling and field evaluations in order to select the best strategies and their cost effectiveness [16].

ii. Lymphatic Filariasis and Ivermectin

A vector-borne illness called lymphatic filariasis is brought on by the parasitic worms *Wuchereria bancrofti*, *Brugia malayi*, and *Brugia timori* (*Brugia* spp.) (Figure 8.3). By 2018, it was endemic in 72 countries [30] and can result in both acute and chronic infections. The World Health Assembly in 1997 made a decision to eliminate the disease as a public health issue by 2020 and the Global Programme to Eliminate Lymphatic Filariasis (GPELF) was launched in 2000 (Molyneux). More than 40 million people were affected by lymphatic filariasis when GPELF was initiated and exhibited various levels of incapacitation and disfigurement mainly due to lymphoedema and elephantiasis (advanced form of lymphoedema), and

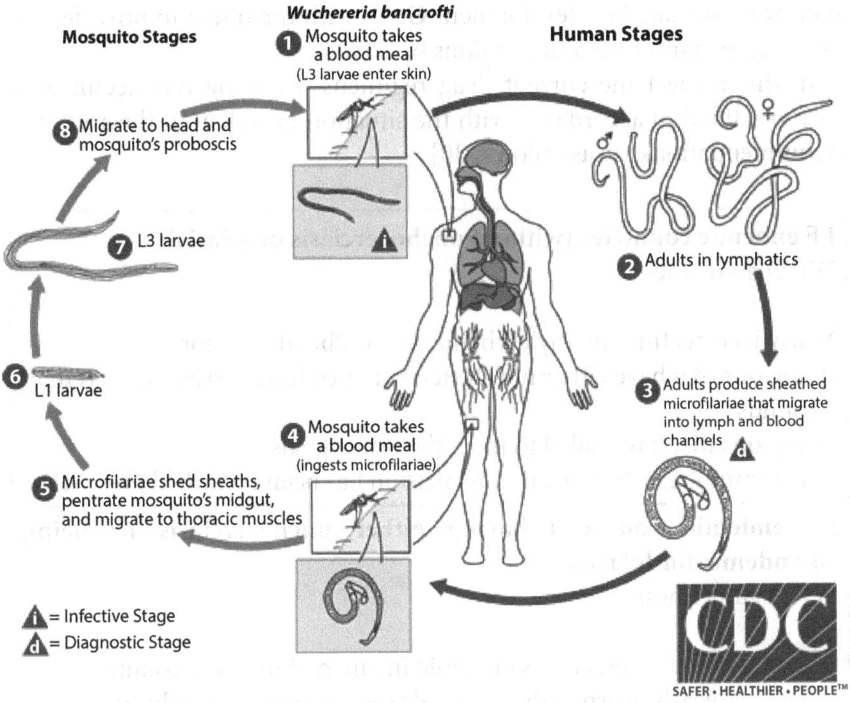

Figure 8.3 Life cycle of the parasitic worms *Wuchereria bancrofti*. Source: https://www.cdc.gov/parasites/lymphaticfilariasis/biology_w_bancrofti.html.

hydrocele. The focus of the program was (i) to curb the spread of LF infection via mass drug administration (MDA) (ii) the reduction in discomfort via morbidity management and disability prevention (MMDP). By 2018 major progress has been achieved on the reduction of transmission and spread of LF but management of disability remains limited [31].

The most widely used method of treating LF involves administering a mixture of the three anthelminthic medications, including ivermectin (150-200 g/kg), albendazole (400 mg) and diethylcarbamazine (6 mg/kg). These drugs are given for at least a span of five years, in conjunction with the end of reproductive lifespan of the adult worms [30].

By 2017, 20 of the 72 endemic countries have achieved their elimination goals, discontinued MDA programs and are currently under surveillance [31]. 22 of the remaining 52 countries still remain endemic and had not started the MDA till 2017, therefore could not achieve the 2020 elimination goals. Currently the WHO has revised its elimination of LF targets to align with Sustainable Development Goals with the intent to provide care to all patients in LF endemic regions by 2030.

In this respect the current drug regimens involving ivermectin have been modified in accordance with the situation prevalent in the area. The recommendations are as follows [30]:

LF endemic countries (without onchocerciasis or loiasis): WHO guidelines: Yearly ivermectin + diethylcarbamazine + albendazole for -> regions that have either not started MDA or have carried out less than 4 rounds -> regions that have failed to meet the target goals -> communities where local transmission has been identified post-MDA
LF endemic countries having either onchocerciasis or being co-endemic for loiasis): WHO guidelines: i. Onchocerciasis is endemic in any region of the country: yearly ivermectin+ albendazole instead of yearly ivermectin + diethylcarbamazine + albendazole. ii. Onchocerciasis is endemic in region of the country: yearly ivermectin+ albendazole instead of twice yearly ivermectin+ albendazole, except in regions where ivermectin is already being delivered for onchocerciasis biannually. iii. Regions where co-endemicity of LF is present with loiasis and no ivermectin was distributed for either onchocerciasis or LF, WHO recommends albendazole biannually instead of albendazole yearly in these regions.

iii. Trichuriasis and Ivermectin

Soil-transmitted helminths (STHs) infections caused by the *Ascaris lumbricoides*, hookworm and *Strongyloides stercoralis* affect nearly 2 billion people particularly in the tropics. These infections cause abdominal pain and discomfort, dysentery, diarrhea, inflammation, impaired immune system, faulty nutrition uptake and mental and physical retardation particularly in children and reduction in working capacity in adults [32].

Earlier the favored treatment for these infections was the drug albendazole that was administered under MDA programs. Albendazole, even though well tolerated exhibits differential results for the three types of worms. It is highly effective against *A. lumbricoides* (Egg reduction rate of >98% and high cure rates of 96%), has a lower efficacy against hookworm infections (Egg reduction rate of > 90% and high cure rates of 80%) but a very low efficacy against *Trichuris trichiura* (Egg reduction rate of > 50% and high cure rates of 31%) [33].

Results of clinical trials carried out using a combination of ivermectin and albendazole show a greater superiority in terms of high cure and egg reduction rates over only albendazole. These will help in formulating better STH control strategies with an overall reduction in morbidity due to these infections. The WHO also includes ivermectin-albendazole in its Model List of Essential Medicines for treating the intestinal helminths [34].

iv. Strongyloidiasis and Ivermectin

Strongyloidiasis, caused by the Strongyloides stercoralis is one of the most neglected tropical diseases despite affecting more than 600 million people worldwide. Its health burden has been severely underestimated, and it is

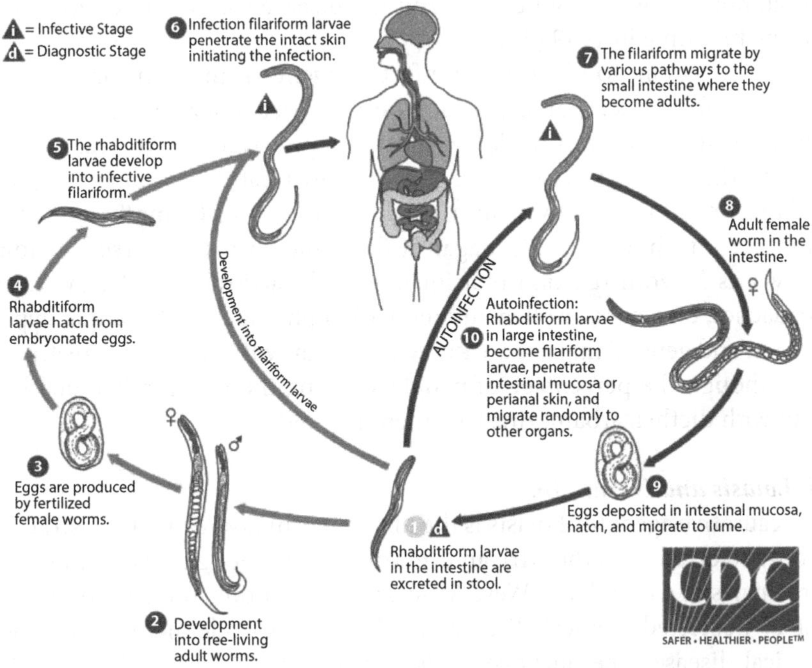

Figure 8.4 Life cycle of parasitic worms *Strongyloides stercoralis*. Source: CDC DPDx: (https://www.cdc.gov/dpdx/).

only in 2020 that WHO has included it in its global roadmaps and laid emphasis on its prevention and control [31].

The disease is asymptomatic in around 30% of the people infected while the rest exhibit a number of symptoms such as diarrhea, constipation, pain in the abdomen, flatulence, obstruction in various parts of the intestine, gastrointestinal bleeding and malabsorption of nutrients. It can also cause respiratory symptoms when the larva migrate (Figure 8.4). The favored drug for treatment of *strongyloidiasis* has been ivermectin, administered at a dose of 200 mcg/kg of body weight daily for a minimum of two days [35]. Considering the latest evidence that supports the combination of ivermectin and albendazole over ivermectin alone it is preferred that this combination is used for the treatment of STHs along with strongyloidiasis in a single MDA program [31].

v. Mansonelliasis and Ivermectin

Mansonelliasis in humans is caused by the three worms *Mansonella ozzardi*, *M. streptocerca* and *M. perstans*. The three worms differ from each other in anatomy, vectors, clinical symptoms and regions of endemicity [36]. Transmission is by dipteran vectors belonging to the families Ceratopogonidae and Simuliidae and the disease was identified in Brazil for the first time in 1949 [37].

It is important to identify the infective agent for mansonelliasis, as all the three forms need different medications. Mebendazole is most effective against *M. perstans* while ivermectin is preferred for *M. ozzardi* and ivermectin+ diethylcarbamazine are used to treat *M. streptocerca* [36]. Ivermectin is usually given in a single dose of 0.14-0.2mg/kg and it is not clear if it shows any effect against the adult worms. Adverse reactions known as Mazotti reaction that include chills, arthralgia, lethargy, fever, headaches, dizziness, swollen and tender lymph nodes, tachycardia, hypotension and general malaise are known to occur with ivermectin treatment even though the percentage of people showing the reaction is much less than with diethylcarbamazine treatment [38, 39].

vi. Loiasis and Ivermectin

The causative agent for Loiasis is the filarial worm *Loa loa* that is also commonly referred to as the African eye worm. Even though it is prevalent in large parts of Central and West Africa, it is still not included in the WHO list of neglected tropical diseases (NTD). Its co-endemicity with other tropical diseases like onchocerciasis and lymphatic filariasis leads to the potential risks when treated with ivermectin. These risks include potentially fatal encephalopathy that can derail the mass drug administration programs in these areas. This reaction is mostly shown in individuals with very high L. loa micro filarial density in the blood [40].

Analysis of the pharmacovigilance database, Vigibase of WHO in 2017 showed the possibility of an association between ivermectin use and both the neurological as well as non-neurological suspected serious adverse drug reactions (sADRs). Ivermectin is known to be a safe drug and is therefore used widely in mass drug administration campaigns but it is also imperative to be aware of its potential sADR effects and take these into account while implementing MDA programs [40].

8.2.3 Challenges of Ivermectin Use as an Anthelminthic

Even though ivermectin is known to be tolerated well by humans and its use results in minimal side effects, latest studies involving analysis of Vigibase, the pharmacovigilance database of WHO has brought to light both neurological and non-neurological side effects of its uses. These side effects are not just associated with L. loa co-endemicity but are also prevalent without it in some individuals. Apart from this, Mazzotti reaction is known to occur with ivermectin treatment that is not shown with DEC treatment. It can also cause neurotoxicity in humans when used in combination with drugs that block CYP3A4 and in the presence of mdr-1 gene polymorphism [41].

Long term use of any drug ultimately results in the development of resistance by the organism, it is meant to target and the same seems to be occurring with the long term use of ivermectin to control anthelminthic diseases. It is now time to look into the reducing efficacy of the drug, development of resistance and the gaps in its procurement and availability that hinder the MDA programs being carried out in a number of countries for use of ivermectin to control of the neglected tropical diseases (NTDs) [41].

Research needs to be promoted into the use of combination therapies instead of a single drug so as to improve the efficacy of the treatment. New guidelines need to be formulated for the usage of the combination therapies and potential risks need to be analyzed prior to the shift to these therapies [42]. Since ivermectin lacks the ability to kill adult worms the time required for it to be effective usually lasts for a large number of years dependent upon the life time of the adult female worm. Therefore, it is also important to focus on developing drugs that are effective against adult worms also in order to reduce the treatment time significantly.

Cost also plays a major role in the acceptance and use of a drug particularly when it is being used in some of the most financially strapped nations of the world. New manufactures should be fast tracked for pre-qualification process so as to ensure adequate availability of ivermectin in enough amounts so as to meet its growing demand [27, 42].

8.3 Insecticidal Activity of Ivermectin

Talking about the discovery, application, efficacy and mode of action of ivermectin, it can be seen that there is lots of information available in literature which reflects that the ivermectin, a class of avermectins, are wide-spectrum pesticides with unique effectiveness to control arthropods which have been reported, as parasites of livestock, harmful to plants or fabrics, and a nuisance to man [43].

It was 1976, when Merck & Co. Inc. synthesized a complex of eight closely related natural products, named as avermectins, from a bacterial culture of Streptomyces avermitilis MA- 680 (NRRL8165) which was originally isolate from a soil sample by the Kitasato Institute at Kawana, Ito City, Shizuoka Prefecture, Japan (Figure 8.5). These were claimed as most potent anthelmintic, insecticidal and acaricidal compounds [44].

Nomenclature of avermectins was done using different combination of letters and numbers (e.g. Ah B2, B1a, B2a, etc.), with manufacturers' codes (e.g. MK 936 and MK 933), proper names such as abamectin and ivermectin *etc.* [44]. Out of these naturally occurring compounds, avermectin B1a has been found to be most potent insecticide against the larval stages

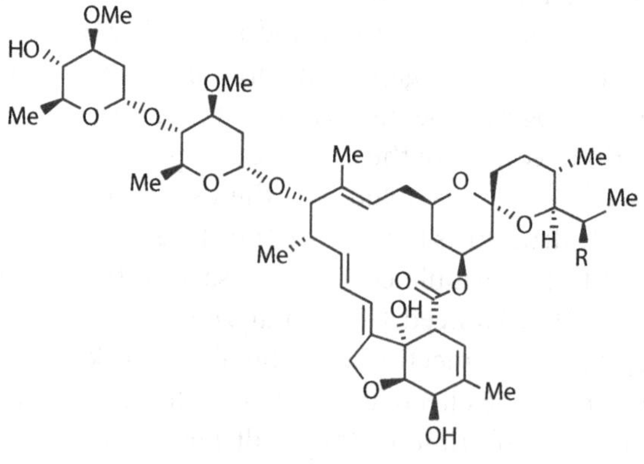

Figure 8.5 Ivermectin (22,23-dihydroavermectin B1). Source: *DrugBank Accession NumberDB00602.*

of various insect pests of orders Coleoptera, Hemiptera, and Lepidoptera,. Therefore, avermectin Bla was considered as an insecticide. It was avermectin Bla with its minor homologue Blb, which is present in 80%:20%, proportions and this is known as avermectin B1; MK 936 (abamectin).

Later on, scientist were able to synthesize a new compound by reduction of the avermectin B1 mixture, which yields 22,23-dihydroavermectin B1 and this product was extensively utilized for the treatment of livestock as well as man [2]. This modified avermectin B1 was named as MK 933 or ivermectin.

Ivermectin was mainly used in humans for the treatment of worm infestations, like strongyloidiasis, ascariasis, trichuriasis, and enterobiasis). But in 2020 during COVID-19 pandemic, ivermectin gained the attraction due to its off-label use for the prophylaxis and treatment of COVID-19. However, further studies are still ongoing in search of strong evidences for ivermectin as a treatment line for COVID-19.

8.3.1 Mode of Action

Ivermectin is a derivative of insecticide class avermectins, that were obtained from naturally occurring soil bacterium, *Streptomyces avermitilis* [45]. These avermectins are also called macrocyclic lactones and claimed for their insecticidal and miticidal activity [46].

They act on GABA receptors as chloride channel agonist at neuromuscular junction and keeps the chloride channels permanently opened that leads to the prolonged influx of Cl^- ions into neurons which results in disruption of nerve impulses and desensitization of the cell functioning. This causes the paralysis in insects ultimately resulted into complete inhibition of feeding activity causing death due to the starvation. It takes approximately 4-5 days to complete its action on insect population.

Ivermectin have been observed to kill the insect both by ingestion [47] and contact (by translaminar activity) in insects [46].

Basically the GABA (*Gama aminobutyric acid*), is a neurotransmitter, when released form presynaptic terminal knob, it reached to the postsynaptic membrane and binds with the ligand gated channels. Now these channels get opened for chloride ion influx, this Cl^- influx cause IPSP (inhibitory post synaptic potential) in postsynaptic membrane and inhibit the nerve excitation [48].

Insect possesses these GABA receptors trans-membrane proteins both in central and peripheral nervous system and they have both metabotropic and ionotropic receptors which are linked with G-protein-coupled receptors and the chloride channels, respectively. However, in insects the ionotropic GABA receptors are the main target of the insecticides in insects.

Ivermectin, when, enter into insect's body, it inhibit the GABA receptors and causes the influx of Cl^- ions in an irreversible manner [48] and due to continuous inflow of Cl^- ions at the postsynaptic nerve membrane. Ultimately this condition leads to the loss of sensitivity and paralysis in the insect.

8.3.2 Overview of Ivermectin as an Insecticide

Egerton *et al.* [49] were the first to demonstrate the anthelminthic activity against gut nematodes in cattle when they applied this chemical through oral and subcutaneous administrations. It has been reported to have efficacies against a number of arthropod parasites including grubs (*Hypoderma bovis, H. lineatum, Dermatobia hominis*), sucking lice (*Haematopinus eurysternus, Linognathus vituli, Solenopotes capillatus*), mange mites (*Sarcoptes scabei, Psoroptes ovis*) and screw worms (*Chrysomya bezziana*) [50].

Ivermectin application showed an effective action against larval, nymphal and adult stages of sucking lice, apparently because of their host blood feeding habit. The injections of ivermectin have controlled the dipteran *C. bezziana* and *D. hominis* infestations. The ticks like *S. scabei* and *P. ovis* has successfully been controlled by single subcutaneous injection of ivermectin whereas mite species, *Chorioptes bovis* required two dose of ivermectin [50].

However, the oral administration of ivermectin didn't show any promising effect against mites. The injection of this chemical has shown its activity against ticks, such as *Boophilus microplus, B. decoloratus* and the soft tick, *Ornithodoros.* In these ticks showed mortality and lesser feeding and engorgement, low fecundity and fewer viable eggs after feeding on ivermectin-treated cattle [50].

The potential role of ivermectin as a control against various insects which are either vector of any disease or pest itself has been demonstrated by various methodologies against different order of insects. In the present chapter we will discuss them in order wise manner including their relevance in animal livestock, agricultural crops or human.

8.3.3 Methods of Application to Animals and Plants

Various methods have been applied to demonstrate the activity of ivermectin as per its effectiveness, best possible administration and applicability.

i. Livestock

Most of the research workers treat livestock for parasites control by applying injection (subcutaneous or intramuscular) or oral administration (by pasting or oral drenching) or Pour-on (topical application) method [2]. For *in vitro* tests, these were added to the diet or rearing medium.

ii. Agricultural Crops

The agricultural practices involve different strategies as required for testing, these application methods have been used in both laboratory and field tests. Researches use spraying, dip, semisynthetic diet/or natural diet mixed/fortified with insecticides, topical application, various trap methods *etc*., consisting a range of concentrations applied at various times of the day or week until the crop is examined.

iii. Human

They were treated through oral administrated and an standard dose per kg body generally given in case of human. This dose may further be given at different times in day and for a certain duration in the course of treatment or till the success full treatment is not achieved.

8.3.4 Disease Vector Control

Ivermectin is claimed as a highly effective and toxic to a broad range of insects. At sub-lethal doses, ivermectin inhibits feeding and disrupts mating behavior, oviposition, egg hatching and development. A comprehensive data is available in literature that shows the anti-insect properties of the avermectins against almost all the insects tested. However, in the present chapter we will focus mainly of insect pests or vectors associated with livestock and humans.

i. Mosquito

Ivermectin has been promising controlling agents against the mosquitoes which are vector of dreadful diseases like malaria, dengue, and filaria. There are several reports which demonstrate the effective management of mosquitoes for example when *Aedes aegypti* and *A. albopictus* adult mosquitoes were allowed to feed on spiked blood meal impregnated with different concentrations (0, 2, 4, 6, 8 and 10 ppm) of ivermectin for 24 and 48 h interval and observed for their behavioral separation as well as for any mortality. When the pupae collected from these mosquitoes which were fed on ivermectin mixed blood meal and they were separated using standard sieving, resulted in reduction of female progeny as compared to the percentage of males pupae of *A. aegypti* (73%) and *A. albopictus* (69%). It was observed that blood meal spiked with 8 ppm of ivermectin when provided for 24h and 48h interval, identified as the most effective way of achieving 100% female elimination for both Aedes species [51].

There two mosquitoes species, *Aedes vexans,* a vector of *Dirofilaria immitis* (dog heartworm), also spread myxomatosis (a deadly rabbit viral disease) and *Tahyna* virus, a less observed *Bunyaviridae virus*, which affects humans in Europe by causing a fever which may lead to the encephalitis or meningitis; and *Psorophora confinnis*, a vector of dog heartworm, *Dirofilaria immitis* [52]. These mosquitoes when released to feed on Ivomec (ivermectin) injection insecticide-treated steers in stable traps, and observed for their mortality. Ivomec showed its maximum activity in terms of mortality in both mosquito species as compared to control, after 48h of treatment. However, the mortality rate was higher in *P. confinnis* [52].

Similar observation were reported by Pampiglione [53] in *Anopheles stephensi, Culex quinquefasciatus* and *A. aegypti*, when these mosquitoes species had fed on mice injected with ivermectin. He observed variable mortality rate among three mosquito species i.e., 100% in *A. stephensi* followed by 50% in mortality in *C. quinquefasciatus* and *A. aegypti* when these were subjected to 36h feeding on treated mice at a rate of 2.8 mg ivermectin/kg body weight [52].

In another study on *C. quinquefasciatus* mosquito, when ivermectin at a concentration of 1.5 ppb was investigated for the behavior and mortality effect by rearing their 3rd and 4th instar larvae on treated diet. It was observed that ivermectin caused paralysis in the larvae as well as killed 73.38% population, further the histological observation showed alteration in mobilization of material stored in the fat body and a marked reduction in fecundity in the adult stage Strong and Brown [43] suggested that the fat body of larval and adult Dipteran has important functions related to the development stage of insect. Ivermectin, in this context may alter the proteins synthesis in larvae and adults indicating a compromised immune state against some bacteria, yeast and viruses, whereas in case of adult females, this alteration in fat body may disturb the production of yolk polypeptides ultimately affecting egg laying and fecundity in female mosquito [54]. In addition, mortality as well as paralysis caused by the ivermectin have been noticed in the adults of *Aedes fluviatilis, Aedes aegypti, Anopheles quadrimaculatus* in addition to other mosquitos' species [54].

Further to combat malaria by suppressing the population of mosquitoes was supported by the study of [55]. In this study, they investigated toxic effects of this endectocides (ivermectin) on the survival and fecundity of *An. arabiensis*. They recorded 77% mortality in mosquitoes who fed on subcutaneous ivermectin injected cattle as compared to the control group. Moreover the mosquitoes that could survive the treatment, exhibited a drastically reduced fecundity up to 90% after feeding on ivermectin treated diet. Similar finding were reported for the survival of *An. Arabiensis* by Pooda *et al.* [56] they also observed a marked reduction in mortality of 75% and 45% after 3rd and 4th weeks of treatment, respectively. In another study, the survival and fecundity of *An. arabiensis* was declined by 52.5% and 64.6%, respectively. Here, in both the studies, ivermectin was applied as a subcutaneous injection method in the cattle [55].

In a mass drug administration program when Anopheles gambiae mosquitoes, the causative agent of malaria in South East Asian and African countries, were assessed for their survival by feeding on volunteers who had received oral administration of ivermectin or nothing. It was observed that the mosquitoes which fed on volunteers, who were given ivermectin the previous day, their mean survival was reduced to 2.3 days, as compared to 5.5 days, the survival rate of those who were allowed to feed on control group. The mortality rate escalated to 73%, 84%, and 89% on 2nd, 3rd, and 4th day, respectively, in the ivermectin group. However, there is no significant difference in the mortality in mosquitoes fed on either group after 14 days of treatment. This shows that ivermectin, as systemic insecticide, possesses safe and significant short-term insecticidal properties. This can be a useful tool to control malaria in humans specially in those areas where mosquitoes are exophagic or exophilic, and has shown drug resistance [57].

Ivermectin as a drug has been proven to be safe and its doses has been standardized and approved in healthy volunteers with very few reactions which are mild in nature. It has been successfully used as anthelminthic in humans and livestock. These qualities can be harnessed in controlling malaria causing vectors as there are limited controlling option available for vector, this becomes more tedious due to exophilic nature of mosquitoes, especially where forest transmission is common and its larval control is less effective or impractical. Further the mechanism of developing resistance against antimalarial drug in these mosquitoes adds another hurdle to control malaria. In these circumstances if we could decrease the longevity and or survival rate of the mosquitoes, then it would reduce the biting rate on human when they are outdoors. This could be an attractive prospect.

Testing on animals and *in vitro* studies have demonstrated that ivermectin has promising effects on the blood-feeding insects when these were allowed to feed animals or blood meals treated with different methods. It has been shown to readily killing adult mosquitoes, involving anophelines. In human trials of MDA for onchocerciasis as well as filariasis, the effect of ivermectin on mosquitoes appears to be sustained in the field [57].

These insecticidal properties of ivermectin suggest that this wonder drug could play a leading role controlling arthropod vector-borne diseases.

The mentioned studies demonstrates that ivermectin, is a drug with a good safety profile, applied in various animals and human, proves to have a major impact on different mosquito species which serve as vector of a number of diseases which affect both human, its pets and livestock which is an important aspect of human livelihood [57].

Though the concept of repurposing ivermectin as a malaria controlling tool originally emerged in 1985, when the in vitro tests revealed that that this drug can kill the malaria vector *A. stephensi* and thereafter evidences were added slowly that shown the potential of this drug. Studies of Papua New Guinea in 1999, randomized controlled trial involving ivermectin's killing effects [57] and recent results using modern techniques of modeling, pharmacology, insectary-based research studies, and several clinical and field MDA trials, have advocated the promising effects of ivermectin and suggested as a premium tool to achieve malaria control [58]. Considering the above facts and findings, the WHO has constituted a technical committee in 2016 for possible uses of ivermectin. It also established Malaria Policy Advisory Committee (MPAC), with a goal to minimize 20% malaria incidence after single round of ivermectin MDA at least 1 month posttreatment of this the standard vector control tools. The malaria-ivermectin Roadmap directs on 1) the product development (e.g., dose and regimen), 2) evidence to aid a global policy recommendation (e.g., safety and efficacy), and 3) the access as well as deployment at scale (e.g., procurement and delivery mode [57].

ii. Fly

Testing on animals and *in vitro* studies have demonstrated that ivermectin has promising effects on blood-feeding insects when these were allowed to feed animals or blood meals treated with different methods. It has been revealed to increase mortality, and reduces the fertility of Tsetse flies, ticks, triatomine bugs, and sandflies.

Ivermectin has a broad antiparasitic activity with long veterinary uses. When ivermectin's activity against *Onchocerca volvulus* was revealed, it was licensed for human the usage and was employed in the MDA programs to control river blindness - it was administered to the 180 million adults as well as children [59].

Leishmaniasis

Ivermectin has been demonstrated as a potent killer of *Phlebotomine* sandfly vectors which transmit *Leishmania* parasites, when they were fed on the rats treated with this insecticide [60], studies showed the toxic impact of ivermectin on blood-feeding sandfly vector, *Phlebotomus papatasi*, where it cause mortality within 1–2 days post treatment. It has been observed that that ivermectin was also able to kill the promastigotes or caused the inability to their infectivity after exposure. Nevertheless, ivermectin was shown to be more effective in killing promastigotes as compared to the conventional drugs such as rifampicin, nystatin and erythromycin [61]. In the case of cutaneous leishmaniasis, *in vitro* studies and subcutaneous inoculation showed that ivermectin kills *Leishmania tropica* parasites, with accelerated skin ulcer healing, further this drug was found to be more effective when compared to the other drugs like pentostam, amphotericin B, rifampicin, metronidazole, berenil, and nystatin. Additionally, when it is combined with an appropriate surgical wound dressing, it shows more vowing curing cutaneous leishmaniasis [59].

African Trypanosomiasis (Sleeping Sickness)
Moreover, Ivermectin has also been proven to be very effective for the treatment option of sleeping sickness (trypanosomiasis) spread by Tsetse flies vector. It has been reported that ivermectin caused mortality within 5 days, in Tsetse flies (*Glossina palpalis*) fed on ivermectin-treated animals. Animal tests has also suggested the pathogen killing ability of ivermectin when mice infected with *Trypanosoma brucei brucei* parasites were treated with subcutaneous dose of ivermectin, enhanced their life span by doubling their survival time. This indicates the scope of examining the applicability of the ivermectin in the treatment of African trypanosomiasis with different aspects [59].

American trypanosomiasis (Chagas disease)
When dogs infected with the *Trypanosoma cruzi* parasites endured a tick infestation, ivermectin cure eliminated the ticks but had no effect on either the dogs or their infection. Triatomine bug vectors of the *T. cruzi* feeding on dogs comparatively soon after the treatment displayed high mortality - which declined hastily as the interval between ivermectin treatment and the blood feed increased [59, 62].

High mortality recorded in hematophagous Muscidae, the horn flies *Haematobia irritans irritans and* stable flies *Stomoxys calcitrans* when these were allowed to feed on heifers treated with the ivermectin injection at a rate of 200 μg/kg body weight [52].

iii. Lice
Lice are the wingless and blood sucking insects that reside either on the head or body surface like pubic region, eyelashes, axillae and

urinogenital regions. Herewith, we will discuss about the nuisance caused by head louse (*Pediculus humanus humanus*), body louse (*Pediculus capitatus*) and crab louse (*Pthirus pubis*) and the preventive effects of ivermectin on these species.

Pediculosis pubis, as well as *phthiriasis palpebrum*, an infestation caused by crab louse, *Pthirus pubis*. During infestation, these parasites can be seen on eyelashes and pubic regions, and in heavy infestation they can inhabits the facial hairs, axillae, chest, and rectal regions. Their serrated claws on the legs facilitate traction on flat, hairless surfaces [63] and also help them move around the entire skin surface. This feature also helps them for transmission among people.

There are various insecticide formulations such as lindane, malathion, and pyrethrin. Treatment of pubic louse during *Pediculosis ciliaris*, also includes mechanical removal with fine forceps, ointment like yellow mercuric oxide, physostigmine and petroleum jelly *etc* [64]. These treatment lines are successful in adults, but in children, as there is less possibility of inhabiting eyelashes because of the temperature and moisture difference, as well as absence of the hairs on most body surface in children, hinders [65] the topical therapy as the insecticide may not reach to all hairy areas.

In this situation, oral administration of ivermectin has been purported to be the best solution to control pubic lice infestation in children. It was observed when 4 children from age group 3 to 10 years, reported the symptoms of pediculosis and on closer examination, numerous nits and motile parasites were found on their eyelashes. When these patients were treated with two 200 μg/kg doses of oral ivermectin (*Stromectol*) weekly. They showed rapid improvement in their condition, they get rid of all adult lice within 2 days, while the nits that were attached to the eyelashes disappeared during the process of growing of eyelid hairs [66].

Ivermectin may be a good alternative to various topical insecticides like malathion and pyrethroid because of their susceptibility of causing resistance in lice populations. Moreover, this repeated and non-judicious use of different products such as insecticidal shampoos, leads to the resistance as well as failure to clear the infestation, further there have been some preliminary epidemiologic evidence like acute leukemia in children, were observed that may be linked to over exposure of the insecticidal shampoos. In addition to this, the infestation can be more prevailing due to poor sanitization, living in close contacts like in classroom, slum area *etc*.

A real time trial was attempted, where the patients were showing topical insecticide resistance in lice and consequently treatment failure. These patients were administrated ivermectin (a dose of 400 μg/kg body weight) on days 1 and 8, for eradicating head-lice infestation [64].

The pre-eminence of ivermectin over failed topical insecticides was observed in the patients on both day 2 and day 8. It was observed that efficacy of orally administrated ivermectin in the patients was more as compared to the patients who were using malathion against lice. The efficacy ivermectin could be observed 8.3% higher than malathion even after 29 days of treatment. Ivermectin administration showed ovicidal activities when the lice were observed with combing procedure.

Further, topical application ivermectin lotion has been reported for the treatment to eradicate permethrin-resistant head lice infestation in children of 2 to 5 years age group [64].

8.3.5 Ivermectin Against Agricultural, Stored Grain Insect Pests and Other Insects

Agriculture has been the mainstay of human life in terms of their livelihood as well as economical support. However, insects pests are the major enemy of human, since the inception of agricultural practices and civilization, as they share a big chunk of food from the agricultural field, granaries, food processing units, as well as destroy household items like furniture, clothes *etc.* which causes a substantial economic loss annually due to their notorious feeding and devastating behavior.

To combat the insect pests human have been relying heavily on synthetic pesticides, however, due to the non-judicious and indiscriminate use of the synthetic pesticides, caused a sever deleterious and harmful effects to environment and human health. Several alternative options of the plant botanicals and bio-natural products were extensively used for the effective management [67].

Various studies suggests that ivermectin, as a naturalyte biocides can be a used as an effective insecticide to target insect pests belonging to different orders as well as possessing different nature of damage. The studies depicted that ivermectin is detrimental against different development stages of insects which is evident from the finding given below.

The toxicity of ivermectin (IVECOP-12) was proved to be potentially effective when it was tested against the lepidopteran pest's larvae of key forest plant defoliators, such as Ailanthus web worm, *Atteva fabriciella* Swederus, bamboo leaf roller, *Crypsiptya coclesalis*, teak leaf skeletonizer, *Eutectona machaeralis*, teak defoliator, *Hyblaea puera* and Albizia foliage feeder, *Spirama retorta etc.* The LC_{50} values and probit analyses revealed that ivermectin was able to kill larvae of these insect pests when they were allowed to feed on treated host plant and also applied topically on the larvae. Its toxicity was highly proved against *H. puera*, followed by *C. coclesalis, S. retorta, E. machaeralis* and *A. fabriciella* [68].

In a study, the insecticidal activities of ivermectin were evaluated and compared with one of biocides, spinosad and a well-known synthetic insecticide, malathion; against three important stored grain insect pests i.e., *Sitophilus oryzae*, *Rhyzopertha dominica*, and *Tribolium castaneum.* The bioassays revealed that ivermectin was able to prevent 100% adult emergence in *S. oryzae*, followed by 81-100% reduction in *R. dominica* and *T. castaneum.* It was observed that Ivermectin even at 50 ppm concentration could kill 96 and 88% larval population in *R. dominica* and *S. oryzae*, respectively, after two weeks of exposure in semisynthetic diet [69].

In another study, where 3% ivermectin dust was used to prevent the menace of *Coptotermes formosanus* is an important termite species, which severely damage housing and construction furniture made of wood, and old trees in residential areas in south of China. The residents were using one of the conventional chemical mirex dust which was ineffective against these termites. In a study when 3% ivermectin dust was applied to eradicate the colonies of *C. formosanus* in residential areas of Hangzhou city, Zhejiang province, China from April 2010 to May 2012. It was found that 20-30 grams of 3% ivermectin dust was applied topically on these termites, which were feeding at 18-80 meter far away from their colony, and allowed to move to their nests, the ivermectin dust could successfully eliminated the colony within 3.5 to 8 months and proved to be good alternative of the conventional mirex dust [70].

Recent advances show that various molecular labs are looking for more effective and novel active domains of ivermectin by creating molecular modification, and are in the process of developing bioactive molecules against insect pests. An example of such hybrid compounds was described by Zhang *et al.* in 2015 [71], when they synthesized a hybrid molecule of ivermectin and milbemycin by domain swapping and resulted in producing 25-ethyl and 25-methyl ivermectin (**1** and **2**), 25-ethyl and 25-methyl ivermectin (derivative **1** and **2**) and tested its insecticidal activity against 2nd instar larvae of armyworm, *Mythimna separata*, a polyphagous pets. These derivatives were compared to their parent compounds milbemycin and ivermectin. This derivative demonstrated its dose-dependent lethal effects in terms of larval mortality and also it was found to be insecticidal than its parent compounds.

This concluded that 25-methyl and 25-ethyl ivermectin obtained by domain swapping can be an promising and novel insecticide in agriculture specially when there is always a threat of developing resistance against insecticides in insects [71].

8.3.6 Ivermectin Usage in Livestock

Livestock is key component of any country's economy. About two third of human population depends on livestock for their livestock. It acts as a considerable source of income for small farm and rural community, especially of developing countries. In addition, livestock adds employment and also contributes in GDP of a country. However, livestock faces various threats of diseases and pests that greatly affect the livelihood of livestock handlers, these pests can have major impact on profit expected from livestock business. Out of various pests and diseases, ectoparasites are one of the important nuisance and this includes a number of insect pests that affects the livestock and livestock farmers both. In the present section, a brief information if provided which reflects the controlling measurements of insects parasites using ivermectin in different forms.

In cattle, ivermectin has successfully been administrated through different methods such as oral, subcutaneous [72–74], and pour-on methods [75], to kill the arthropod ectoparasites of cattle. The ivermectin has been found to be very effective against a number of ectoparasites like, grubs (*H. lineatum, Hypoderma bovis, Dermatobia hominis*), sucking lice (*Linognathus vituli, Haematopinus eurysternus, Solenopotes capillatus*), mange mites (*Psoroptes ovis* and *Sarcoptes scabei,*) and screw worms (*Chrysomya bezziana*).

Earlier studies has shown that ivermectin administrated either orally or subcutaneous, has toxic effects against larval stages of *Hypoderma* spp. Whereas, it has been very effective against all the life stages of lice as they are blood sucker and could be affected due to ingestion of ivermectin injected host blood. The injectable doses have also succeeded in controlling screw worm *C. bezziana* and *D. hominis* infestations.

Pour-on and subcutaneous treatment of ivermectin showed excellent results in controlling mites like *S. scabei, P. ovis* and *C. bovis*. Ivermectin had also shown promising efficacy against hard ticks, like *Boophilus microplus, B. decoloratus*, soft tick, and Ornithodoros as well [50].

i. Ticks and Mites

Pour-on treatment, however, was very effective to control the mites for example, C. *bovis,* if it is used in higher concentration as compared to subcutaneous doses. This could happened due to its presence on the skin and hair cattle and the same was evident with negligible lesions around the area of application [50, 76].

Though the drug was found effective during such treatment methods, but its persistence is again a question of interest in the case of livestock management. Limited number of the studies has been conducted for evaluation of the persistence of ivermectin against mange mites. However, studies conducted [73] demonstrated that a single intramuscular ivermectin dose, protects cattle against *P. bovis* infestation up to 3 weeks. In case of ticks, there is scarcity of any studies and also it is restricted to *B. microplus*, where ivermectin could kill more than 80% population and its persistence was observed for one week when animals were administrated with subcutaneous ivermectin injection [50, 77].

ii. Lice

There are 4 economically important lice species, Haematopinus eurysternus, Linognathus vituli, Solenopotes capillatus and Bovicola (Damalinia) bovis, which were better controlled by topical (pour-on) application of ivermectin as the injectable drug was not successful due to surface feeding (feed on skin and debris) behavior of lice. High efficacy of the topical application suggested a greater exposure of these ectoparasites to high concentration of ivermectin that acted both as contact and feeding poison [78]. Moreover, in case of lice control, ivermectin persistence was as longer as 35-49 days and eradicated 100% B. bovis infestations, when pour-on method was applied cattle [50].

iii. Warble Fly Larvae and Screw Worms

The most prevalent and economically important Warble flies (*H. bovis* and *H. lineatum*) have been successfully prevented by ivermectin injections and topical applications, by killing all the larval population of these flies [74]. Further it was observed that ivermectin doses were persistent and effective till 7–10 days of treatment and successfully killed around 97% of larval population [50]. This showed the potential application of ivermectin in livestock pest management.

The screw worm, *Cochliomyia hominovorax*, is considered as pest of cattle in The New World, it is an obligatory parasite that causes myiasis and can infect almost all the warm-blooded animals, and also humans; though these flies were successfully controlled in the USA after adopting sterile male release program. However, it is still a threat in southern part of USA and Mexico, if there is any disruption in pest management practice. Studies showed that if the calves are administrated with subcutaneous ivermectin injection with 24 h of birth or after castration, then they can be protected from navel or scrotal myiasis [79]. Recently, it was demonstrated that ivermectin and other macrolytes such as doramactin are able to reduce 90% chances of myiasis even after 12 days of the treatment [50, 80].

iv. Hornflies

The hornfly, *H. irritans,* is a blood sucking parasite, cosmopolitan in nature, causes a considerable damage by blood sucking and spread secondary infection in cattle [50]. A conventional method to control this fly is to simply catch and kill but this is no sufficient measure as the developing stages of this fly escape this strategy due to different site of the development habitat such as manure. This is one of the major challenges to control this fly. However, various studies were conducted to evaluate the efficacy and persistence of ivermectin when pour-on technique was used to control the flies, Lysyk and Colwell (1996) [81] reported that topical application of ivermectin could prevent 90% adult populations up to 8-16 days, and the efficacy was still evident by reducing 50% population till 18-26 days. The results were more promising with repeated application of ivermectin and its efficacy was observed even after 35 days of treatment [50, 82].

8.3.7 Environmental Impact of Ivermectin

Ivermectin have been recognized as one of the best available, effective and inexpensive tool of treatment and control of various pests of crops, ectoparasites of livestock and human also. Several findings have demonstrated its efficiency and persistence that helps eradicate several economically important and health affecting pests; however, there has always a concern of its deleterious effects on environment biodiversity.

There are possibilities of environmental contamination by this drug as it has been used extensively in cattle industries and crop management and there are chances of appearance of ivermectin in food chain, farmland system through runoff and leaching process and mixing with dung or dung pat which may affect non-target insects and other biological agents [83].

It has been observed that 6%2-98% of unmetabolized ivermectin dose, in cattle feces and this contaminated dung containing persistent ivermectin may harm coprophagous fauna. It is found in the dung fly *Scathophaga stercoraria*, that after feeding on dung mixed with various concentration of ivermectin greatly affected the larval population and also disrupted the reproductive behavior in adults. Larval duration was lengthened after exposure and it was also observed that ivermectin contaminated dung caused reduced egg-to-adult survival, body size, and the male biased sex ratio and reduced oviposition in adults. These observations strongly appeal that such treatments should be applied after due and accurate eco-assessments [84].

Ivermectin, once contaminated the dung can affect the life cycle pattern of various non-target dung feeding insects like dung beetles, dipteran larvae and some birds which feed on these insects [85]. Also the reduction of these dung fauna can result in decrease in dung degradation which may invite other contaminating factors in environment.

Prolonged use of ivermectin in meat producing animals like goat, sheep, buffalo *etc.* can also cause accumulation of the ivermectin residues in meat that may cause adverse effect on human health also [86].

Undoubtedly, these macrocyclic lactone derivatives have shown their unprecedented efficiency and width of promising effects against a large number of human, veterinary parasites and agricultural plant pests [87]. But, also reports are there that indicated the resistance in certain veterinary parasites after their extensive use. This observation raise the concern of critical understanding about the pace and methods of application of ivermectin as well as the development of resistance mechanism insects so as to manage the resistance mechanism in insects [88].

Conclusions

Although there are various studies that have shown the negative impact of ivermectin against different ecological components such as algae, plants, non-target invertebrates, and vertebrates, including fish *etc.* and the same has been taken into consideration by various nodal agencies like Food and Drug Administration (FDA) [88], but further to strengthen the use of ivermectin and its modified version, needs a detailed study of possible short- or long-term effects on abiotic and biotic components of ecosystems. However, ivermectin have huge potential to control insect fauna that is a nuisance of human and its resources that causes millions of US $, loss annually, which attracts the commercial relevance of ivermectin and makes the use of ivermectin obvious in livestock industry as well as human health related program such mass malaria eradication, combating leishmaniasis, trypanosomiasis *etc.*

Hence it is advisable in case of ivermectin usage as insecticide for effective and strategic insect pest management, its risk and benefit assessment must be done critically before applying it in the crop field, livestock and or on human population.

References

1. Õmura, S. and Crump, A., The life and times of ivermectin — A success story. *Nat. Rev. Microbiol.*, 2, 12, 984–989, Dec. 2004.
2. Campbell, W.C., Ivermectin: An update. *Parasitol. Today*, 1, 1, 10–16, Jul. 1985.
3. Lakwo, T., Oguttu, D., Ukety, T., Post, R., Bakajika, D., Onchocerciasis elimination: Progress and challenges. *Res. Rep. Trop. Med.*, 11, 81–95, Oct. 2020.
4. Stokstad, E. and Vogel, G., Neglected tropical diseases get the limelight in Stockholm. *Science*, 350, 144–145, Oct. 2015.
5. Õmura, S., Ivermectin: 25 years and still going strong. *Int. J. Antimicrob. Agents*, 31, 2, 91–98, Feb. 2008.
6. Martin, R.J., Robertson, A.P., Choudhary, S., Ivermectin: An anthelmintic, an insecticide, and much more. *Trends Parasitol.*, 37, 1, 48–64, Jan. 2021.
7. Sainas, S., Dosio, F., Boschi, D., Lolli, M.L., Targeting human onchocerciasis: Recent advances beyond ivermectin, in: *Annual Reports in Medicinal Chemistry*, vol. 51, pp. 1–38, Elsevier, Cambridge, US, 2018.
8. Sauerbrey, M., Rakers, L.J., Richards, F.O., Progress toward elimination of onchocerciasis in the Americas. *Int. Health*, 10, suppl_1, i71–i78, Mar. 2018.
9. Dadzie, Y., Amazigo, U.V., Boatin, B.A., Sékétéli, A., Is onchocerciasis elimination in Africa feasible by 2025: A perspective based on lessons learnt from the African control programmes. *Infect. Dis. Poverty*, 7, 1, 63, Dec. 2018.
10. Fobi, G. *et al.*, Managing the fight against onchocerciasis in Africa: APOC experience. *PLoS Negl. Trop. Dis.*, 9, 5, e0003542, May 2015.
11. World Health Organization, Progress report on the elimination of human onchocerciasis, 91, 2016 2015.
12. World Health Organization, Progress report on the elimination of human onchocerciasis, 2019 2018.
13. Traore, M.O. *et al.*, Proof-of-principle of onchocerciasis elimination with ivermectin treatment in endemic foci in Africa: Final results of a study in Mali and Senegal. *PLoS Negl. Trop. Dis.*, 6, 9, e1825, Sep. 2012.
14. World Health Organizatio, *Guidelines for stopping mass drug administration and verifying elimination of human onchocerciasis: Criteria and procedures*, World Health Organization, Africa, 2016.
15. Gebrezgabiher, G., Mekonnen, Z., Yewhalaw, D., Hailu, A., Reaching the last mile: Main challenges relating to and recommendations to accelerate onchocerciasis elimination in Africa. *Infect. Dis. Poverty*, 8, 1, 60, Dec. 2019.
16. Boussinesq, M., Fobi, G., Kuesel, A.C., Alternative treatment strategies to accelerate the elimination of onchocerciasis. *Int. Health*, 10, suppl_1, i40–i48, Mar. 20184.
17. Tekle, A., Elhassan, E., Isiyaku, S., Amazigo, U., Bush, S., Noma, M., Impact of long-term treatment of onchocerciasis with ivermectin in Kaduna State, Nigeria: First evidence of the potential for elimination in the operational area

of the African Programme for Onchocerciasis Control. *Parasites Vectors*, 5, 1, 28, 2012.
18. Diawara, L., Traoré, M., Badji, A., Bissan, Y., Doumbia, K., Goita, S., Feasibility of onchocerciasis elimination with ivermectin treatment in endemic foci in Africa: First evidence from studies in Mali and Senegal. *PLoS Negl. Trop. Dis.*, 3, 7, e497, 2009.
19. Evans, D. *et al.*, Status of Onchocerciasis transmission after more than a decade of mass drug administration for onchocerciasis and lymphatic filariasis elimination in central Nigeria: Challenges in coordinating the stop MDA decision. *PLoS Negl. Trop. Dis.*, 8, 9, e3113, 2014.
20. Zarroug, I. *et al.*, The first confirmed elimination of an onchocerciasis focus in Africa: Abu Hamed, Sudan. *Am. J. Trop. Med. Hyg.*, 95, 5, 1037–40, 2016.
21. World Health Organization, "World Health Organization & African Programme for Onchocerciasis Control (APOC), *Guidelines for revising ivermectin treatment boundaries within the context of onchocerciasis elimination*, African Programme for Onchocerciasis Control., WHO/MG/15.21, 2015.
22. Mackenzie, C.D., Homeida, M.M., Hopkins, A.D., Lawrence, J.C., Elimination of onchocerciasis from Africa: Possible? *Trends Parasitol.*, 28, 1, 16–22, Jan. 2012.
23. Whittaker, C., Walker, M., Pion, S.D.S., Chesnais, C.B., Boussinesq, M., Basáñez, M.-G., The population biology and transmission dynamics of Loa loa. *Trends Parasitol.*, 34, 4, 335–350, Apr. 2018.
24. Kim, Y.E., Remme, J.H.F., Steinmann, P., Stolk, W.A., Roungou, J.-B., Tediosi, F., Control, elimination, and eradication of river blindness: Scenarios, timelines, and ivermectin treatment needs in Africa. *PLoS Negl. Trop. Dis.*, 9, 4, e0003664, Apr. 2015.
25. Colebunders, R. *et al.*, From river blindness control to elimination: Bridge over troubled water. *Infect. Dis. Poverty*, 7, 1, 21, Dec. 2018.
26. Senyonjo, L. *et al.*, Factors associated with ivermectin non-compliance and its potential role in sustaining Onchocerca volvulus transmission in the West Region of Cameroon. *PLoS Negl. Trop. Dis.*, 10, 8, e0004905, Aug. 2016.
27. Bundy, D.A.P., Dhomun, B., Daney, X., Schultz, L.B., Tembon, A., Investing in onchocerciasis control: Financial management of the African programme for onchocerciasis control (APOC). *PLoS Negl. Trop. Dis.*, 9, 5, e0003508, May 2015.
28. Dissak-Delon, F.N. *et al.*, Barriers to the National Onchocerciasis Control Programme at operational level in Cameroon: A qualitative assessment of stakeholders' views. *Parasitol. Vectors*, 12, 1, 246, Dec. 2019.
29. NTD Modelling Consortium Onchocerciasis Group, The World Health Organization 2030 goals for onchocerciasis: Insights and perspectives from mathematical modelling. *Gates Open Res.*, 3, 1545, Sep. 2019.
30. World Health Organization, *Lymphatic filariasis: Managing morbidity and preventing disability: An aide-mémoire for national programme managers*,

2nd ed, World Health Organization, Geneva, 2021, Accessed: Aug. 01, 2022. [Online]. Available: https://apps.who.int/iris/handle/10665/339931.
31. World Health Organization, *Guideline: Alternative mass drug administration regimens to eliminate lymphatic filariasis*, World Health Organization, Geneva, 2017, Accessed: Aug. 01, 2022. [Online]. Available: https://apps.who.int/iris/handle/10665/259381.
32. Patel, C. *et al.*, Efficacy and safety of ivermectin and albendazole co-administration in school-aged children and adults infected with Trichuris trichiura: Study protocol for a multi-country randomized controlled double-blind trial. *BMC Infect. Dis.*, 19, 1, 262, Dec. 2019.
33. Moser, W., Schindler, C., Keiser, J., Efficacy of recommended drugs against soil transmitted helminths: Systematic review and network meta-analysis. *BMJ*, 358, j4307, 2017.
34. Keller, L. *et al.*, Long-term outcomes of ivermectin-albendazole versus albendazole alone against soil-transmitted helminths: Results from randomized controlled trials in Lao PDR and Pemba Island, Tanzania. *PLoS Negl. Trop. Dis.*, 15, 6, e0009561, Jun. 2021.
35. Ganesh S. and Cruz R.J., Strongyloidiasis: A multifaceted disease. *Gastro. Hepato.*, 7, 3, 194–196, Mar. 2011.
36. Downes, B. and Jacobsen, K., A systematic review of the epidemiology of mansonelliasis. *Afr. J. Infect. Dis.*, 4, 1, 7–14, Jun. 2010.
37. de Almeida Basano, S. *et al.*, Phase III clinical trial to evaluate ivermectin in the reduction of Mansonella ozzardi infection in the Brazilian Amazon. *Am. J. Trop. Med. Hyg.*, 98, 3, 786–790, Mar. 2018.
38. Ramadhan, Z.I.R., *et al.*, Neglected co-infection of Malaria: A brief explanation of the parasite - Mansonella ozzardi, p. 5. https://www.researchgate.net/profile/Ranindya_Cipta/project/Neglected-Co-infection-of-Malaria-A-Brief-Explanation-of-The-Parasite-Mansonella-Ozzardi-2/attachment/5bfaa88ecfe4a7645504aa87/AS:696851724509184@1543153806166/download/Mansonellosis+Fix.pdf?context=ProjectUpdatesLog
39. Turcios, R.M. *et al.*, Temporal and geographic trends of rotavirus activity in the United States, 1997–2004. *Pediatr. Infect. Dis. J.*, 25, 5, 451–454, May 2006.
40. Campillo, J.T., Boussinesq, M., Bertout, S., Faillie, J.-L., Chesnais, C.B., Serious adverse reactions associated with ivermectin: A systematic pharmacovigilance study in sub-Saharan Africa and in the rest of the World. *PLoS Negl. Trop. Dis.*, 15, 4, e0009354, Apr. 2021.
41. Chandler, R.E., Serious neurological adverse events after ivermectin—Do they occur beyond the indication of onchocerciasis? *Am. J. Trop. Med. Hyg.*, 98, 2, 382–388, Feb. 2018.
42. Freeman, M.C. *et al.*, Challenges and opportunities for control and elimination of soil-transmitted helminth infection beyond 2020. *PLoS Negl. Trop. Dis.*, 13, 4, e0007201, Apr. 2019.

43. Strong, L. and Brown, T.A., Avermectins in insect control and biology: A review. *Bull. Entomol. Res.*, 77, 3, 357–389, 1987, https://doi.org/10.1017/S0007485300011846.
44. Fisher, M.H., Recent advances in avermectin research. *Pure Appl. Chem.*, 62, 7, 1231–1240, Jan. 1990.
45. Burg, R.W. *et al.*, Avermectins, new family of potent anthelmintic agents: Producing organism and fermentation. *Antimicrob. Agents Chemother.*, 15, 3, 361–367, Mar. 1979.
46. Jansson, R.K. and Dybas, R.A., Avermectins: Biochemical mode of action, biological activity and agricultural importance, in: *Insecticides with Novel Modes of Action*, pp. 152–170, Springer, Berlin, Heidelberg, 1998, [Online]. Available: https://doi.org/10.1007/978-3-662-03565-8_9.
47. Turner, M.J. and Schaeffer, J.M., Mode of action of ivermectin, in: *Ivermectin and Abamectin*, W.C. Campbell, (Ed.), pp. 73–88, Springer New York, NY, 1989, [Online], Available: https://doi.org/10.1007/978-1-4612-3626-9_5.
48. Chen, I.-S. and Kubo, Y., Ivermectin and its target molecules: Shared and unique modulation mechanisms of ion channels and receptors by ivermectin: Modulation mechanisms of ivermectin targets. *J. Physiol.*, 596, 10, 1833–1845, May 2018.
49. Egerton, J.R., Brokken, E.S., Suhayda, D., Eary, C.H., Wooden, J.W., Kilgore, R.L., The antiparasitic activity of ivermectin in horses. *Vet. Parasitol.*, 8, 1, 83–88, 1981, https://doi.org/10.1016/0304-4017(81)90020-0.
50. Vercruysse, J. and Rew, R.S. (Eds.), *Macrocyclic lactones in antiparasitic therapy*, CABI Pub, Oxon, UK; New York, NY, 2002.
51. Gunathilaka, N., Ranathunge, T., Udayanga, L., Wijegunawardena, A., Gilles, J.R.L., Abeyewickreme, W., Use of mechanical and behavioural methods to eliminate female Aedes aegypti and Aedes albopictus for sterile insect technique and incompatible insect technique applications. *Parasitol. Vectors*, 12, 1, 148, Dec. 2019.
52. Loftin, K.M., Byford, R.L., Craig, M.E., Steiner, R.L., Evaluation of cattle insecticide treatments on attraction, mortality and fecundity of mosquitoes. *J. Am. Mosq. Control Assoc.*, 12, 1, 17–22, 1996.
53. Pampiglione, S., Majori, G., Petrangeli, G., Romi, R., Vermectins, MK-933 and MK-936, for mosquito control. *Trans. R. Soc. Trop. Med. Hyg.*, 79, 6, 797–799, 1985.
54. Alves, S.N., Serrão, J.E., Mocelin, G., de Melo, A.L., Effect of ivermectin on the life cycle and larval fat body of Culex quinquefasciatus. *Braz. Arch. Biol. Technol.*, 47, 3, 433–439, Jul. 2004.
55. Makhanthisa, T.I., Braack, L., Lutermann, H., The effect of cattle-administered ivermectin and fipronil on the mortality and fecundity of Anopheles arabiensis Patton. *Parasitol. Vectors*, 14, 1, 349, Dec. 2021.
56. Pooda, H.S. *et al.*, Administration of ivermectin to peridomestic cattle: A promising approach to target the residual transmission of human malaria. *Malar. J.*, 14, 1, 496, Dec. 2015.

57. Chaccour, C., Lines, J., Whitty, C.J.M., Effect of ivermectin on anopheles gambiae mosquitoes fed on humans: The potential of oral insecticides in Malaria control. *J. Infect. Dis.*, 202, 1, 113–116, 2010, https://doi.org/10.1086/653208.
58. Chaccour, C., Hammann, F., Rabinovich, N.R., Ivermectin to reduce malaria transmission I. Pharmacokinetic and pharmacodynamic considerations regarding efficacy and safety. *Malar. J.*, 16, 1, 161, Dec. 2017.
59. Crump, A., Ivermectin: Enigmatic multifaceted 'wonder' drug continues to surprise and exceed expectations. *J. Antibiot. (Tokyo)*, 70, 5, 495–505, May 2017.
60. Mascari, T.M., Mitchell, M.A., Rowton, E.D., Foil, L.D., Ivermectin as a rodent feed-through insecticide for control of immature sand flies (Diptera: Psychodidae). *J. Am. Mosq. Control Assoc.*, 24, 2, 323–326, Jun. 2008.
61. Distelmans, W., D'Haeseleer, F., Mortelmans, J., Efficacy of systemic administration of ivermectin against tsetse flies. *Ann. Soc. Belg. Med. Trop.*, 83, 119–125, 1983.
62. Dias, J.C.P., Schofield, C.J., Machado, E.M., Fernandes, A.J., Ticks, ivermectin, and experimental Chagas disease. *Mem. Inst. Oswaldo Cruz*, 100, 8, 829–832, Dec. 2005.
63. Roberts, R.J., Head lice. *N. Engl. J. Med.*, 346, 1645–50, 2002.
64. Chosidow, O. *et al.*, Oral ivermectin versus malathion lotion for difficult-to-treat head lice. *N. Engl. J. Med.*, 362, 10, 896–905, Mar. 2010.
65. Dodd, C., Interventions for treating headlice, in: *Cochrane Database of Systematic Reviews*, The Cochrane Collaboration (Ed.), p. CD001165, John Wiley & Sons, Ltd, Chichester, UK, 2001.
66. Burkhart, C.N. and Burkhart, C.G., Oral ivermectin therapy for *Phthiriasis Palpebrum. Arch Ophthalmol.*, 118, 1, 134–135, Jan. 2000.
67. Tarkeshwar, Das, A., Thakur, S., Ruchika, K., Choudhary, K., Kapinder, Nanoparticles: Applications in medicine and agriculture. *Int. J. Sci. Res.*, 12, 3, 28–33, 2022, http://dx.doi.org/10.29322/IJSRP.12.03.2022.p12306.
68. Roychoudhury, N. and Mishra, R.K., Toxicity of ivermectin against the defoliator larvae of some major forest insect pests-a critical review. *Pestology*, 44, 4, 34–38, 2020.
69. Ahmed, S. and Nasr, H., Evaluation of spinosad and ivermectin against three of stored grain insects in wheat grains. *J. Plant Prot. Pathol.*, 6, 4, 713–726, Apr. 2015.
70. Zhao, J., Dong, Y., Yu, B., Zhang, Z., Mo, J., Ivermectin dust for the control of coptotermes formosanus in residential areas. *Sociobiology*, 59, 4, 1365–1374, 2012.
71. Zhang, J., Yan, Y.-J., An, J., Huang, S.-X., Wang, X.-J., Xiang, W.-S., Designed biosynthesis of 25-methyl and 25-ethyl ivermectin with enhanced insecticidal activity by domain swap of avermectin polyketide synthase. *Microb. Cell Factories*, 14, 1, 152, Dec. 2015.

72. Guillot, F.S. and Meleney, W.P., The infectivity of surviving Psoroptes ovis (Hering) on cattle treated with ivermectin. *Vet. Parasitol.*, 10, 1, 73–78, 1982, https://doi.org/10.1016/0304-4017(82)90009-7.
73. Meleney, W.P., Control of psoroptic scabies on calves with ivermectin. *Am. J. Vet. Res.*, 43, 329–331, 1982.
74. Benz, G.W., Roncalli, R.A., Gross, S.J., Use of ivermectin in cattle, sheep, goats, and swine, in: *Ivermectin and Abamectin*, W.C. Campbell (Ed.), pp. 215–229, Springer-Verlag, New York, 1989, [Online]. Available: https://doi.org/10.1007/978-1-4612-3626-9_15.
75. Lonneux, J.F., Nguyen, T.Q., Losson, B.J., Efficacy of pour-on and injectable formulations of moxidectin and ivermectin in cattle naturally infected with Psoroptes ovis: Parasitological, clinical and serological data. *Vet. Parasitol.*, 69, 319–330, 1997.
76. Jenkinson, D.M., Hutchison, G., Jackson, D., McQueen, L., Route of passage of cypermethrin across the surface of sheep skin. *Res. Vet. Sci.*, 41, 2, 237–241, 1986.
77. Cramer, L.G., Bridi, A.A., Amaral, N.K., Gross, S.J., Persistent activity of injectable ivermectin in the control of the cattle tick Boophilus microplus. *Vet. Rec.*, 122, 25, 611–612, 1988.
78. Polley, L.R., Wagner, B.A., Ward, T.I., Campbell, J.R., Effect of topical ivermectin and moxidectin for naturally acquired *Damalinia bovis* infestations in cattle treated under winter conditions in Canada. *Vet. Rec.*, 143, 3, 80–81, Jul. 1998.
79. Benitez Usher, C. *et al.*, Prophylactic use of ivermectin against cattle myiasis caused by Cochliomyia hominivorax. *Vet. Parasitol.*, 72, 215–220, 1997.
80. Anziani, O.S. *et al.*, Persistent activity of doramectin and ivermectin in the prevention of cutaneous myiasis in cattle experimentally infested with Cochliomyia hominivorax. *Vet. Parasitol.*, 87, 2–3, 243–247, Jan. 2000.
81. Lysyk, T.J. and Colwell, D.D., Duration of efficacy of diazinon ear tags and ivermectin pour-on for control of horn fly (Diptera: Muscidae). *J. Econ. Entomol.*, 89, 6, 1513–1520, Dec. 1996.
82. Uzuka, Y. *et al.*, Chemical control of Haematobia irritans with 0.5% topical ivermectin solution in cattle. *J. Vet. Med. Sci.*, 61, 3, 287–289, 1999.
83. Souza, A., Ferreira-Châline, R., Châline, N., Korasaki, V., Beiroz, W., Louzada, J., Does Neem trigger the same response as Ivermectin? Dung beetle behaviour and physiology. *PeerJ Prepr.*, preprint, Mar. 2018.
84. Mahdjoub, H., Blanckenhorn, W.U., Lüpold, S., Roy, J., Gourgoulianni, N., Khelifa, R., Fitness consequences of the combined effects of veterinary and agricultural pesticides on a non-target insect. *Chemosphere*, 260, 126–271, 2020, https://doi.org/10.1016/j.chemosphere.2020.126271.
85. Iqbal, A., Usman, M., Abubakar, M., Mini review: Current tick control strategies in Pakistan are possible environmental risks. *Iraqi J. Vet. Sci.*, 31, 2, 81–86, Dec. 2017.

86. Jackson, H.C., Ivermectin as a systemic insecticide. *Parasitol. Today*, 5, 5, 147, 1989, https://doi.org/10.1016/0169-4758(89)90079-3.
87. Campbell, W.C. (Ed.), *Ivermectin and abamectin*, Springer New York, New York, NY, 1989.
88. Hu, J., Liang, P., Shi, X., Gao, X., Effects of insecticides on the fluidity of mitochondrial membranes of the diamondback moth, *Plutella xylostella*, resistant and susceptible to avermectin. *J. Insect Sci.*, 8, 3, 1–9, Jan. 2008.

9

Potential Applications of Ivermectin (IVM) in Dermatology

Vikrant Jayant and Rashid Ali*

Department of Chemistry, Jamia Millia Islamia, Jamia Nagar, Okhla, New Delhi, India

Abstract

Ivermectin (IVM) is a type of unsurpassed semisynthetic and broad-spectrum anti-parasitic drug, which possess remarkable potency as compared to the previously reported compounds. Introduction of IVM is a most important breakthrough in the treatment of several parasitic dermatoses in animals. It is highly useful as veterinary medicine, and significant work against two important phyla of the parasites for instance nemathelminthes or nematodes (round-worms) and arthropods (mites, insects, and ticks), has been reported. Over the past several years, the usage of IVM in various parasites such as bacteria, fungus, virus, protozoa, arthropods, nematodes, etc., has uncovered their excellent potential in the field of skin parasitology. Importantly, in the branch of veterinary sciences, their resistance of IVM in dermatology is now-a-days widespread, but the exact mechanisms for the resistances are still unknown, and will inspire the researcher to find out the exact causes for the real one. In this chapter, we presented the applications of IVM in dermatology, which will make a valuable platform for the prodigious impact of IVM drug in the treatment of severe skin related issues and other, if any.

Keywords: Ivermectin, avermection, dermatology, doses, diseases, skin infection

**Corresponding author*: rali1@jmi.ac.in

Rashid Ali and Shahid ul-Islam (eds.) Chemistry and Biological Activities of Ivermectin, (199–230)

9.1 Introduction

Ivermectin (IVM), a white to slightly yellowish-white crystalline solid, is a non-hygroscopic synthetic derivative of a broad range of anti-parasitic class of macrocyclic lactones. This vital compound is almost insoluble in water but soluble in CH_3OH as well as 95% C_2H_5OH. As can be seen from the inspection of the Figure 9.1, IVM has several potential applications. Notably, IVM (MK-0933), a 22,23-dihydroderivative of avermectin B1, is basically a mixture of mainly two chemical compounds for instances, a 22,23-dihydroavermectin B1a (80%–90%) with a molecular formula $C_{48}H_{74}O_{14}$ and the 22,23-dihydroavermectin B1b (10%–20%), having the molecular formula of $C_{47}H_{72}O_{14}$. The chemical representation of IVM (B1a and B1b) is shown in the Figure 9.2. First time IVM was isolated by Satoshi Ōmura using the process of fermentation of soil microorganism, for instances actinomycete *Streptomyces avermitilis* (*S. avermitilis*) from a golf course in Japan in the late 1960s, which produced the active ingredient avermectin. Similar type of macrolides, for example emamectin, doramectin, milbemycin D, eprinomectin, abamectin, moxidectin is also of high value compounds for different purposes. At present, IVM is recognized as

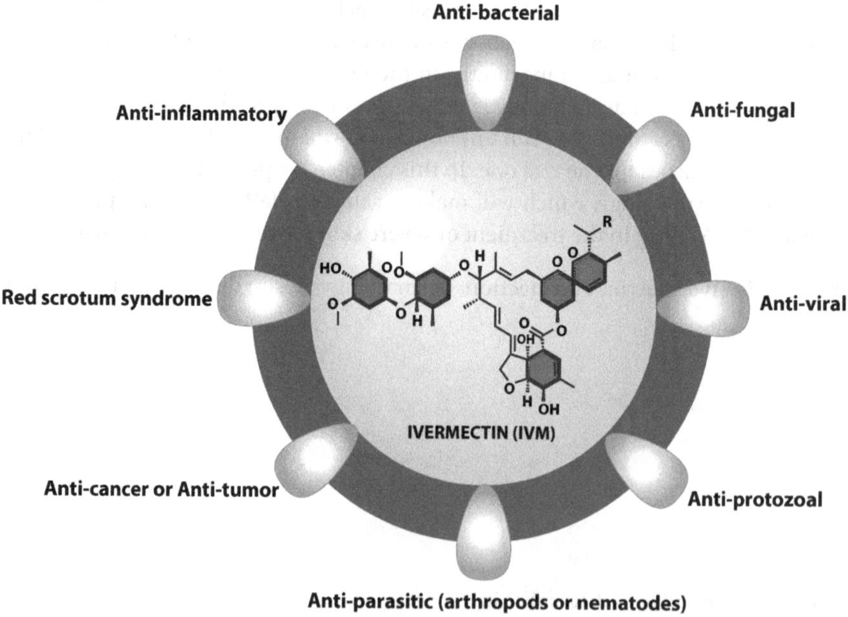

Figure 9.1 Systematic representation of diverse potential applications of ivermectin (IVM).

a 'Wonder Drug' because of its great expedient effects documented on the animal as well as human beings health. The brand product name of IVM is "*Stromectol*" - an oral anti-parasitic agent, approved for a proper treatment of "worm infestations". Basically, IVM was first discovered in the 1970s during a veterinary medication screening effort at Merck pharmaceuticals, and was approved in 1981 for the commercial use in veterinary sciences for the parasitic diseases in domestic pets and animals under the brand name "*Mectizan*". Since then, IVM has been used to treat several parasitic diseases year-by- year, and showed a great impact against other severe parasitic infections. The anti-parasitic drug IVM, also possess a broad nature of the spectrum due to its noteworthy anti-inflammatory as well as anti-helmintic properties. Importantly, IVM is a miracle medication because of its long-lasting effects on the global health with around a half-century of history, and having received the Nobel Prize in *physiology or medicine* in the year 2015.

Dermatology is that branch of medicine which deals with the skin and related issues - found creepy from both the medical as well as surgical perspectives. Basically, the *Dermatologist,* is a specialist in the field of medical arena who deals with the diseases related to the nails, hair, skin, and cosmetic problems as well. Interestingly studies accounted that, a dermal route of administration is fancied over a contemporary and/or an oral route for IVM due to its superior bioavailability [1].

Recently, uses of IVM in the dermatological part against few crucial parasitic infestations in both humans as well as animals with cutaneous tropism, was immensely studied as the drug possess a great attentiveness in the treatment of the patients. There are several indications of IVM drug to be used in dermatology as: rosacea, cutaneous larva currens, human body lice, head lice, pediculosis, onchocerciasis, scabies, red scrotum syndrome (RSS), demodicosis, loasis, filariasis, myiasis, cutaneous larva migrans, strongyloidosis, and dermatitis *etc.* Moreover, oral IVM is also effective against a broad range of ectoparasites (insects and acarine) as well as endoparasites (nematodes) of animals and human beings. Thus, at current condition, the curiosity of dermatologists in IVM raised, due to encouraging results - perceived during the treatment of ectoparasites. The economic impact of IVM is very difficult to assess, thus IVM provides a highly effective easy-to-use therapy for many parasitic skin diseases of small and exotic animals [2].

IVM was initially tested on humans in 1982 by Aziz *et al.* against the nematodes *Onchocerca volvulus* (*O. volvulus*), and currently regarded under the world-class medication with the help of numerous therapeutic uses. Since 1987, it has widely been applied among the people to treat

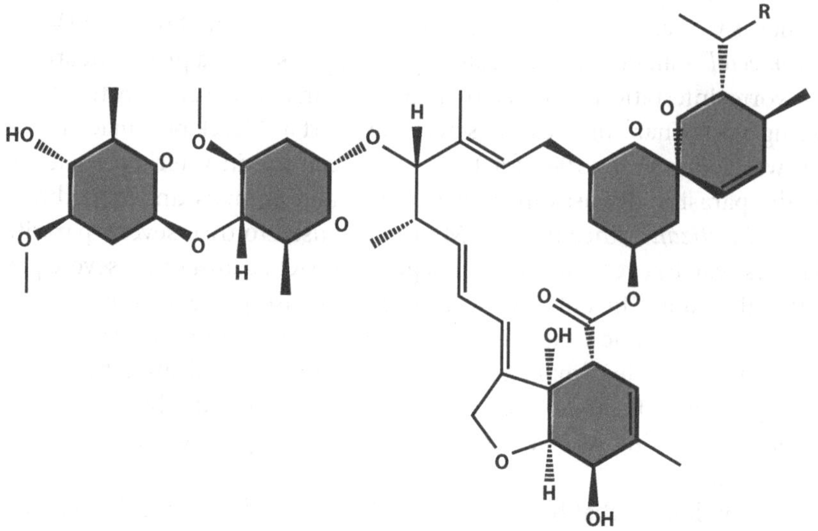

Avermectin (22,23-dihydroavermectin) B_{1a}, R=CH_2CH_3, $C_{48}H_{74}O_{14}$ = 875.106 g/mol
Avermectin (22,23-dihydroavermectin) B_{1b}, R=CH_3, $C_{47}H_{72}O_{14}$ = 861.079 g/mol

Figure 9.2 The chemical structures of the two components of ivermectin (IVM), namely, 22,23-dihydroavermectin B1a and 22,23-dihydroavermectin B1b.

endemic onchocerciasis in the nations of Africa as well as Latin America. Importantly, due to its broad-spectrum nature, IVM is continuously evolving in the direction of anti-bacterial, anti-viral, anti-fungal, anti-protozoal, anti-inflammatory and anti-cancer, and anti-parasitic besides, it is able to reduce the effects of syndrome RSS. More importantly, IVM drug has gained a great attention of the scientific community during the COVID-19 pandemic as well. This medication has a multi-modal activity, and is also: cheap, convenient, safe, much effective and widely accessible. Moreover, currently it appeared in a form of valuable alternative for the complete eradication of resistant of head lice's infestation in children or kids, and it also work quite efficiently while treating intestinal worms as well as many skin conditions such as itching and small bumps on the skin. Though, in the past decades, several older drug molecules with recognized dermatologic benefits are being revealed. However, these medications have been used safely for the treatment of a large number of skin issues in small animals, though Food and Drug Administration (FDA), has not approved the majority of these medications for the usage of animals [3]. Over the past four decades, IVM has transformed from a "blockbuster" medicine to cure the human kind ectoparasitic disorders and nematodal infestation. Remarkably, IVM is considered to be a safe, non-toxic, and effective therapeutic agent - helpful in the treatment of common human infestations, while previously reported drugs causes the resistance for the management of general parasitic skin diseases. In this particular chapter, diverse potential applications of the topical IVM in the area of dermatology, has been revealed. Moreover, this chapter also demonstrate severe skin related problems which have been resolved from time to time by using the IVM drug [4]. We hope this chapter will for sure inspire to the newcomers as well as to those who are devoted their own research to the application part of IVM specifically in the dermatology, and can provide a new option with their useful information to the research community around the globe.

9.2 Mechanism of Action, Toxicity, and Side Effects of IVM

Basically, IVM drug selectively binds to the glutamate-gated chloride channels in the invertebrates. This type of medication works very well by interfering with invertebrate's muscle cells and the nerve cells. In general sense, IVM's mode of action involves the binding to glutamate chloride ion channels preferentially, and that condition is able to increase cell permeability to the chloride ions. In consequence, the condition facilitates the hyperpolarization of the muscle as well as nerve cells – leading finally to the parasitic death instantly. It encourages the secretion of inhibitory neurotransmitter such as γ-aminobutyric acid (GABA) from the presynaptic nerve terminals and supports its binding to the postsynaptic receptors too. Which in turn, increases the permeability subtly to the chloride ions influx, and hence resulting hyperpolarization of the muscle, and nerve cells will ultimately causes the parasite's paralysis and death. It acts pleasantly on the both endoparasites as well as ectoparasites by inhibiting the conduction process of nerve impulse in intermediary neurons and in nerve-muscle synapses, respectively. Alternately, we can say that the inhibitory neurotransmitter (GABA) is released, cause for the hyperpolarization of postsynaptic membrane and the prolonged hyperpolarization ultimately leads to the paralysis and death. The GABA blocks neurotransmission from interneurons to motor neurons in nematodes, while it blocks from motor neurons to muscle cells in arthropods. Due to the IVM's limited blood-brain barrier (BBB) penetration, most mammals do not experience the GABA-agonist effects of IVM. The IVM is considered relatively safe, as no genotoxicity was observed; but at a very high dose, it exhibited embryotoxicity in animals. Despite being uncommon, drug overdose, either acute or chronic, can cause the toxicity. The toxicity of IVM manifests with a lot of symptoms such as diarrhea, muscle fasciculation, drooping of lips, breathing difficulty, increased salivation, bilateral mydriasis, ataxia, absent menace reflex and rarely encephalopathy recumbency, depression, reduced pupillary reflex and death. Typically, within the first 36 h following injection, clinical symptoms usually progress. In cases of dermatological conditions, the side effects are exceptionally minimal for the IVM drug when it is taken in a form of therapeutic doses, like nausea, muscle aches, dizziness, abdominal pain, headache, tremors, edema, sleepiness, dizziness, and malaise *etc.* These types of side effects are nearly uncommon with topical IVM but may include a burning sensation, dandruff, and dry skin as well. It is important to seek medical assistance right immediately if the side effects are severe or observed persistently [5]. On the basis of past accomplishment, the few important dosing schedule of IVM drug for a common dermatological conditions are tabulated in the Table 9.1 [5b].

Table 9.1 The dosing schedule of IVM for a common dermatological conditions.

S. no.	Diseases	Suggested treatment regimens
1.	**Scabies**	*The classical scabies* Oral IVM - two doses of 200 μg/kg, 7 days apart (not FDA-approved, used when FDA-approved treatment has failed or could not be tolerated) Crusted scabies: three regimens reliant on severity Days 1, 2, and 8 or Days 1, 2, 8, 9, and 15 or Days 1, 2, 8, 9, 15, 22, and 29
2.	**Pediculosis**	*Pediculosis capitis* Topical IVM 0.5% lotion Oral IVM - two doses (400 μg/kg), 7 days apart *Phthirus pubis* Oral IVM - two doses (250 μg/kg), 7 day apart Severe cases: oral IVM - two doses (400 μg/kg), 7 days apart *Phthiriasis palpebrarum* Oral IVM – two doses (200 μg/kg), 7 days apart

(*Continued*)

Table 9.1 The dosing schedule of IVM for a common dermatological conditions. (*Continued*)

S. no.	Diseases	Suggested treatment regimens
3.	**Onchocerciasis**	Oral IVM 200 μg/kg - every 6 months as long as the infected person has evidence of skin or eye infection
4.	**Demodicosis**	Oral IVM of 250 μg/kg (single dose)
5.	**Rosacea**	Topical 1% IVM cream QD for 3 months
6.	**Furuncular myiasis**	Albendazole 400 mg (single dose) with 150-200 μg/kg of IVM (single dose)
7.	**Myiasis**	*Furuncular myiasis*: 1% IVM cream for 2 h followed by manual removal of larva Migratory myiasis: oral IVM 200 μg/kg (three doses) *Wound myiasis:* 1% ivermectin in propylene glycol solution for 2 h followed by washing with saline solution
8.	**Cutaneous larva migrans**	Oral IVM 200 μg/kg (single dose) Second dose - during relapse
9.	**Strongyloidosis**	Oral IVM 200 μg/kg for 1-2 days *Hyperinfection syndrome*: oral IVM 200 μg/kg/day until stool and/or sputum investigations are negative for two weeks along with stoppage of immunosuppressive therapy. Cutaneous larva currens: oral IVM 200 μg/kg for the two consecutive days.
10.	**Loasis**	*Microfilaremia*: <1000-2000 mf/ml – oral IVM 150 μg/kg (single dose), if co-infection with *Onchocerca volvulus*; treatment repeated every three months. *Microfilaremia*: 2000-8000 mf/ml - oral IVM 150 μg/kg (single dose), repeated monthly until microfilaremia levels are <2000 mf/ml. *Microfilaremia*: 8000–30,000 mf/ml – oral IVM 150 μg/kg (single dose), under close monitoring. *Microfilaremia*: >30,000 mf/ml – oral IVM (150 μg/kg), for 5 days under supervision in the hospital.

9.3 Motivational Approach of IVM in the Treatment of Skin

Skin is a complex organ which provides a strong barrier against external infections, and protects against germs, enables touch sensations and regulates body temperature. In these days, skin diseases are associated with major global health issues because of a large number of population affected by it, in this regard, IVM plays a noteworthy role to resolve the skin related issues. Generally, skin disorders are minor but some may indicate a more serious matter as well. The epidermis, dermis, and hypodermis are the three primary layers of the skin, and are very much sensitive toward skin cancer, acne, wrinkles, rashes *etc.* Skin conditions have a wide range of symptom including ulcers, peeling skin, dry or cracked skin, and a loss of skin pigment *etc.* Some chronic-type skin related issue appears since the birth, whereas suddenly some other severe type disorder seems in later stage of life. So far, several foreign agents have successfully been reported for the skin related matters, such as bacteria, fungus, virus, protozoans, and arthropod *etc.* Their associated problem could be resolved by the treatment with IVM. On the other front, stress causes hormonal imbalance which could trigger the skin disorders. Moreover, the Sun radiation can also induces many different skin disorders out of which few are very common and harmless, whereas others are rare or life-threatening such as skin cancer. Herewith are the contributions of IVM drug in skin-related issues caused by the bacterial infection, fungal infection, viral infection, and protozoal infection.

9.3.1 Anti-Bacterial Agent

Staphylococcus aureus (*S. aureus*) is a kind of pathogenic bacteria, a cause for the large number of diseases both in animals as well as humans. This bacterium is being called as one of the most pathogenic organisms. IVM nicely works against these parasites, and possibly micromolar concentrations of the amount of drug at the tissue levels is sufficient to get a therapeutic effect. Recently, methicillin made resistance against the treatment of *S. aureus*, hence quest for the alternate anti-staphylococcal agent require an urgent attention. Interestingly, in 2018, Zhao's research group studied *in vitro* activity of IVM against *S. aureus* clinical isolates, where different anthelmintics including levamisole, albendazole, and IVM were tested against the bacterium *S. aureus* to determine their minimum inhibitory concentrations. The outcomes showed that IVM but not levamisole or albendazole exhibited a potent anti-staphylococcal activity at the concentrations of 6.25 μg/ml and 12.5 μg/ml against the two isolates. As per said information, obtained from the pharmacology/toxicology of IVM, these novel outcomes displayed that IVM possess an unusual anti-bacterial effect against the certain type of *S. aureus* isolates. On the other side, it was not able to inhibit the growth of all type of *Staphylococci* in the limitation of IVM. These authors extended the broad-spectrum range of pharmacological effects of IVM in a more sophisticated way, but its requisite is potentially appealing further to develop as an anti-staphylococcal agent [6].

Chlamydia is a common type of infection, spread through Sexually Transmitted Infection (STI), and the bacteria responsible for spreading this particular disease *Chlamydia trachomatis (C. trachomatis)* which primarily infect mucosal epithelial cells, and their infection multiply by the medium of intercourse, such as anal sex or oral sex. Because *Chlamydia* frequently has no symptoms, a lot of people who carries it, unknowingly spread it to the other people - their infection can only be stopped by routine screenings. On the other hand, Pettengill *et al.* demonstrated that IVM is able to prevent the infection of human cervical epithelial cells caused by a well-known bacterial pathogen *C. trachomatis*, and this pathogen clinically provide an envisioned for the ocular infections and STI in which the minimal concentrations of IVM drug is responsible for the inhibitory action. The concentration of 5 μM IVM added one hour post infection (hpi), inhibit meaningfully the growth of that contagious infection. Additionally, under the treatment of 1 μM concentration of IVM at 1 hpi is able to slightly reduce the size of inclusions, while 5 μM concentration of IVM significantly reduces the dimensions of inclusions in cells, and surprisingly 10 μM concentration of IVM completely inhibit the inclusion development process [7].

Traditionally, IVM is believed to be inactive against all the bacteria (pathogens). In 2013, Ameen and Drancourt studied the anti-tuberculous activity of the IVM drug [8]. In a separate work, in the same year, Lim *et al.* has reported for first time, the anti-mycobacterial activity of the IVM. In their studies, most of the avermectin's family members for instance IVM, selamectin, and moxidectin, worked nicely against the species (mycobacterial), containing extensively drug-resistant (XDR) and multidrug-resistant (MDR) clinical strains of *Mycobacterium tuberculosis (M. tuberculosis)*. Thus, based on the previous studies, IVM also permitted for the clinical as well as veterinary uses including the tuberculosis treatment. Moreover, bactericidal or bacteriostatic nature of the avermectins, was shown by a survival kinetic experiment for the drug incorporating IVM, selamectin, and moxidectin. Noticeably, almost all types of the avermectins had been proved for the bactericidal description to reduce the initial viability of bacterium *up to* 6 order of magnitude which means - *the limits of CFU detection*. Thus, all the tested avermectins disclosed an inspiring bactericidal activity against the MDR-type of strain, signifying a remarkable potential for the treatment of MDR and XDR TB patients. A recent report of avermectins (e.g. IVM, selamectin and moxidectin), revealed the inhibition growth of different type of *M. tuberculosis* strains *in-vitro* at different concentrations including 2 to 8 μg/ml [9].

Mycobacterium ulcerans (*M. ulcerans*), cause for the infection of subcutaneous tissue known as Buruli ulcer (BU), and WHO recommended 8-weeks course of antibiotic (*streptomycin or rifampicin*) for the treatment. Because of the application in the treatment of river blindness, IVM is affordable and widely used in the west part of Africa. Daily streptomycin injections and adverse effects such as ototoxicity - are the issues with this regimen, thereby needs to be revolutionized. In 2015, Stinear's research team reported the IVM activity against *M. tuberculosis*, besides *in-vitro* efficacy of IVM and moxidectin drugs on the *M. ulcerans*, where both of them were able to significantly inhibit the growth of *M. ulcerans* at a minimum inhibitory concentrations (4-8 μg/ml), and consequently a specific dose-dependent reduction was observed in both culture-based and the bioluminescence assays [10].

9.3.2 Anti-Fungal Agent

Ringworm or dermatophytosis, is a zoonotic common type of skin and hair infection in animals/humans caused by *Trichophyton verrucosum* (*T. verrucosum*). Dermatophytes are the surface parasites of keratinized tissue or hair of livestock, and are communicable easily to the human beings. The clinical usefulness of IVM drug on the bovine dermatophytosis was evaluated by Kirmizigul *et al.*, and the findings illustrated that IVM was successful in the proper treatment of the dermatophytosis infection for the animals. During their studies, they cogitated in total 30 cattle - treated them as 20 in group-I, and 10 cattle for the control (group-II) of the age between 3-14 months, while the entire practice related to the farm management of animals was not changed during their studies. A single lengthy injection of IVM drug containing the amount of the dose around 0.2 mg/kg administered subcutaneously to the animals of the group-I, whereas animals in the group-II treated under the control without IVM dosage. Thus, total 18 animals (90%) from the group-I recovered in 45 days, on the other hand in the group-II, no recovery was observed during the same period of time [11]. Therefore, this study revealed that IVM is widely used in form of anti-parasite medicine, which automatically regulate as well as activate the immune system of the body. Another study was performed by the Dawood's research group using IVM in a treatment of cattle dermatophytosis. In this exercise, they administered a single dermal injection of IVM of dose around 200 microgram/kg to the affected animals. After the drug administration, recovery rate was observed after 10 days, started by appearance of the hair and scales dropping. These observation show that IVM can be used fruitfully in the treatment of dermatophytosis in bovine [12].

On the other hand, recently Kumar *et al.* studied the anti-fungal activity of IVM in ringworm infection in the Malnad Gidda cow. During their studies, the skin scraping collected from the cow's skin - showing bounded erythematous greyish-white crusty lesions onto the body. The direct microscopic analysis of skin's scrapings using lactophenol cotton blue visible to macroconidia of trichophyton species, and the four doses of IVM administered subcutaneously at the interval of seven days, consequently, 80% improvement was observed after the administration of IVM in that condition. Thus, this study resulted that IVM provide a better result in the treatment of dermatophytosis in cattle as a substitute of local/parenteral anti-fungal drugs [13].

9.3.3 Anti-Viral Agent

IVM drug play a crucial role in the mechanism of several biological information, therefore the drug is able to serving as a potential candidate in the effective treatment of a wide range of viruses such as cutaneous bovine papillomatosis and bovine/ovine capripoxviruses *etc.* In this regard, Feyisa, described a successful treatment of cutaneous bovine papillomatosis by using IVM in two cattle for example crossbred heifer and calf. Cutaneous bovine papillomatosis is an infectious type of neoplastic viral disease of animals - characterized visually by the presence of multiple skin tumors or wart lesion, particularly on the head part of the body (faces). Even though, some cases regress on their own, others may progress to a malignant state, and requires the animal to be killed, which is problematic. The cases were identified as being caused by bovine cutaneous papillomatosis based on the historical record as well as clinical examination. The animals were given IVM treatment just after the surgically removal of mass warts, and followed-up for two months, resulting the end of warts. Pleasingly, IVM was found as a viable option for treating of bovine papillomatosis [14]. Moreover, detailed analysis of the IVM effect on the bovine warts was studied by Jameel *et al.* During their studies a benign tumor that is cutaneous warts in bovine, was induced by the virus related to the family of papillomaviruses. The IVM drug successfully administrated subcutaneously in the treatment of endoparasites and ectoparasites in animals in addition to the bovine warts, and the amount of dose was around 300 μg/Kg [15].

By the World Organization for Animal Health, capripoxvirus (CaPVs) diseases were considered as reportable, these includes: sheeppox virus (SPPV), lumpy skin disease virus (LSDV), and goatpox (GTP). They can further be differentiated easily on the basis of moderately severe and occasionally serious infections in the sheep/cattle. To control the infections, most efficient method is the vaccination, and it is always demanded to find out a suitable anti-viral drug, having the capability to work against these type of diseases. To this line, very recently, Yeşilbağ and co-workers planned to scrutinize the anti-viral efficiency of the IVM drug at the different phases of *in-vitro* replication of SPPV as well as LSDV. Remarkably, IVM inhibits strongly *in-vitro* replication stage of LSDV, inhibiting around 99.82%, and SPPV inhibit about 99.87%. While, LSDV was treated with a IVM dose of 2.5 μM, fewer infectious virions were observed, and the amount of infectious virions decreased at the attachment and replication stages when only SPPV was treated with 2.5 μM IVM. The IVM drug affected the process of replication cycle of capripoxviruses in an impressive way in the post-entry stages rather than in the pre-entry stages. Thus, based on the studies, IVM was found to be more effective statistically on LSDV than SPPV - during virus attachment stages as well as penetration stages ($P < 0.05$). And the IVM drug was able to inhibit the mechanism of capripoxviruses including (LSDV/SPPV) at the several stages of the proliferation [16].

Nose is the foremost protruding site of the face from which the throat and lung get infections, thus the usage of the mask(s) may be advantageous to protect the nasal passages. There are several therapeutic strategies that reduce the rate of infection from the virus in the nose, for instance antiviral nasal sprays and nasal irrigation. Most of the viruses infect the airway-lining cells, known as "*epithelial cells*" for example influenza viruses, SARS-CoV-2 virus, or coronavirus (COVID-19). Keeping the earlier experience of IVM in mind, in 2021, Aref *et al.* facilitated the therapeutic efficacy with the help of IVM mucoadhesive nanosuspension intranasal spray for the patient's (*total 114 patients*) treatment effected with a mild COVID-19. They divided them into the two groups where group A including 57 patients, administrated the IVM mucoadhesive nanosuspension nasal spray twice in a day plus the Egyptian protocol for the mild COVID-19, on the other front, the group B also having 57 patients just followed the Egyptian protocol but no IVM, the group A was found to be much beneficial with IVM, thereby displaying the role of it [17].

9.3.4 Anti-Protozoal Agent

Malaria is a protozoan infection besides life-threatening illness, caused by a malarial parasite - transmitted to the human beings through a bite of female mosquito known as *Anopheles.* Generally, malarial infection is associated with the genus *Plasmodium,* which is of four types, i.e., *P. falciparum, P. ovale, P. malariae* and *P. vivax.* The symptoms of malaria include paroxysms of fever, and the length of which depends upon the particular *Plasmodium* species. There are three main successive stages - originated in the malaria paroxysm. The first stage has a duration of 15-60 minute, a cold stage including the symptoms such as shivering and a feeling of cold. Then the next stage comes for 2-6 hour, a hot stage including fever, sometimes reaches the temperature around 41°C with flushed often headache, dry skin, nausea, and vomiting. Finally, the duration of 2-4 hour, a sweating stage, during this stage the fever drops quickly, and the patient immediately undergoes heavy sweating. Notably, serious consequences, such as death, can result especially after infection with *P. falciparum.*

On the basis of past studies, IVM has exhibited the ability to kill malaria vectors efficiently. A recent cluster-randomized clinical trial using IVM, repeated in every 3 weeks during a rainy season showed a decrease (20%) in the number of malarial episodes in children.

In 2020, Singh *et al.*, revealed that IVM and analogues inhibit the growth of *P. falciparum* erythrocyte *in vitro.* On the basis of their studies of useful trial by the continual administration of IVM (150-200 μg/kg), it was noticed that a regularly repeated mass administrations of IVM at the time of malaria transmission season for the 18-week, able to reduce the malaria - among the children or adults without increasing the side effects [4, 18].

Babesiosis infection is caused by the microscopic parasites (*Babesia*) which infect the red blood cells (RBCs). The parasitic infection is usually transmitted by a certain tick bite. Although, many people who are infected by *Babesia* do not show any type of symptoms, except few skin bruising, yellowish skin/eyes, and severe headache *etc.* Similarly, Theilerioses is a collection of tickborne disease usually caused by a certain type of protozoan parasites associated with the *Genus Theileria.* A huge number of Theileria species, occurs in the domestic animals as well as wild ungulates, found in the tropical and subtropical regions. Other one is equine piroplasmosis, a type of blood-borne protozoal infection found in horses associated with the causing agent Theileria/Babesia. The research based on chemotherapy against the pathogen Babesia as well as Theileria, has become urgent because of the appearance of the resistance of the parasite to the already existing drugs. Noticeably, IVM drug is the world's largest endectocide, having the capability of killing a wide variety of parasites/vectors, residing both inside as well as outside the body. A study was documented related to the efficacy of the IVM on the growth of Babesia and Theileria *in vitro* as well as *in vivo*. It affected both the parasites Babesia and Theileria, but the viability assay indicated that the IVM drug is much more effective against the Babesia parasites than Theileria parasites [19].

On the other hand, Giardiasis is a type of infection that occurs mostly in the small intestine, and spread through the contact with infected people. The causing agent of the disease is a microscopic parasite commonly known as *Giardia lamblia*. Along similar lines, Cryptosporidiosis, is an infection caused by the parasite *cryptosporidium* that causes diarrhea. Once the parasite enters into the body, it passes through the digestive tract, and infects the fecal matter, and their infection infect to the other person when they consumed contaminated food or water with the stool. One can also get infected, if touches the things or surfaces contaminated with stool. In 1996, Youssef's research group studied the effect of IVM drug on the combined intestinal protozoal infection like cryptosporidiosis and giardiasis. They perceived that IVM has a great tendency for the broad spectrum anti-parasitic agent in addition to be an inspiring drug for the treatment of both the parasitic infection as well as infestation. In their investigations, the activity of IVM against the two common protozoal infection, giardiasis and cryptosporidiosis, was successfully recorded in a rat model. The results indicated that the IVM therapy is so much effective or impressive against both the parasites when the amount of dose of 200 micrograms/kg was used in the study [20].

9.4 Role of IVM with Good Anti-Parasitic Properties Against the Infection of Skin

On the skin, parasites typically are the form of tiny insects or worms that make shelter into the skin, and reside over there. Skin is the most sensitive part of the body which is always exposed to the environment, thereby entrance for several parasites. Because of increasing travel and immigration parasite infestations, the systemic illnesses in people result in cutaneous lesions which are becoming more common in dermatology clinical practices. The IVM drug has proved to be tremendously advantageous in a suitable treatment of a wide range of arthropod and nematode type of parasitic infections.

9.4.1 Arthropods

Scabies are itching type of skin complications caused by a very small burrowing mite known as, *Sarcoptes scabiei* var. *hominis*. On the basis of the earlier studies, it could be inspected that scabies are contagious in nature, and can spread rapidly through close physical contact in the problematic areas of a family, overcrowding, child care groups, poor sanitation school classes and nursing home. Therefore, the doctors always advises to frequently treat entire families or contact groups due extremely contagious nature of the scabies. It is severe and highly spreadable form of infection that could be transmitted *via* an airborne route. Pleasingly, at present a lot of effective treatments are available for the scabies. In this context, Rezaee's research group reported a comparable studies of the efficacy and safety of permethrin 5% lotion with oral IVM for the treatment of scabies [21].

In comparison to the topical treatments, the oral IVM is a more efficient and affordable option for treating scabies infection. Based on the evidences, we can say that oral IVM is useful for the safe as well as effective treatment of scabies; even if the IVM drug is not FDA-approved for this purpose. Globally, permethrin drug is being considered for the most effective treatment of the scabies. Oral IVM recommended for that type of patients who generally have failed while treating with permethrin drug, and/or for those patients who don't tolerate FDA-approved topical medications. In these particular cases, two doses of oral IVM (200 μg/kg) are generally given for treating the classic scabies. Though, currently IVM is not FDA-approved for the treatment of scabies, but still it shows superiority over the most topical agents including lindane, not permethrin. Interestingly, the IVM drug was analyzed to be well-tolerated in case of infants where the healing rate 80%, was observed. Although the IVM drug is not ovicidal but in a case of the adult stages of the mite, it works very effectively. Sometimes, a single dose may be insufficient to treat the infestation, hence multiple doses needs to be repeated for 1-2 weeks. A large scabies epidemics in nursing homes including other facilities as well as nodular scabies, crusted clinical manifestations, infestation in immunocompromised hosts, and nodular scabies (often which is resistant to the topical treatment), are all great indicators for the oral IVM. If the scabies are crusted type, more frequent dosage e.g. 2-3 doses for 1-2 week, are recommended. In addition to the oral IVM, topical 1% IVM (used for the whole skin over the course of one night duration, after that repeated for a week), is also involved in the category of an effective anti-scabietic treatment. The center for disease control and prevention (CDC), recommended a combined treatment using topical scabicide plus oral IVM (200 μg/kg). Notably, in case of crusted scabies, multiple doses of oral IVM are administrated [22].

In the year 2017, Sanz-Navarro *et al.* studied 23 patients affected with the human scabies, and treated with the oral IVM drug; out of them, only 10 patients received a single oral dose (200 μg/kg) and 13 patients received an amount of oral dose which is closer to the 400 μg/kg. Noticeably, a second, or even a third dose was administered only in those cases when the treatment become failure. The first 10 patients needed at least two (80%) or three (20%) doses of IVM drug for the entire resolution of infection, whereas the remaining cases were resolved with a single amount of oral dose of 400 μg/kg. Thus, oral IVM drug has a highly effective way for the proper treatment of human scabies, and a single amount of oral dose of around 400 μg/kg has been proven for the high efficacy and good tolerance [23].

On the other hand, quite recently, in 2021, Vázquez-Osorio *et al.* has noticed another interesting case of scabies in a 14-year-old girl, affected with the superficial epidermolytic ichthyosis. She faced a very dramatic problem associated with the generalized scaling as well as hyperkeratosis both from one year of the age with severe pruritus. The basis of scabies mites detection, in addition to the molecular genetic analysis, consequently showed a very rare pathogenic variant in the keratin namely, 2(KRT2) gene, confirmed by a proper diagnosis of superficial epidermolytic ichthyosis, and the proper treatment with oral IVM drug led to the complete reduction of the symptoms [24].

Epidermolysis bullosa is a collection of very rare diseases of skin in which skin seems to be more fragile and easily blister. Rubs or bumps on the skin leads to the tears, sores, and blisters in the skin which can appeared anywhere on the body. Interestingly, in 2004, Angelo *et al.* studied on a 14-year-old female candidate, suffering from the generalized bullous eruption, appeared on the skin since her birth. On the basis of the observations, the bullae spread asymmetrically all over the cutaneous surface with the time and slowly swapped into the erosions form, resolved with the scarring areas. By means of a clinical trials, a diagnosis of the recessive dystrophic epidermolysis bullosa was made, and the microscopic examination displayed the presence of numerous mites and ova. For this purpose, a single dose of IVM (200 μg/hg) was administered, resulting in the initial improvement, thereby IVM represents an effective drug for severe forms of dermatoses [25].

Cutaneous myiasis is another type of infestation interconnected in the human beings caused by the fly larvae, known as *Musca domestica*. To this regard, Hasan's research group has studied numerous modes of treatment including oral IVM, topical IVM, and irrigation with IVM drug in an appropriate way. The best way to treat this particular disease was through the manual extraction along with the administration of two oral doses of IVM drug (6 mg for the each) within 24 hours. Another example is nasal and nasopharyngeal myiasis - a rare condition causes by the flies, found frequently in slum and rural areas where the unhygienic conditions are prevalent. Notably, the conventional treatment of myiasis includes the removal of all the larvae following manual protocol, either directly or by the endoscopic assistance. In this case, Hasan and his research team has concluded that oral IVM drug is safer as well as more effective way for the treatment of both nasal and nasopharyngeal myiasis. In this particular study, they found that the amount of dose of oral IVM (6 mg stat, and then repeated after 24 h) was highly effective in the treatment of nasopharyngeal myiasis. These workers concluded that the IVM drug in oral formulation is more economical, and can be administered orally with decent compliance (only 2 doses) and minimal local/systemic rare adverse effects. The IVM drug can also be used in a case of drowsy patients (*via* nasogastric tube) and thus all these features make the IVM an 'ideal' drug to combat the nasal and nasopharyngeal myiasis too [26].

On the other hand, the maggot infestation, also known as orbital myiasis, is an uncommon illness for which only few reports have publicly been made. To this context, Puthran *et al.* has used IVM drug successfully as a non-invasive treatment for the orbital myiasis (safely removal of maggots) - particularly to those which are deeply embedded in the tissues/skin of orbit [27].

The research studies on myiasis, in 2018 by Tay *et al.*, described an another interesting case of nasal myiasis in a 97-year-old Peruvian farmer with undiagnosed mucocutaneous leishmaniasis which occur due to *Musca domestica*. In this case, at the initial stage, the removal of fly larvae was done by manual extraction with the help of a toothed forceps. Later on, a nasal irrigation with IVM solution was found useful for the patient's self-expulsion of around 50 larvae within 15 minutes. They also secured a course using appropriate amount of oral IVM drug, after a CT scan, showing clear sinuses. They declared that IVM irrigation is the simplest and the effective way of proper treatment for nasal myiasis [28]. In a separate study, Costa *et al.* demonstrated one of the unique rare case of the rhino-orbital myiasis in a 55-year-old man caused by *Cochliomyia hominivorax*, using oral dose of IVM drug (200 μg/kg) [29].

An infection of the hairy areas of the body or clothing with lice eggs, larvae, or adults is known as *pediculosis*. This bug feeds on human blood in its crawling stages, causes intense itching. The IVM was found effective as well as curable against the permethrin-resistant head lice *in-vitro* condition, and was able to eliminate the parasites partially refractory to malathion *in-vivo* condition as well. The treatment of *pediculosis capitis* using IVM lotion (0.5%) is entirely US-FDA approved. The single 10-minutes usage of topical 0.5% of IVM lotion on the dry hair for 15 days, makes the 79% of the part utterly louse-free. In the cases of patients associated with the pyrethroid-resistant head lice infestations, an RCT (*randomized control trials*), found that a single oral dose of IVM, approx. 400 μg/kg of body weight, repeated for the every 7 days, provided good results [30]. Noticeably, overdose of IVM drug causes the development of resistance.

Rosacea is considered as a chronic inflammatory skin disease having different phenotypes (variety of signs and symptoms of the skin). There are lots of evidences associated with commensal *Demodex mite* to papulopustular rosacea. It is believed that, a well-known treatment with topical metronidazole, azelaic acid, and tetracyclines are associated by anti-inflammatory effects. These treatments are linked to the repeated relapses, and failure in the therapeutic process, therefore novel beneficial treatment options is/are highly needed. To this regard, in 2018, Ebbelaar *et al.* studied the topical IVM drug in an appropriate treatment of papulopustular rosacea, where topical IVM cream (1.0%) - a new effective option having tendency for both anti-inflammatory and acaricidal activity against *Demodex mites*, opened the door to a more etiologic tactic, and its application was adopted by the clinical guidelines. On the basis of past accomplishments, the amount of 1% cream of IVM drug is now accepted by US-FDA for the genteel treatment of inflammatory rosacea. The IVM drug not only targets *Demodex folliculorum* but also decreases the inflammation associated with the condition. The result analysis showed that within 2-weeks it control the things by improving the patient satisfaction and quality of life. Furthermore, it was noticed that IVM 1% cream is superior than metronidazole 0.75 % cream for reducing a number of inflammatory lesions, and react to the point of almost clear response and patient satisfaction. Thus, based on the studies, IVM drug is also well-tolerated and advantageous for the treatment of perioral dermatitis [31].

On the other front, a separate unique issue explained to the *Demodex* infestation (demodicosis) of the eyelids where the major problem was associated with the substantial ocular morbidity as well as with chronic blepharitis in the patients of all ages. In this context, in 2022, Helm evaluated the new treatment of ocular *Demodex* infestation associated with *Demodex* blepharitis using a topical 1% IVM cream which is effective for a months-long reduction in the clinical signs of *Demodex* infestation [32]. Whereas in a separate work, Schaller's research group evaluated the efficacy and safety of combining 1% IVM cream and doxycycline (40-mg) modified-release capsules (30-mg immediate-release and 10-mg delayed-release beads) IVM *versus* DMR, and a placebo for genteel treatment of severe rosacea. Specifically, in the case of severe rosacea, the combining IVM and DMR were found to be able to improve the response rates, faster responses, and increase patient satisfaction too [33].

9.4.2 Nematodes

The parasitic worm *Onchocerca volvulus* causes the neglected tropical disease (NTD) onchocerciasis, also familiar as the river blindness. IVM is a novel well-tolerated as well as operative-drug for the infection caused by *Onchocerca volvulus* (onchodermatitis). The *Simulium blackflies* that repeatedly bite a person can spread this matter disease. Onchocerciasis, in Africa is controlled as a public health based on yearly mass IVM delivery. It had been observed that a single dose of IVM approx.150 μg/kg, provided a significant progress in dermatitis condition in the first 3 months [34].

Notably, Lymphatic filariasis is an acute type of disease allied with the parasitic infection *W. bancrofti*, one of three different types of nematodes for instance *Wuchereria bancrofti, Brugia malayi*, or *Brugia timori*. In both children as well as adults, the microscopic worms enter into the human body *via* mosquito transmission which can live *up to* 5-7 years inside the lymphatic system. The major symptoms includes: lymphadenitis, orchitis, acute lymphangitis of the legs, and elephantiasis. Interestingly, studies has revealed that a single dose of IVM drug is always microfilaricidal, working very efficiently against the *W. bancrofti* for a duration of 3 months. Whereas, larger doses of the IVM drug approx. 150–400 μg/kg, is sufficient for the no return of microfilaremia worm. More importantly, a balanced combination of IVM drug added to the standard therapy with the drugs such as diethylcarbamazine (DEC) and albendazole shows a greater reduction rate in the population of microfilaremia worm. A triple drug treatment using an oral dose of IVM drug (200 μg/kg), DEC (6 mg/kg), and albendazole (400 mg) was found to be fairly better option for the effective treatment of *brugian filariasis* than the albendazole and DEC [35]. On the other hand, the administration of a single oral dose of IVM (150 μg/kg), leads to quickly microfilaricidal effect and most of the microfilariae cleared within a week of therapy period, and the skin affected by the microfilarial worm resolved up to the 95%–99% after the one month treatment. Interestingly, this method also prevent the formation of new microfilariae by adult female worms [36].

Strongyloidiasis is the infection caused by roundworm/nematode, called *Strongyloides stercoralis (S. stercoralis)*, and rashes on the skin may appear immediately after the contact with the *S. stercoralis* roundworm. Gastrointestinal IVM drug is considered to be the first-line treatment for both the acute as well as chronic strongyloidosis, and a single amount of dose of the IVM drug is enough for non-disseminated infections with a cure rate of 83%. Two successive days for the oral IVM therapy were found effective or curable in treating the cutaneous form of strongyloidiasis known as larva currens [37].

Cutaneous larva migrans (CLM), shows the clinical sign of creeping eruption, and considered under the infectious syndrome caused by the multiple types of hookworms most commonly transmitted by the animal feces - depositing eggs in the soil, where the larvae enter into the human body through direct contact with the skin. Cutaneous larva migrans is caused by the cutaneous penetration of the larvae form of the animal hookworms (typically known as *Ancylostoma braziliense*). In this condition, the IVM drug should be given in an amount of one or two doses approx. 150 to 200 μg/kg once daily to get the relief. Interestingly, the topical use of IVM drug was found to be more beneficial in the localized cases [38].

Human gnathostomiasis is caused by the parasitic worms (nematodes) interrelated to the genus Gnathostoma. Based on the survey, the disease is most commonly observed in the South-East Asia, in some areas of Africa, and the part of the South or central America too. At the primary stage the people become infected directly by eating under cooked or raw fresh water fish, frogs, reptiles, eels, and birds. The two most typical symptoms of the illness observed in the people are: 1) higher level of the eosinophils in the blood level; 2) the migratory swellings under the skin. The amount of dose of IVM (200 μg/kg/day) is very effective as well as curable for 2 days duration in the gnathostomiasis treatment with a successful cure rate of 100% [39].

Ascariasis is a type of roundworm infection on the skin, and the ascaris species are very large nematodes (roundworms) that parasitize the human intestine. Interestingly, IVM drug is quite safe and curable for both children as well as adults – affected with the ascaris infection, using IVM doses (100 to 200 μg/kg) [40].

Enterobiasis is a kind of pinworm infection which occurs mostly by a very small, thin, white type of roundworm *Enterobius vermicularis*.

The worm infection is able to affect all the people of diverse community, normally found in children. The pinworm infection of *E. vermicularis* could be cured by 85% IVM (50 to 200 μg/kg). Moreover, in these cases, the combination of IVM with albendazole drug was also found to be worthy choice for the treatment [41].

9.5 Importance of IVM as an Anti-Cancer or Anti-Tumor Agent in Curing the Skin

Melanoma is the category of most aggressive cancer associated with the skin which induces the metastasis, and both melanoma as well as non-melanoma skin cancers incidences are sharply rising. Exposure to the ultraviolet-B-radiation (UVB), induces the basal and squamous skin cancer cells. Sometime, radioactive and UV-radiation are also responsible for the malignancy of the skin, which later generates the tumor. Moreover, several other environmental issues are also associated with the case of melanoma cancer, basal cell carcinoma cancer (BCC), and squamous cell carcinoma cancer (SSC). Interestingly, Gallardo *et al.* identified that IVM shows strong anti-melanoma activity, and able to potentiate for the anti–BRAF V600 inhibitor. It has been found that the specific part of macrolide of the IVM drug holds *in-vitro* cytotoxicity against the BRAF V600 mutated melanoma cells lines. Noticeably, the serine/threonine p21-activated protein kinase 1 (PAK1) is the main aim of IVM drug which has true license for its anti-melanoma activity [42].

In a separate study, Chen *et al.*, noticed that IVM possess a great potential for the anti-tumor activity, working effectively against the esophageal squamous cell carcinoma (ESCC) cells *in-vitro* and in the nude mice. The effect of IVM drug inhibits the colony formation which induces the apoptosis process with the help of a mitochondrial dependent mode in the ESCC cells. Interestingly, the amount of IVM drug also disallows the invasion, ESCC cell migration, besides protein levels of MMP-2 and MMP-9. The specific amount of IVM dose strongly inhibits the expression of PAK1 protein by gain-of-function/loss-of-function in the experiments, and showed that PAK1 protein molecules played a conclusive role in the IVM-mediated inhibitory effects on the ESCC cells. Often, the combination of IVM with chemotherapeutic drug agents for instances 5-fluorouracil and *cis*-platin, induces the apoptosis of ESCC cells. Fascinatingly, *in-vivo* experiments also confirmed that the IVM drug successfully suppress the lungs metastasis of the ESCC cells in addition to the growth of tumor cells [43].

On the other hand, it has been exposed by Tang *et al.* that the IVM drug is able to inhibit the process of proliferation of numerous tumor cells by regulating the pathways associated with multiple signaling, and also specified that the IVM drug acts as an anti-cancer agent [44]. In 2017, González's research group showed the role of IVM in the breast cancer cells. They revealed that IVM drug targeted the *stem cell* population specifically in the MDA-MB-231 human breast cancer cells, and also observed that the amount of IVM drug inhibits the viability of CSC-enriched populations in comparison to the total cells population [45].

Recently, a synergistic anti-tumor effect of dichloroacetate and IVM drug was studied by Ishiguro *et al.* In that study, they mentioned the case of three patients in which the combination of omeprazole plus tamoxifen, dichloroacetate, and IVM intensely relieved the symptoms associated with cancer, and sarcoma progression. Only the amount of IVM drug was shown in the above combination administrated for Case-I, 12 mg IVM per day (on Day 1), Case-II, 12 mg IVM per day (on Days 1 and 4), Case-III, 12 mg IVM twice a week [46].

In 2019, Jiang *et al.* revealed that a specific amount of IVM drug reverses the drug resistance in the cancer cells with the help of EGFR/ERK/Akt/NF-κB pathway. The outcomes from their studies revealed that the amount of IVM drug boosted the anti-cancer efficacy of the chemotherapeutic agents substantially to the tumor cells, particularly in a case of the drug-resistant cells [47]. On the other front, Wang's research group demonstrated that the amount of IVM drug regulates as well as control the expression of proteins such as cleaved-PARP, Bax, Bcl-2, PARP, and Caspase-3, in the apoptosis pathway by increase the production of reactive oxygen species (ROS), and consequently inhibit the process of cell cycle in the stage of S-phase to inhibit the proper colorectal cancer cells at that stage. Based on results of the studies, they revealed that IVM drug may have an abundant potential as a new anti-cancer drug for the clear-cut treatment of the human colorectal cancer (CRC) and other severe cancers too [48].

9.6 Social Value of IVM in the Medical Care of Red Scrotum Syndrome (RSS)

Red scrotum syndrome (RSS), is still a very rare well-demarcated scrotal erythema, almost chronic disease associated with the unknown etiology, mostly affecting males in their second half life. It is analyzed by erythema with sharp borders, pain/hyperalgesia, itching, and burning sensation.

Generally, this type of disease is associated with the continued administration of topical steroids such as topical corticosteroids, and mostly people over the age of 50 are affected with it. Because of its chronic nature, it is very difficult to treat and directly impact on the reduction in quality of life and psychological distress. The RSS shows few symptoms like redness of the anterior part of scrotum, severe itching, tenderness, pain, pruritus, hyperalgesia, dysesthesia or scrotal erythema. In regard to this, Martinez *et al.* successfully testified three cases where they reported the proper treatment of this particular disorder with the oral IVM drug. Remarkably, oral IVM is a highly operative, safer and curable in RSS disease as it does not require any type of laboratory work-up just before the administration of the drug [49].

9.7 Utility of IVM as an Anti-Inflammatory Drug in the Treatment of Skin-Related Issues

Immune system related to any physical body, play an energetic role in retaining health of all the organ or tissues of the body. The immune system of the body interacts with invaders, for example microorganisms, pathogens, parasites, any foreign substances and cancer cells which initiate inflammation to attack these invaders. Typically, reaction associated with the immune system protects the physical body, and also promote the healing process. Sometime the reactions of immune system undergoes to the misdirection or overreaction at healthy tissues, causing the intense inflammation/swelling and impairments. These types of unusual immune responses are also associated with the hypersensitivity of the body like allergies, asthma, and skin disease (patches) *etc.* The immune system of anybody consists of a large number of special cells (e.g. B-cells, and T-cells), tissues, and organs system that effort collectively to protect us. Many skin disorders associated with the immune system's reaction, and only few affect directly to the dermal parts without interfering the immune system. Thus, immunomodulating effects of the IVM drug offer an unconventional methodology for the accurate treatment of skin disease which involves the immunosuppression. Based on the earlier studies, IVM drug works effectively to control the human innate immune system by proper interaction between antigen and antibody. Few important hypersensitivity reactions of the skin and their related disorders as well as their control by IVM drug are mentioned below:

9.7.1 Allergy

Allergy associated with the skin become irritated or harmless when the immune system reacted to something (foreign substances) known as an allergic reaction. Many different type of allergens causes the allergic reaction such as the rashes, itching, redness, bumps, burning, hives, and swelling *etc.* The most common allergic skin condition is dermatitis, eczema, urticaria, hives, and angioedema *etc.* In other words, a skin allergy is a skin condition where the immune system interacts strongly with an allergen, and the exact reason behind the allergy depends upon the condition and the allergen as well. Generally, the treatment for skin allergy varies depending on the problem and type of allergen, but usually contains two main approaches: first one is to avoid the allergen, if it is known, and second one uses medicines, creams, and other strategies to relieve the itching, swelling, or pain.

At present, IVM is widely used in both human as well as veterinary medicine to treat the parasitic infections. Recent reports suggested that IVM could also have anti-inflammatory properties. In this context, Vocanson's research team investigated the activity of IVM drug in a murine model of atopic dermatitis (AD) induced by frequent exposure to the allergen *Dermatophagoides farina* in standard cellular immunology assays. The outcomes of the experiment showed that topical IVM enrich allergic skin inflammation by decreasing the priming and activation of allergen-specific T-cells, besides the production of inflammatory cytokines. But on the other side, IVM have no major role on the functions of dendritic cells *in vivo* as well as *in vitro*. Notably, IVM directly impaired activation of T-cells, proliferation and cytokine production and follow the polyclonal and antigen-specific stimulation. The drug treatment performed by applying 10 μl of IVM (0.1%) or BMV (0.01%) to the site of allergen exposure, worked effectively. Thus, overall results showed that IVM drug is enriched with a topical anti-inflammatory properties for the treatment of T-cells mediated skin inflammatory diseases [50].

As there is a lack of literature related to the link between parasites and allergic diseases. In this regard, Rapur *et al.* studied the effect of IVM on an allergy-type manifestations in occult strongyloidiasis where immunomodulatory effects of the helminths have nicely been discussed. In their studies, they considered a number of patients with allergy-type symptoms and coexisting strongyloides infection. They exposed the effect of strongyloides eradication therapy employing IVM drug by considering 84 patients [51].

Rosacea is also considered under the chronic inflammatory illness, differentiated on account of various symptoms e.g. telangiectasia, flushing (transient erythema), persistent erythema, inflammatory papules, skin irritation (burning and stinging), and pustules (inflammatory lesions) *etc.* These type of symptoms may appeared solely and/or in combination. Sometime, the symptoms of rosacea disease give rise to the facial hypersensitivity. Few environmental factors that are also responsible for it are: sunlight, temperature change, stress, spicy foods, and heavy exercise *etc.* It also occurs due to the high levels of proteins involved in inflammatory pathway, or because of hyper-responsive immune system. Remarkably, the amount of topical IVM cream (10 mg/g) and the amount of systemic oral anti-inflammatory doxycycline approx. 40 mg modified-release have been approved scientifically for the medical treatment of papulopustular rosacea (PPR). Related to this, Steinhoff's research group proposed the topical IVM cream as an inspiring new competitor for the first-line medical treatment to inflammatory lesions (skin issues) of rosacea, the IVM drug can also be used in form of combination with the systemic doxycycline modified-release to deliver best medical treatment methodology which include all the inflammatory PPR pathways [52]. Additionally, there are two reports related to IVM drug, administered either by topically or systemically, making an effort to the anti-inflammatory effect in the murine models of allergic inflammation (atopic dermatitis and asthma), while systemic effect was accomplished via 2 mg/kg IVM dose [53].

9.7.2 Psoriasis or Crusted Scabies

Based on the literature studies about psoriasis/crusted scabies, Goyal and Wong studied the case of 67-year-old female affected with the symptom, non-itchy erythematous scaly - on the fingertips/nail thickening. The crusted scabies, indicated an inappropriate host response to the Sarcoptes scabei mite, causing hyperinfestation with a huge number of mites. Moreover, the similar type of situation was also observed in the persons affected with the condition of immunosuppression like leukemia, human T-cell lymphotropic virus 1 infection, and human immunodeficiency virus (HIV) infection. The effective treatment of crusted scabies disease, particularly in the nail improvement was very problematic, thus an aggressive therapy with two doses of IVM approx. 200 μg/kg for 2 weeks in combination with keratolytics and topical permethrin was suggested. Overall, this type of medical treatment cleared the nail changes as well as the crusted skin lesions [54].

9.7.3 Asthma

Asthma is an inflammatory kind of disease associated with the lungs, caused by the increasing number of inflammatory cell infiltration into the airways and inoperable respiratory function. To treat inflammatory diseases, Yan *et al.* used IVM drug in asthma treatment. In this study, they investigated the inhibitory effects of IVM drug on the allergic asthma symptoms in the mice. The overall studies displayed that IVM have an anti-inflammatory properties both in the lung tissue as well as in systemic level [55, 56]. In another report, Kadaria's research group disclosed the anti-inflammatory activity of IVM, where the patient was diagnosed with "hyperinfection syndrome" or "disseminated disease" due to *Strongyloides stercorali (SS)* including the symptom like abdominal pain, shortness of breath where most of the symptoms probably exacerbated by the prednisone. During the treatment these workers discontinued the steroids due to exacerbation properties, and the patient was given IVM drug to get better condition [57].

In the context of the effect of IVM drug on the immune response in mice, studied by Blakley and Rousseaux, where assess to the effect of IVM on immune function (antibody production), male CD-1 mice were inoculated with an antigen the day after SC administration of IVM (0.2 mg/kg of body weight or 20 mg/kg). Response evaluated 5 days after inoculation of the antigen. Antibody production against the sheep RBC, a T-lymphocyte, macrophage-dependent response, was enhanced by IVM treatment ($P = 0.00049$). In contrast, antibody production against dinitrophenyl-ficoll, a T-lymphocyte-independent, macrophage-dependent response, was not altered by IVM treatment. Overall results indicated that the immunostimulatory properties of IVM are associated with altered function of T-lymphocytes, in particular, T-helper lymphocytes. The immunomodulating effects of IVM may provide an alternative approach for treatment of disease problems involving immunosuppression [58].

9.8 Conclusions

IVM is effective in preventing a variety of diseases, which means it could have a positive impact on the health of the communities where it is being used. Herewith, we have shown the wide importance of IVM drug in the arena of dermatology. Evidences based on the literature, strongly supports its tremendous usage in the clinical applications, specifically in skin related diseases. Based the studies, IVM is revolutionizing the management of nematodal and ectoparasitic infection – making the IVM unprecedentedly unique, multifaceted 'wonder' drug of the past or present, and for sure will be exceptional from the future perspective as well. This chapter will open the door for the new comers, besides repurposing of IVM to control or cure novel diseases in forthcoming future.

Acknowledgements

The authors gratefully thank the DST-SERB, New Delhi for financial support (Project File no. ECR/2017/000821). V.J. is thankful to DST-SERB for the project fellowship. R.A. is also thankful to Jamia Millia Islamia, New Delhi for providing the necessary research facilities.

References

1. Crump, A., Ivermectin: Enigmatic multifaceted 'wonder'drug continues to surprise and exceed expectations. *J. Antibiot.*, 70, 5, 495–505, 2017.
2. Sharun, K., Shyamkumar, T.S., Aneesha, V.A., Dhama, K., Pawde, A.M., Pal, A., Current therapeutic applications and pharmacokinetic modulations of ivermectin. *Vet. World*, 12, 8, 1204, 2019.
3. Laing, R., Gillan, V., Devaney, E., Ivermectin–old drug, new tricks? *Trends Parasitol.*, 33, 6, 463–72, 2017.
4. Mathachan, S.R., Sardana, K., Khurana, A., Current use of ivermectin in dermatology, tropical medicine, and COVID-19: An update on pharmacology, uses, proven and varied proposed mechanistic action. *Indian Dermatol. Online J.*, 12, 4, 500, 2021.
5. (a). Mundell, A.C., New therapeutic agents in veterinary dermatology. *Vet. Clin. North Am.: Small Anim. Pract.*, 20, 6, 1541–56, 1990. (b). Gowtham, S. and Karthikeyan, K., Wonder drug for worms: A review of three decades of ivermectin use in dermatology. *Indian J. Dermatol. Venereol. Leprol.*, 85, 6, 674–8, 2019.
6. Ashraf, S., Chaudhry, U., Raza, A., Ghosh, D., Zhao, X., *In vitro* activity of ivermectin against *Staphylococcus aureus* clinical isolates. *Antimicrob. Resist. Infect. Control.*, 7, 1, 1–6, 2018.
7. Pettengill, M.A., Lam, V.W., Ollawa, I., Marques-da-Silva, C., Ojcius, D.M., Ivermectin inhibits growth of *Chlamydia trachomatis* in epithelial cells. *PloS One*, 7, 10, e48456, 2012.
8. Ameen, S.M. and Drancourt, M., Transparency declarations. *Microbiol. Rev.*, 13, 534–46, 2000.
9. Lim, L.E., Vilchèze, C., Ng, C., Jacobs Jr., W.R., Ramón-García, S., Thompson, C.J., Anthelmintic avermectins kill Mycobacterium tuberculosis, including multidrug-resistant clinical strains. *Antimicrob. Agents Chemother.*, 57, 2, 1040–6, 2013.

10. Omansen, T.F., Porter, J.L., Johnson, P.D., van der Werf, T.S., Stienstra, Y., Stinear, T.P., *In-vitro* activity of avermectins against Mycobacterium ulcerans. *PloS Negl. Trop. Dis.*, 9, 3, e0003549, 2015.
11. Kirmizigul, A.H. and Erkilic, E.E., Clinical effectiveness of ivermectin on bovine dermatophytosis. *Kafkas Univ. Vet. Fak. Derg.*, 18, 3, 523–526, 2012.
12. Jameel, G.H., Ivermectin activity in treatment of cattle dermatopyhtosis. *Diyala Agric. Sci. J.*, 7, 1, 30–40, 2015.
13. Kumar, N., Tajunnisa, M., NS, H.R., Thippan, M., Evolution of antifungal activity of ivermectin as novel alternative therapies for dermatophytosis in Malnad Gidda cow. *Int. J. Chem. Stud.*, 6, 6, 293–294, 2018.
14. Feyisa, A.F., Cutaneous bovine papillomatosis (Warts) treatment outcome using: Ivermectin: A case of crossbred heifer and calf. *J. Vet. Sci. Technol.*, 9, 544, 2, 2018.
15. Jameel, G.H., Jawad, S.J., Ebraheem, T.A., Majeed, W.A., Study of Ivermectin effect on the bovine warts. *Diyala J. Pure Sci.*, 7, 2, 121–131, 2011.
16. Toker, E.B., Ates, O., Yeşilbağ, K., Inhibition of bovine and ovine capripoxviruses (Lumpy skin disease virus and Sheeppox virus) by ivermectin occurs at different stages of propagation *in vitro*. *Virus Res.*, 310, 198671, 2022.
17. Aref, Z.F., Bazeed, S.E., Hassan, M.H., Hassan, A.S., Rashad, A., Hassan, R.G., Abdelmaksoud, A.A., Clinical, biochemical and molecular evaluations of ivermectin mucoadhesive nanosuspension nasal spray in reducing upper respiratory symptoms of mild COVID-19. *Int. J. Nanomed.*, 16, 4063, 2021.
18. (a). Singh, L., Fontinha, D., Francisco, D., Mendes, A.M., Prudêncio, M., Singh, K., Molecular design and synthesis of ivermectin hybrids targeting hepatic and erythrocytic stages of plasmodium parasites. *J. Med. Chem.*, 63, 1750–62, 2020. (b). Foy, B.D., Alout, H., Seaman, J.A., Rao, S., Magalhaes, T., Wade, M., Parikh, S., Soma, D.D., Sagna, A.B., Fournet, F., Slater, H.C., Efficacy and risk of harms of repeat ivermectin mass drug administrations for control of malaria (RIMDAMAL): A cluster-randomised trial. *Lancet*, 393, 10180, 1517–26, 2019.
19. Batiha, G.E., Beshbishy, A.M., Tayebwa, D.S., Adeyemi, O.S., Yokoyama, N., Igarashi, I., Evaluation of the inhibitory effect of ivermectin on the growth of Babesia and Theileria parasites *in vitro* and *in vivo*. *Trop. Med. Health*, 47, 1, 1–2, 2019.
20. Youssef, M.Y., Essa, M.M., Sadaka, H.A., Eissa, M.M., Rizk, A.M., Effect of ivermectin on combined intestinal protozoal infection (giardiasis and cryptosporidiosis). *J. Egypt Soc. Parasitol.*, 26, 3, 543–53, 1996.
21. Ranjkesh, M.R., Naghili, B., Goldust, M., Rezaee, E., The efficacy of permethrin 5% *vs.* oral ivermectin for the treatment of scabies. *Ann. Parasitol.*, 59, 4, 189–94, 2013.
22. (a). Chandler, D.J. and Fuller, L.C., A review of scabies: An infestation more than skin deep. *Dermatology*, 235, 2, 79–90, 2019. (b). Rosumeck, S., Nast, A., Dressler, C., Ivermectin and permethrin for treating scabies. *Cochrane Database Syst. Rev.*, 4, 4, CD012994, 2018.

23. Sanz-Navarro, J., Feal, C., Dauden, E., Treatment of human scabies with oral ivermectin. Eczematous eruptions as a new non-reported adverse event. *Actas Dermo-Sifiliogr. (English Edition)*, 108, 7, 643–9, 2017.
24. Vázquez-Osorio, I., Moreiras-Arias, N., Pérez-Feal, P., Sainz-Gaspar, L., Ortiz-Cabrera, N.V., Hernández-Martín, A., Scabies in a 14-year-old girl with superficial epidermolytic ichthyosis. *Pediatr. Dermatol.*, 39, 2, 305–6, 2022.
25. Angelo, C., Pedicelli, C., Provini, A., Annessi, G., Zambruno, G., Paradisi, M., Successful treatment of Norwegian scabies with ivermectin in a patient with recessive dystrophic epidermolysis bullosa. *Minerva Pediatr.*, 56, 3, 353–7, 2004.
26. Sayeed, A., Ahmed, A., Sharma, S.C., Hasan, S.A., Ivermectin: A novel method of treatment of nasal and nasopharyngeal myiasis. *Indian J. Otolaryngol. Head Neck Surg.*, 71, 3, S2019–S2024, 2019.
27. Puthran, N., Hegde, V., Anupama, B., Andrew, S., Ivermectin treatment for massive orbital myiasis in an empty socket with concomitant scalp pediculosis. *Indian J. Ophthalmol.*, 60, 3, 225–227, 2012.
28. Tay, S.Y., Ramasamy, B.R., Watson, D.A., Montoya, M., Treatment of nasal myiasis with ivermectin irrigation. *Case Rep.*, *2018*, bcr–2017, 2018.
29. Costa, D.C., de Tarso Ponte Pierre-Filho, P., Mac Cord Medina, F., Mota, R.G., Carrera, C.R., Use of oral ivermectin in a patient with destructive rhino-orbital myiasis. *Eye*, 19, 9, 1018–20, 2005.
30. (a). Foucault, C., Ranque, S., Badiaga, S., Rovery, C., Raoult, D., Brouqui, P., Oral ivermectin in the treatment of body lice. *J. Infect. Dis.*, 193, 3, 474–6, 2006. Pariser, D.M., Meinking, T.L., Bell, M., Ryan, W.G., Topical 0.5% ivermectin lotion for treatment of head lice. *N. Engl. J. Med.*, 367, 18, 1687–93, 2012.
31. Ebbelaar, C.C.F., Venema, A.W., Van Dijk, M.R., Topical ivermectin in the treatment of papulopustular rosacea: A systematic review of evidence and clinical guideline recommendations. *Dermatol. Ther.*, 8, 3, 379–387, 2018.
32. Helm, C.J., Treatment of ocular Demodex infestation with topical ivermectin cream. *Am. J. Ophthalmol. Case Rep.*, 26, 101551, 2022.
33. Schaller, M., Kemény, L., Havlickova, B., Jackson, J.M., Ambroziak, M., Lynde, C., Gooderham, M., Remenyik, E., Del Rosso, J., Weglowska, J., Chavda, R., A randomized phase 3b/4 study to evaluate concomitant use of topical ivermectin 1% cream and doxycycline 40-mg modified-release capsules, versus topical ivermectin 1% cream and placebo in the treatment of severe rosacea. *J. Am. Acad. Dermatol.*, 82, 2, 336–43, 2020.
34. Campillo, J.T., Chesnais, C.B., Pion, S.D., Gardon, J., Kamgno, J., Boussinesq, M., Individuals living in an onchocerciasis focus and treated three-monthly with ivermectin develop fewer new onchocercal nodules than individuals treated annually. *Parasitol. Vectors*, 13, 1, 1–1, 2020.
35. (a). Dembele, B., Coulibaly, Y.I., Dolo, H., Konate, S., Coulibaly, S.Y., Sanogo, D., Soumaoro, L., Coulibaly, M.E., Doumbia, S.S., Diallo, A.A., Traore, S.F.,

Use of high-dose, twice-yearly albendazole and ivermectin to suppress Wuchereria bancrofti microfilarial levels. *Clin. Infect. Dis.*, 51, 11, 1229–35, 2010. (b) Brown, K.R., Ricci, F.M., Ottesen, E.A., Ivermectin: Effectiveness in lymphatic filariasis. *Parasitology*, 121, S1, S133–46, 2000.

36. Basáñez, M.G., Pion, S.D., Boakes, E., Filipe, J.A., Churcher, T.S., Boussinesq, M., Effect of single-dose ivermectin on Onchocerca volvulus: A systematic review and meta-analysis. *Lancet Infect. Dis.*, 8, 5, 310–22, 2008.
37. (a). Buonfrate, D., Requena-Mendez, A., Angheben, A., Muñoz, J., Gobbi, F., Van Den Ende, J., Bisoffi, Z., Severe strongyloidiasis: A systematic review of case reports. *BMC Infect. Dis.*, 13, 78, 1–0, 2013. (b). Buonfrate, D., Salas-Coronas, J., Muñoz, J., Maruri, B. T., Rodari, P., Castelli, F., Zammarchi, L., Bianchi, L., Gobbi, F., Cabezas-Fernández, T., Requena-Mendez, A., Multiple-dose versus single-dose ivermectin for Strongyloides stercoralis infection (Strong Treat 1 to 4): a multicentre, open-label, phase 3, randomised controlled superiority trial. *Lancet Inf. Dis.*, 19, 11, 1181-90, 2019.
38. Gerbig, A.W. and Kempf, W., Topical treatment of cutaneous larva migrans with ivermectin 1%. *Int. J. Dermatol.*, 59, 1, e21–2, 2020.
39. Nontasut, P., Claesson, B.A., Dekumyoy, P., Pakdee, W., Chullawichit, S., Double-dose ivermectin *vs* albendazole for the treatment of gnathostomiasis. *Southeast Asian J. Trop. Med. Public Health*, 36, 3, 650–2, 2005.
40. Conterno, L.O., Turchi, M.D., Corrêa, I., de Barros Almeida, R.A., Anthelmintic drugs for treating ascariasis. *Cochrane Database Syst. Rev.*, 4, 1–101, 2020.
41. Lloyd, A.E., Honey, B.L., John, B.M., Condren, M., Treatment options and considerations for intestinal helminthic infections. *J. Pharm. Technol.*, 30, 4, 130–139, 2014.
42. Gallardo, F., Teiti, I., Rochaix, P., Demilly, E., Jullien, D., Mariamé, B., Tilkin-Mariamé, A.F., Macrocyclic lactones block melanoma growth, metastases development and potentiate activity of anti–BRAF V600 inhibitors. *Clin. Skin Cancer*, 1, 1, 4–14, 2016.
43. Chen, L., Bi, S., Wei, Q., Zhao, Z., Wang, C., Xie, S., Ivermectin suppresses tumour growth and metastasis through degradation of PAK1 in oesophageal squamous cell carcinoma. *J. Cell. Mol. Med.*, 24, 9, 5387–401, 2020.
44. Tang, M., Hu, X., Wang, Y., Yao, X., Zhang, W., Yu, C., Cheng, F., Li, J., Fang, Q., Ivermectin, a potential anticancer drug derived from an antiparasitic drug. *Pharmacol. Res.*, 163, 105207, 2021.
45. Dominguez-Gomez, G., Chavez-Blanco, A., Medina-Franco, J.L., Saldivar-Gonzalez, F., Flores-Torrontegui, Y., Juarez, M., Díaz-Chávez, J., Gonzalez-Fierro, A., Dueñas-González, A., Ivermectin as an inhibitor of cancer stemlike cells. *Mol. Med. Rep.*, 17, 2, 3397–403, 2018.
46. Ishiguro, T., Ishiguro, R.H., Ishiguro, M., Toki, A., Terunuma, H., Synergistic anti-tumor effect of dichloroacetate and ivermectin. *Cureus*, 14, 2, e21884, 2022.

47. Jiang, L., Wang, P., Sun, Y.J., Wu, Y.J., Ivermectin reverses the drug resistance in cancer cells through EGFR/ERK/Akt/NF-κB pathway. *J. Exp. Clin. Cancer Res.*, 38, 1, 1–8, 2019.
48. Zhou, S., Wu, H., Ning, W., Wu, X., Xu, X., Ma, Y., Li, X., Hu, J., Wang, C., Wang, J., Ivermectin has new application in inhibiting colorectal cancer cell growth. *Front. Pharmacol.*, 12, 2145, 2021.
49. Martinez, J.D., Soria Orozco, M., Cardenas-de la Garza, J.A., Oral ivermectin for the treatment of red scrotum syndrome. *J. Dermatol. Treat.*, 33, 2, 1102–3, 2022.
50. Ventre, E., Rozières, A., Lenief, V., Albert, F., Rossio, P., Laoubi, L., Dombrowicz, D., Staels, B., Ulmann, L., Julia, V., Vial, E., Topical ivermectin improves allergic skin inflammation. *Allergy*, 72, 8, 1212–21, 2017.
51. Rampur, L., Jariwala, S.P., Hudes, G., Rosenstreich, D.L., de Vos, G., Effect of ivermectin on allergy-type manifestations in occult strongyloidiasis. *Ann. Allergy Asthma Immunol.*, 117, 4, 423–8, 2016.
52. Steinhoff, M., Vocanson, M., Voegel, J.J., Hacini-Rachinel, F., Schäfer, G., Topical ivermectin 10 mg/g and oral doxycycline 40 mg modified-release: Current evidence on the complementary use of anti-inflammatory rosacea treatments. *Adv. Ther.*, 33, 9, 1481–501, 2016.
53. DiNicolantonio, J.J., Barroso-Arranda, J., McCarty, M., Ivermectin may be a clinically useful anti-inflammatory agent for late-stage COVID-19. *Open Heart*, 7, 2, e001350, 2020.
54. Goyal, N.N. and Wong, G.A., Psoriasis or crusted scabies. *Clin. Exp. Dermatol.: Continuing Professional Development (CPD), 33*, 2, 211–2, 2008.
55. Yan, S., Ci, X., Chen, N.A., Chen, C., Li, X., Chu, X., Li, J., Deng, X., Anti-inflammatory effects of ivermectin in mouse model of allergic asthma. *Inflammation Res.*, 60, 6, 589–96, 2011.
56. (a). Portmann-Baracco, A., Bryce-Alberti, M., Accinelli, R.A., Antiviral and anti-inflammatory properties of ivermectin and its potential use in covid-19. *Arch. Bronconeumol.*, 56, 12, 831, 2020.
57. (a). Alsharif, A., Sodhi, A., Murillo, L.C., Headley, A.S., Kadaria, D., Wait!!! No steroids for this asthma. *Am. J. Case Rep.*, 16, 398, 2015. (b). Salam, R., Sharaan, A., Jackson, S.M., Solis, R.A., Zuberi, J., Strongyloides hyperinfection syndrome: A curious case of asthma worsened by systemic corticosteroids. *Am. J. Case Rep.*, 21, e925221-1, 2020.
58. Blakley, B.R. and Rousseaux, C.G., Effect of ivermectin on the immune response in mice. *Am. J. Vet. Res.*, 52, 4, 593–5, 1991.

10

Antiviral Uses of Ivermectin

Shabnam Kawoosa[1], Rabiah Bashir[1], Sheeba Nazir[1], Zubaid-ul-Khazir Rather[2*], Gulam Nabi Yatoo[1,2] and Nisar Ahmad Khan[1†]

[1]*Department of Pharmaceutical Sciences, School of Applied Sciences and Technology, University of Kashmir, Srinagar, J&K, India*
[2]*NPCL-Lab Department of Chemistry National Institute of Technology, Srinagar, J&K, India*

Abstract
Ivermectin is known as a wonder drug because of its potential to cure many diseases and effectively combats microorganisms, including certain viruses. Ivermectin has the potential to inhibit/kill the proliferation of microbial, viral, and malignant cells rendering it effective in the treatment of infections and cancers. In this chapter, mechanism of action of ivermectin and its anti-viral effects against RNA viruses such as Zika, Dengue, Yellow fever, West Nile, Hendra, Newcastle, Avian influenza A, Porcine Reproductive and Respiratory Syndrome, Human immunodeficiency virus type 1 and some DNA viruses such as Equine herpes type 1, BK polyomavirus, Pseudorabies, Porcine circovirus 2, and Bovine herpesvirus are discussed. The significant role of ivermectin in several biological pathways makes it a prospective drug for the treatment of a wide range of viruses, including COVID-19.

Keywords: Ivermectin, COVID-19, DNA virus, RNA virus, antiviral drugs

10.1 Introduction

Replication of virus occurs only in live cells of a life form and is an infectious agent having size in the sub-microscopic range. All forms of life such as microorganisms including bacteria and archea, plants and animals are

**Corresponding author*: rather.zubaid007@gmail.com
†*Corresponding author*: nakhan2008@gmail.com

Rashid Ali and Shahid ul-Islam (eds.) Chemistry and Biological Activities of Ivermectin, (231–246)

infected by the viruses [1]. In 1892, Dmitri Ivanovsky described a pathogen that infected tobacco and Martin Beijerneck in 1898 discovered the tobacco mosaic virus. Although millions of viruses are present in the environment, only 9000 species of viruses have been described in detail [2]. Viruses are the most numerous biological entities and are found in almost every ecosystem [3, 4]. Virology, a subspeciality of microbiology is the study of viruses. Original virus is duplicated thousands of times by the host cell when infected by the viruses. They live as independent particles called virions when they are outside or while infecting a cell. Virions consist of 3 parts (i) long molecules of RNA or DNA encoding the protein architecture by which the virus acts is the genetic material of virus (ii) capsid the protein coat surrounding and protecting the genetic material and occasionally (iii) covering of lipids outside. Virus particles have shapes ranging from simple helical to more complex structures and icosahedral forms. Virions in some species are very minute that are unable to see by using optical microscope having size of 1/100th of most bacteria.

The origin theory of viruses is not clear; some may have evolved from bacteria while others might have evolved from fragments of DNA moving amid cells called as plasmids. Viruses play a key role in gene transmission, increasing genetic variety in a manner akin to sexual reproduction [5]. Some biologists believe that viruses constitute a type of life since they have genetic material, can reproduce, and have evolved through natural selection despite not having cell structure, one of the essential characteristics of life. Because they possess some but not all of the characteristics of life viruses are now referred to as replicators and organisms on the border of life [6].

Viral spread occurs by many ways. One method of spread is by means of vectors disease bearing organisms. For instance, alphids (insects that consume plant sap) and blood sucking insects spread virus from plant to plant. SARS-CoV, influenza virus, smallpox, chicken pox, and measles viruses spread in atmosphere by sneezing and coughing. Common viruses that cause viral gastroenteritis are norovirus and rotavirus which transmit by the faeco-oral means passed by food, water or by hand to mouth contact. Less than 100 particles of norovirus are necessary for human infection [7]. HIV is spread through sexual interaction and coming into contact with infected blood. A virus' host range refers to all the cells that it can infect Range can be broad or narrow depending on the capability of a virus to infect more or less species respectively. When animals are infected by the virus an immune response is provoked that helps to eliminate the virus. Vaccines aid in triggering an immune response that confers acquirement of artificial immunity to the specific viral infection. Certain viruses eliminate these immune responses like those that cause viral hepatitis, HIV, HPV infections and as a result chronic infection occurs.

Viruses require host cells for making new products as they do not have their own metabolism therefore they do not have the ability to reproduce naturally outside the host cell [8]. Cell division is used by the accepted forms of life for reproduction but viruses spontaneously assemble within cells.

Ivermectin is an anthelmintic and antiparasitic drug that has a broad spectrum of efficacy against animal nematode worms and ectoparasites. In addition to treating onchocerciasis in humans, ivermectin is also used to treat strongyloidiasis, ascariasis, trichuriasis, and enterobiasis. It is included in mass treatment programs in endemic areas. Ivermectin is an analogue of avermectin. In invertebrate muscle and nerve cells of microfilaria binding with glutamate gated Cl ion channels occurs selectively thereby increasing the movement of Cl ions across the cell membrane; resulting in hyper polarization of the cell. This process leads to paralysis of the parasite and eventually death. It also disrupts the GABA mediated CNS transmission acting as GABA agonist. Intrauterine development of microfilaria is also disturbed by this inhibiting their release from uteri of gravid female worms [9].

10.2 Mechanism of Action of Ivermectin

Ivermectin is a macrocyclic lactone, consisting of 22, 23-dihydroavermectin B1a (80%) and B1b (20%) exhibiting a wide spectrum anti-parasitic action [10]. Ivermectin was proven to be the most reliable, safe and potent semi-synthetic analogue compared to whole class of avermectins, which were first found by Professor Satoshi Mura in 1975 while studying the fermentation byproducts of an actinomycete, Streptomyces avermitilis [11].

Its primary mechanism of action is to bind to chloride channel proteins that are glutamate-controlled and are characteristic of certain classes of invertebrates, increasing movement of this electrolyte across the cell membrane. Thus results in the commencement of hyperpolarization in the outer protective membrane of target cells; that causes blockage of neuronal signal conductance within them, leading to paralysis and cell mortality [12]. Its low cost, superior effectiveness, and safety, as well as its pronounced helminth tropism, have all contributed to its long-standing success on the market since 1981; due to which it has been listed on the 21st edition of the WHO's Essential Medicines in 2019. It is a versatile molecule having distinct qualities that make it an intriguing candidate in both basic and applied research. In addition to antibacterial [13, 14], antiviral [15, 16], and anticancer activities, it has shown beneficial results in the management of some chronic diseases [17, 18]; due to the multi range cellular targeting by ivermectin Figure 10.1 [19].

The FDA gave the medication approval as a wide-spectrum antiparasitic medicine [20]. Ivermectin has the ability to treat a variety of ailments since it possesses antibacterial, antiviral, and anti-cancer effects. Ivermectin has exhibited inhibitory action against different groups of viruses, containing either DNA or RNA as genetic materials [21]. Some examples of RNA viruses are SARS-coronavirus, Dengue virus, Newcastle virus, Zika virus, Hendra virus, Yellow fever virus, and West Nile virus. Some DNA viruses include bovine herpes virus (BHV), BK polyomavirus, and Equine herpes virus type 1 (EHV-1) [20].

Ivermectin, however, correspond to be a potent antiviral and hence valuable as potential curative agent for novel coronavirus linked disease.

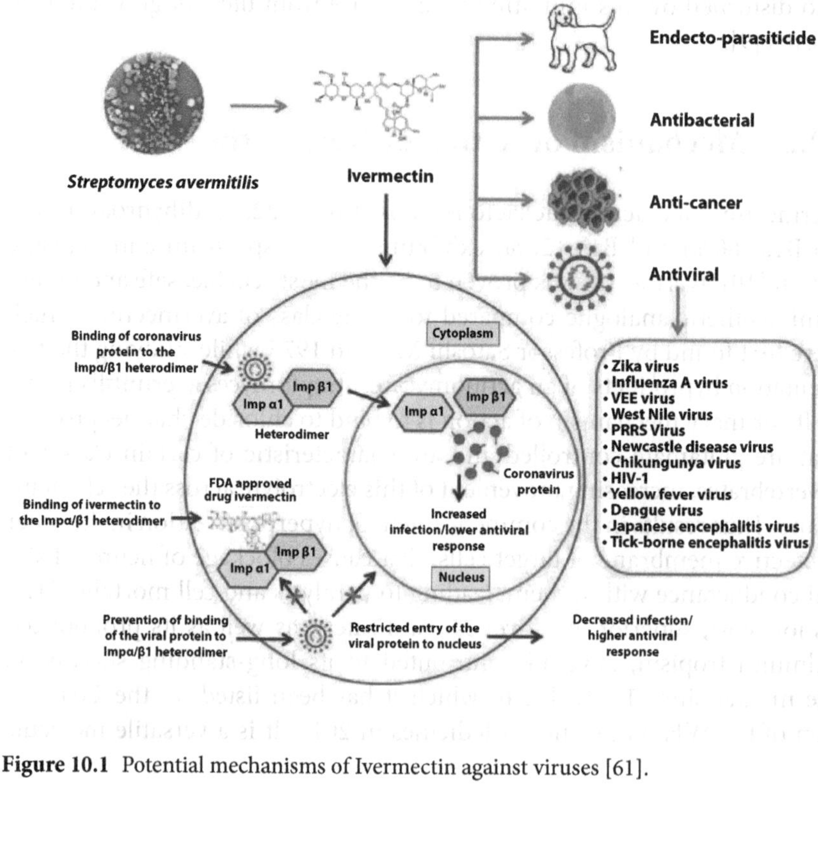

Figure 10.1 Potential mechanisms of Ivermectin against viruses [61].

Its potential role as an ionophore agent has only lately been suggested but hasn't been adequately described [15]. The structure of ionophore is perfectly designed in two sites; the hydrophilic pocket of the structure has a specific affinity to bind some ions whereas the hydrophobic layer allows the assembled complex to penetrate the cell membrane; thereby leading to modification of hydro-electrolyte dynamics [22]. The respiratory route and ATP generation in the mitochondria of eukaryotes have long been studied using certain chemical compounds commonly referred to be decoupling agents viz 2, 4-dinitrophenol. Also their antibacterial action has long been recognized [23]. Additionally, their potential use as antiviral medications [24, 25] and anticancer chemotherapeutic medicines has been proposed [26]. Monensin A and valinomycin, two of the most significant ionophores, are polyether and depsipeptide antibiotics, respectively. It is clear from their internal structures that they contain numerous oxygen atoms (having lone pairs of electrons), which are crucial for attachment with cations and transferring them across phospholipid cellular membrane. The two components of the ivermectin formula initially appear to lack the chemical characteristics necessary for a substance to be classified as an ionophore, as well as those that were discussed above. However, it is conceivable that two ivermectin molecules working in a head-tail manner could produce a complex structure to be utilized as such. The same molecules may interact naturally or by the binding of certain plasma transport proteins, particularly albumin [27], which would have the function of putting the molecules in the proper orientation to achieve the intended configuration. As can be seen, this results in the formation of an interior cavity because the oxygen atom acting as Lewis base that can attach to positively charged atoms (Lewis acid). Since acid catalysis is typically not achievable at elevated temperatures (*in vitro* and *in vivo*), ether bonds (-O- bridges) are difficult to form. To stabilize the new structure, it's expected that the -OH groups will be crucial for establishing chemical bonds (hydrogen bonds) between the functional groups. However, it is impossible to rule out the emergence of further weak and strong interactions of different types. Otherwise, certain cations could join the two molecules in the manner suggested, producing, and stabilizing the final structure themselves. Examples of this are already known from the literature [28]. Therefore, the complex's external component would already possess the hydrophobic characteristics needed to carry ions across the viral envelope. The osmotic lysis that follows the recapture of water indicates an ionic imbalance between the internal and exterior environments. This would enable the virus to be neutralized at initial phase of infection before adhering to the host cell, enabling them to enter and access the metabolic activities of those cells to produce additional viral particles. This theory, however, would only prove true for viruses lacking protein capsids that demonstrate the endurance to the osmotic pressure, albeit to a lesser degree than the cell walls of bacteria and fungus [29]. However, coronavirus only exhibits a phospholipid membrane to safeguard the genetic information and has the few proteins it needs to exit infected cells [30]. The potential of the viral membrane may also be impacted by this unusual method of electrolyte uptake, endangering its stability and effectiveness. The viral proteins in this region are no different. Additionally, as a result of the shift in concentration of particular cations, several crucial viral enzymes required for replication, such as RNA-dependent RNA polymerases (RdRp) [31] which are already pharmaceutical targets, may be inhibited.

10.3 Anti-Viral Effects Against Various DNA and RNA Viruses

In a freshwater crayfish (*Cherax quadricarinatus*) model, the first *in vivo* anti-viral study of ivermectin showed its action against parvo viruses [32]. The wide-spectrum endo/ectoparasiticide has demonstrated potent antiviral effects against variety of DNA and RNA viruses, including the SAR-coronavirus, HIV-1, Zika virus (ZKV), Chikungunya virus (CGV), influenza A virus (Inf AV), Yellow fever virus (YFV), Venezuelan equine encephalitis virus (VEEV), West Nile virus (WNV), porcine reproductive and respiratory syndrome virus, Newcastle disease virus (NCDV) and dengue virus (DENGV).

10.3.1 RNA Viruses

10.3.1.1 COVID-19

Murine hepatitis virus (MHV) and SARS-CoV-2 correspond to the *Betacoronavirus* genus; they possess structural familiarity in terms of gene sequence; therefore MHV has been exploited as the substitute model for SARS-CoV-2 infections. Recently Tan and his colleagues in 2021 studied the mono and combinational therapy of anti-COVID drugs, *viz.* remdesivir, chloroquine, ivermectin, and doxycycline, which were tested for inhibition of live virus and replication of viral RNA utilizing an *in vitro* model of MHV infection of RAW264.7 macrophages. Remdesivir monotherapy (6 μM) produced the effective results by suppressing the replication of viral RNA and live virus by around 2-log10 and 1-log10, respectively. The treatment with ivermectin displayed the highest selectivity index. By significantly reducing the amount of active virus by 7-log10 and viral RNA by 2.5-log10 in infected macrophages, remdesivir and ivermectin demonstrated a highly effective synergism. Additionally, the levels of IL-6, TNF-a, and leukaemia IF were observed to be minimal in this combinational therapy. These findings necessitate further research regarding the mechanism of this synergism, in conjunction with upcoming *in vivo* investigations and clinical trials for SARS-CoV-2 management [33]. An *in vitro* experiment found that a single dose of ivermectin could suppress SARS-CoV-2 replication in 24-48 h, presumably by constraining IMPαβ1 mediated nuclear transport of viral proteins. After 24 hours post infection, the amount of viral RNA produced from affected cells and cell-linked to viral RNA were reduced considerably by greater than 90% and 99%, respectively. Interestingly, ivermectin treatment of SARS-CoV-2-infected cells for 48 h showed a 5000 times decline of viral RNA verses the control group. But, no more diminution in concentration of viral RNA was found after 72 hours [34]. Moreover, the study found no evidence of ivermectin toxicity in any of the groups during observation period. SARS-CoV-2 binds specifically to the importin-α (IMP-α) and importin-β (IMP-β) heterodimer and is thereafter transported via the nuclear pore complex (NPC). The complex is finally released within the nucleus. The SARS-CoV-2 protein subsequently stimulates the host shut-off, which suppresses immune response of host and enables the virus to proliferate. Ivermectin interacts with the importin-/β complex and destabilizes the importin-/β heterodimer to stop SARS-CoV-2 protein from entering into the nucleus and thereby facilitating the effective host immune response. Ivermectin has gained more attention in medical research because of the preclinical research indicating its ability to suppress SARS-CoV-2. The Food and Drug Administration (FDA) reportedly addressed it by releasing a formal letter within which it was emphasized that investigation was currently at a preliminary phase and there is a necessity for additional phases of clinical trials to evaluate whether ivermectin is efficient in treating SARS-CoV-2. This is crucial because the study may encourage consumers to engage in the practice of self-medication [35]. Despite the disagreement; this work represents a significant advancement in our knowledge of the positive effects of ivermectin against SARS-Coronavirus infection.

In a recent investigation, the combination therapy of ivermectin and hydroxychloroquine was suggested to prevent or treat COVID-19 which showed a synergistic interaction by restriction of viral infiltration and multiplication. Nevertheless, pharmacokinetic findings demonstrated that a greater concentration/dose was needed to reproduce the antiviral action. In humans, it is therefore quite challenging to obtain the required inhibitory concentration. Ivermectin's effectiveness in SARS-CoV-2 patients has to be established by conducting further randomized clinical trials [36–41].

10.3.1.2 Zika Virus

The anti-viral properties of Ivermectin were validated in an *in vitro* experiment performed by Barrow *et al.* using Huh-7 cells that were Zika-(ZIKMEX_1_7) [42]. In another *in vivo* experiment on bone marrow cells of male Sprague–Dawley rat, treatment of ivermectin (of 0.4 kg−1) with aged garlic extract (AGE) (300,600 and 1200 mg kg−1) was administered and the results revealed a decrease in the cytotoxicity [43]. It might be stated that increased dosages of ivermectin could be administered in conjunction with AGE for treatment of ZKV in mice, allowing researchers to explore the antiviral activities of ivermectin in larger concentrations. The impact of ivermectin on number of ZKV- infected cell lines has also been evaluated by researchers in an *in vitro* investigation [44, 45]. The drug was possibly able to pass the epithelial barrier of intestine when administered orally and achieve an optimum blood concentration; however there was minimal damage in epithelial cells; while no hepatotoxicity was observed. Additionally, *in vitro* tests revealed that the treatment exhibited temperature-dependent stability and did not pass through the placental barrier [46].

The researchers demonstrated that the IMP α/β1 heterodimer can be dissociated by ivermectin. The fact that ivermectin could specifically bind to the IMP's armadillo connect repeat domain and alter its structural conformation, and that might be the cause of the inhibition of binding to IMPβ1. They arrived at the conclusion that ivermectin could prevent IMPα from recognizing NLS-containing proteins like NS5 in a cell environment. Ivermectin impairs NLS recognition and nuclear targeting [47, 48].

10.3.1.3 Dengue Virus

Kylie *et al.* demonstrated that Ivermectin has been shown to limit the growth of DNGV at high concentrations (25–50 μM) using infected human cervical adenocarcinoma cells (Hela) by impeding the IMP α/β1-dependent movement of viral proteins from cytoplasm to the nucleus of the host cell [49]. Another study was conducted on Huh-7 cells infested with four distinct DNGV serotypes, which on treatment with ivermectin inhibited the nuclear transport mediated via IMP α/β. Ivermectin's potential to be used as an antiviral medication in the management of dengue was mentioned by the authors [50]. Researchers established that the ivermectin effectively inhibited NS5 resulting in a 60% drop in NS5 protein concentrations in the nucleus after 7 hours of treatment. These outcomes are in line with some other research that demonstrated ivermectin reduces DNGV reproduction by preventing NS5 interaction with IMP α/β transporter [44].

10.3.1.4 Foot-and-Mouth Disease Virus

The FMDV is an RNA virus with 8500 nucleotides and a positive-sense single strand that causes several epidemics in livestock worldwide. A research was conducted to investigate the antiviral activity of ivermectin against various FMDV serotypes employing a BHK-21 cell line. In comparison to the attachment and entry stages, the viral intensity was lowered more during the replication stage. This work demonstrated ivermectin's *in vitro* anti-FMDV efficacy and suggested its use to treat FMDV [51].

10.3.1.5 Hendra Virus

The effectiveness of ivermectin against Hendra virus HEV was studied *in vitro*. The virus's capacity to block the host's type-one interferon response by generating the polycistronic p gene contributes to part of its primary pathogenicity. Researchers found that HEV dynamically travels through the IMP- α1 between the nucleus and the cytoplasm. With mammalian cells, the findings revealed that ivermectin prevented HEV infection and even a single, non-optimized concentration of 10 μM could suppress the virus approximately 5 times without cytotoxicity. Ivermectin was found to be beneficial in preventing HEV infection by halting the transmission of virus via IMP α1/β1 [52].

10.3.1.6 Newcastle Virus

Azeem and his coworkers studied the cytotoxic impact of ivermectin as well as its possible antiviral activity against NCV. *In vivo* and *in vitro* studies were conducted simultaneously on a 9-day-old chick embryo and a chick primary fibroblast cell line respectively. Ivermectin was evaluated in this study at concentrations of 6.25, 12.5, 25, 50, 100, and 200 μg ml^{-1}. The results showed that the drug exhibited cytotoxic effects at concentrations of 100 μg ml^{-1} or above. But at 50 μg ml^{-1} or less concentrations, it was safe with no drug induced cytotoxicity, and only moderate to weak antiviral potential could be found [53].

10.3.1.7 Avian Influenza A Virus

An investigation was conducted employing Avian influenza A-infected chicken hepatocellular carcinoma cells. Treatment with 10 μM ivermectin completely stopped the nuclear transmission of numerous viral RNA complexes [54].

10.3.1.8 Human Immunodeficiency Virus Type 1

Researchers conducted *in vitro* tests to determine how ivermectin affected the transport of the nuclear protein of the HIV-1 virus. The outcomes demonstrated that ivermectin decreased the NLS-containing protein binding by IMP α/β and inhibited such interactions at lower levels (IC50: 4.8 μM). Ivermectin is a nuclear transport inhibitor via IMP/α/β; however, it has no impact on nuclear transfer by IMP β1 alone. It entirely prevents the nuclear import of the HIV-1 active integrase protein, a key component of the pre-integration complex [55]. In a study using infected human cervical adenocarcinoma cells (Hela), Kylie *et al.* demonstrated that ivermectin has an inhibitory effect on HIV-1 proliferation at high doses (25–50 μM). It is accomplished by impeding the IMP α/β1-dependent movement of viral proteins from the cytoplasm to the nucleus of the host cell. Ivermectin prevented HIV-1 integrase from aggregating together in nucleus, as per the study [49].

10.3.2 DNA Viruses

10.3.2.1 Equine Herpesvirus Type 1 (EHV-1)

The anti-viral potential of Ivermectin against different DNA viruses was investigated in a wide number of studies. Two distinct strains of EHV-1, viz Rac-H and Jan-E were investigated at various dosages employing primary murine neurons to determine the anti-viral potential of ivermectin. The results showed that ivermectin lowered the proliferation of Jan-E; whereas Rac-H remained unaffected. The results showed that various EHV-1 strains use distinct receptors to infiltrate the nucleus. The results of the study indicate that IMP α/β along with the other receptors plays a role in nuclear import in the EHV-1 [56].

10.3.2.2 *Pseudorabies Virus (PRV)*

PRV swine virus is an enclosed double-stranded DNA virus that belongs to the alpha-herpesviridae subfamily and infects pigs for the whole of their lives, The DNA polymerase enzyme of pseudorabies virus is composed of two subunits classified as UL30 and UL42 AL [57]. It is discovered that the UL42 subunit has a bipartite NLS that is mediated by IMP-α/β- and transports both subunits into the cell nucleus. The *in vitro* and *in vivo* antiviral effects of Ivermectin were studied by Lv *et al.* The study revealed that ivermectin prevented DNA polymerase subunit UL42 from entering the nucleus; consequently, as the dose of drug was increased, fewer UL42 were detected in the nucleus by employing the western blot method. The drug prevented UL42 from moving from the cytoplasm to the nucleus via the NLS, but it had no effect on UL42 expression in the cytoplasm. Ivermectin exhibited the drastic decrease of viral replication in the brain and kidney of all animals in the PRV-infected mouse model, and this decline was particularly pronounced in the kidneys that serves as the principal organ engaged in the metabolism of ivermectin. It was found that the drug might be used as a possible antiviral treatment against PRV [58].

10.3.2.3 *BK Polyomavirus (BKPyV)*

As reported by Wagstaff *et al*, ivermectin was demonstrated to be able to particularly block the nuclear transfer pathway through IMP α/β [49]. Bennet and his coworkers looked into the impact of ivermectin on BKPyV - infected renal proximal tubule epithelial cells; after treatment with10 μM ivermectin; a descriptive examination utilizing RTPCR revealed a decrease in the quantity of the early protein T Antigen mRNA. The study indicated the restricted entrance into nuclease that led to decreased gene expression of viral nucleic acids. Ivermectin's inhibitory impact demonstrates that active NPC transfer allows polyomavirus to enter the nucleus [59].

10.3.2.4 *Porcine Circovirus 2 (PCV2)*

In PK-15 cells infected with PCV2 (Circoviridae family), the inhibitory effect of ivermectin on virus multiplication was examined. Ivermectin did not have harmful effects at concentrations of 50 or 100 μg ml^{-1} at 24 or 48 h after treatment, but at 200 μg ml^{-1}, it dramatically reduced cell viability (*P* value 0.05). The study demonstrated a decline in therapeutic efficacy at subsequent time points as was mentioned in other research [60].

10.4 Conclusion

It could be concluded that ivermectin has antiviral effects against a wide variety of RNA and DNA viruses. Because ivermectin has been found to be efficacious in primary infection as per experimental investigations, it is postulated that the intake of ivermectin may result in prevention of infection in the initial stages, which needs further human investigations and clinical trials to confirm this claim. Because of its antiviral activity, ivermectin may play an important role in multiple vital biological processes, making it a possible candidate for the treatment of various viruses, including COVID-19. However it must be evaluated through clinical studies, and calls for more research. Ivermectin may possess broad-spectrum antiviral capabilities that could enable it to treat additional viral diseases and make it useful in future epidemic management.

References

1. Koonin, E.V., Senkevich, T.G., Dolja, V.V., The ancient virus world and evolution of cells. *Biol. Direct*, 1, 1, 29, 2006.
2. Breitbart, M. and Rohwer, F., Here a virus, there a virus, everywhere the same virus?. *Trends Microbiol.*, 13, 278, 2005.
3. Lawrence, C.M., Menon, S., Eilers, B.J., Bothner, B., Khayat, R., Douglas, T., Young, M.J., Structural and functional studies of archeal viruses. *J. Biol. Chem.*, 284, 12599, 2009.
4. Edwards, R.A. and Rohwer, F., Viral metagenomics. *Nat. Rev. Microbiol.*, 3, 504, 2005.
5. Canchaya, C., Fournous, G., Chibani-Chennoufi, S., Dillmann, M.L., Brüssow, H., Phage as agents of lateral gene transfer. *Curr. Opin. Microbiol.*, 6, 417, 2003.
6. Rybicki, E.P., The classification of organisms at the edge of life, or problems with virus systematics. *S. Afr. J. Sci.*, 86, 182, 1990.
7. Robilotti, E., Deresinski, S., Pinsky, B.A., Norovirus. *C.M.R.*, 28, 134, 2015.
8. Wimmer, E., Mueller, S., Tumpey, T.M., Taubenberger, J.K., Synthetic viruses: A new opportunity to understand and prevent viral disease. *Nat. Biotechnol.*, 27, 1163, 2009.
9. Ikeda, T., Pharmacological effects of ivermectin, an antiparasitic agent for intestinal strongyloidiasis: Its mode of action and clinical efficacy. *Nihon Yakurigaku Zasshi*, 122, 527, 2003.
10. Gonzalez Canga, A., Prieto, A.M.S., Liebana, M.J.D., Martinez, N.F., Vega, M.S., Vietiez, J.J.G., The pharmacokinetics and interactions of ivermectin in humans–a mini-review. *AAPS J.*, 10, 42, 2008.

11. Crump, A., Ivermectin: Enigmatic multifaceted 'wonder' drug continues to surprise and exceed expectations. *J. Antibiot. (Tokyo)*, 70, 495–505, 2017.
12. Geary, T.G., Ivermectin 20 years on: Maturation of a wonder drug. *Trends Parasitol.*, 21, 530, 2005.
13. Lim, L.E., Vilchèze, C., Ng, C., Jacobs, W.R., Jr, Ramón-García, S., Thompson, C.J., Anthelmintic avermectins kill Mycobacterium tuberculosis, including multidrug-resistant clinical strains. *Antimicrob. Agents Chemother.*, 57, 1040, 2013.
14. Ashraf, S., Chaudhry, U., Raza, A., Ghosh, D., Zhao, X., *In vitro* activity of ivermectin against Staphylococcus aureus clinical isolates. *Antimicrob. Resist. Infect. Control*, 7, 27, 2018. doi: 10.1186/s13756-018-0314-4
15. Juarez, M., Schcolnik-Cabrera, A., Dueñas-Gonzalez, A., The multitargeted drug ivermectin: From an antiparasitic agent to a repositioned cancer drug. *Am. J. Cancer Res.*, 8, 317, 2018.
16. Intuyod, K., Hahnvajanawong, C., Pinlaor, P., Pinlaor, S., Anti-parasitic drug ivermectin exhibits potent anticancer activity against gemcitabine-resistant cholangiocarcinoma *in vitro*. *Anticancer Res.*, 39, 4837, 2019.
17. Ashraf, S. and Prichard, R., Ivermectin exhibits potent antimitotic activity. *Vet. Parasitol.*, 226, 1, 2016.
18. Ventre, E., Rozières, A., Lenief, V., Albert, F., Rossio, P., Laoubi, L., Dombrowicz, D., Staels, B., Ulmann, L., Julia, V., Vial, E., Jomard, A., Hacini-Rachinel, F., Nicolas, J.F., Vocanson, M., Topical ivermectin improves allergic skin inflammation. *Allergy*, 72, 1212, 2017.
19. Rizzo, E., Ivermectin, antiviral properties and COVID- 19: A possible new mechanism of action. *Naunyn Schmiedebergs Arch. Pharmacol.*, 393, 1153, 2020.
20. González Canga, A., Sahagún Prieto, A.M., Diez Liébana, M.J. *et al.*, The pharmacokinetics and interaction of ivermectin in humans: A mini-review. *AAPS J.*, 10, 42, 2008.
21. Gotz, V., Influenza A viruses escape from MxA restriction at the expense of efficient nuclear vRNP import. *Sci. Rep.*, 6, 23138, 2016.
22. Freedman, J.C., Chapter 4 - Ionophores in planar lipid bilayers, in: *Cell Physiology Source Book*, 4th edn., N. Sperelakis (Ed.), pp. 61–66, Academic Press, Elsevier, Amsterdam, 2012.
23. Bakker, E.P., Ionophore antibiotics, in: *Mechanism of action of antibacterial agents*, Antibiotics vol. 5:1, F.E. Hahn (Ed.), Springer, Berlin, Heidelberg, 1979.
24. Krenn, B.M., Gaudernak, E., Holzer, B., Lanke, K., Van Kuppeveld, F.J., Seipelt, J., Antiviral activity of the zinc ionophores pyrithione and hinokitiol against picornavirus infections. *J. Virol.*, 83, 58, 2009.
25. Sandler, Z.J., Vu, M.N., Menachery, V.D., Mounce, B.C., Novel ionophores active against La Crosse virus identified through rapid antiviral screening. *Antimicrob. Agents Chemother.*, 2020.

26. Kaushik, V., Yakisich, J.S., Kumar, A., Azad, N., Iyer, A.K.V., Ionophores: Potential use as anticancer drugs and chemosensitizers. *Cancers (Basel)*, 10, 360, 2018. doi: 10.3390/cancers10100360
27. Klotz, U., Ogbuokiri, J.E., Okonkwo, P.O., Ivermectin binds avidly to plasma proteins. *Eur. J. Clin. Pharmacol.*, 39, 607, 1990.
28. Abbott, B.J., Fukuda, D.S., Dorman, D.E., Occolowitz, J.L., Debono, M., Farhner, L., Microbial transformation of A23187, a divalent cation ionophore antibiotic. *Antimicrob. Agents Chemother.*, 16, 808, 1979.
29. Cordova, A., Deserno, M., Gelbart, W.M., Ben-Shaul, A., Osmotic shock and the strength of viral capsids. *Biophys. J.*, 85, 72, 2003.
30. Sigrist, C.J., Bridge, A., Le Mercier, P., A potential role for integrins in host cell entry by SARS-CoV-2. *Antivir. Res.*, 177, 104759, 2020.
31. te Velthuis, A.J., van den Worm, S.H., Sims, A.C., Baric, R.S., Snijder, E.J., van Hemert, M.J., Zn(2+) inhibits coronavirus and arterivirus RNA polymerase activity *in vitro* and zinc ionophores block the replication of these viruses in cell culture. *PLoS Pathog.*, 6, e1001176, 2010.
32. Nguyen, K.Y., Sakuna, K., Kinobe, R., Owens, L., Ivermectin blocks the nuclear location signal of parvo viruses in crayfish, Cherax quadricarinatus. *Aquaculture.*, 288, 420, 2014.
33. Tan, Y.L., Tan, K.S.W., Chu, J.J.H., Chow, V.T., Combination treatment with remdesivir and ivermectin exerts highly synergistic and potent antiviral activity against murine coronavirus infection. *Front. Cell. Infect. Microbiol.*, 11, 700502, 2021. doi: 10.3389/fcimb.2021.700502
34. Caly, L., Druce, J.D., Catton, M.G., Jans, D.A., Wagstaff, K.M., The FDA-approved drug ivermectin inhibits the replication of SARS-CoV-2 *in vitro*. *Antivir. Res.*, 178, 104787, 2020.
35. FDA letter to stakeholders: Do not use ivermectin intended for animals as treatment for COVID-19 in humans, Available online: https://www.fda.gov/animal-veterinary/product-safety-information/fda-letterstakeholders-do-not-use-ivermectin-intended-animals-treatment-covid-19-humans).
36. Patri, A. and Fabbrocini, G., Hydroxychloroquine and ivermectin: A synergistic combination for COVID-19 chemoprophylaxis and treatment? *J. Am. Acad. Dermatol.*, 82, e221, 2020.
37. Momekov, G. and Momekova, D., Ivermectin as a potential COVID-19 treatment from the pharmacokinetic point of view. *MedRxiv*, 34, 469, 2020.
38. Cavalcanti, A.B., Zampieri, F.G., Rosa, R.G., Azevedo, L.C.P., Veiga, V.C., Avezum, A., Damiani, L.P., Marcadenti, A., Kawano-Dourado, L., Lisboa, T., Hydroxychloroquine with or without Azithromycin in Mild-to-Moderate COVID-19. *N. Engl. J. Med.*, 383, e–119, 2020.
39. Liu, Y., Gayle, A., Wilder-Smith, A.J.R., The reproductive number of COVID-19 is higher compared to SARS coronavirus. *J. Travel Med.*, 2, 27, 2020.
40. Frediansyah, A., Tiwaric, R., Sharun, K., Dhama, K., Harapan, H., Antivirals for COVID-19: A critical review. *Clin. Epidemiol. Glob. Health*, 9, 90, 2020.

41. Khan, S., Dhama, K., Patel, S.K., Pathak, M., Tiwari, R., Singh, B.R., Sah, R., Bonilla-Aldana, D.K., Rodriguez-Morales, A.J., Leblebicioglu., H., Ivermectin, a new candidate therapeutic against SARS-CoV-2/COVID-19. *Clin. Microbiol. Antimicrob.*, 19, 23, 2020.
42. Barrows, N.J., Campos, R.K., Powell, S.T. *et al.*, A screen of FDAapproved drugs for inhibitors of Zika virus infection. *Cell Host Microbe*, 20, 259, 2016.
43. Khalil, A.M. and Abu Samrah, H.M., *In vivo* combined treatment of rats with ivermectin and aged garlic extract attenuates ivermectininduced cytogenotoxicity in bone marrow cells. *Res. Vet. Sci.*, 120, 94, 2018.
44. Ji, W. and Luo, G., Zika virus NS5 nuclear accumulation is protective of protein degradation and is required for viral RNA replication. *Virology*, 541, 124, 2020.
45. Tay, M.Y.F., Fraser, J.E., Chan, W.K.K. *et al.*, Nuclear localization of dengue virus (DENV) 1-4 non-structural protein 5; protection against all 4 DENV serotypes by the inhibitor Ivermectin. *Antivir. Res.*, 99, 301, 2013.
46. Surnar, B., Kamran, M.Z., Shah, A.S. *et al.*, Orally administrable therapeutic synthetic nanoparticle for Zika virus. *ACS Nano*, 13, 11034, 2019.
47. Yang, S.N.Y., Atkinson, S.C., Wang, C. *et al.*, The broad spectrum antiviral ivermectin targets the host nuclear transport importin alpha/beta1 heterodimer. *Antivir. Res.*, 177, 104760, 2020.
48. Kosyna, F.K., Nagel, M., Kluxen, L., Kraushaar, K., Depping, R., The importin alpha/beta-specific inhibitor Ivermectin affects HIFdependent hypoxia response pathways. *Biol. Chem.*, 396, 1357, 2015.
49. Wagstaff, K.M., Sivakumaran, H., Heaton, S.M., Harrich, D., Jans, D.A., Ivermectin is a specific inhibitor of importin alpha/beta-mediated nuclear import able to inhibit replication of HIV-1 and dengue virus. *Biochem. J.*, 443, 851, 2012.
50. Tay, M.Y.F., Fraser, J.E., Chan, W.K.K. *et al.*, Nuclear localization of dengue virus (DENV) 1-4 non-structural protein 5; protection against all 4 DENV serotypes by the inhibitor Ivermectin. *Antivir. Res.*, 99, 301, 2013.
51. Naeem, Z., Raza, S., Afzal, S., Sheikh, A.A., Ali, M.M., Altaf, I., Antiviral potential of ivermectin against foot-and-mouth disease virus, serotype O, A and Asia-1. *Microb. Pathogen.*, 155, 104914, 2021.
52. Atkinson, S.C., Audsley, M.D., Lieu, K.G. *et al.*, Recognition by host nuclear transport proteins drives disorder-to-order transition in Hendra virus V. *Sci. Rep.*, 8, 358, 2018.
53. Azeem, S., Ashraf, M., Rasheed, M.A., Anjum, A.A., Hameed, R., Evaluation of cytotoxicity and antiviral activity of ivermectin against Newcastle disease virus. *Pak. J. Pharm. Sci.*, 28, 597, 2015.
54. Gotz, V., Magar, L., Dornfeld, D. *et al.*, Influenza A viruses escape from MxA restriction at the expense of efficient nuclear vRNP import. *Sci. Rep.*, 6, 23138, 2016.

55. Wagstaff, K.M., Rawlinson, S.M., Hearps, A.C., Jans, D.A., An AlphaScreen(R)-based assay for high-throughput screening for specific inhibitors of nuclear import. *J. Biomol. Screen.*, 16, 192, 2011.
56. Slonska, A., Cymerys, J., Skwarska, J., Golke, A., Banbura, M.W., Influence of importin alpha/beta and exportin 1 on equine herpesvirus type 1 (EHV-1) replication in primary murine neurons. *Pol. J. Vet. Sci.*, 16, 749, 2013.
57. Wang, Y.P., Du, W.J., Huang, L.P. *et al.*, The pseudorabies virus DNA polymerase accessory subunit UL42 directs nuclear transport of the holoenzyme. *Front. Microbiol.*, 7, 124, 2016.
58. Lv, C., Liu, W., Wang, B. *et al.*, Ivermectin inhibits DNA polymerase UL42 of pseudorabies virus entrance into the nucleus and proliferation of the virus *in vitro* and vivo. *Antivir. Res.*, 159, 55, 2018.
59. Bennett, S.M., Zhao, L., Bosard, C., Imperiale, M.J., Role of a nuclear localization signal on the minor capsid proteins VP2 and VP3 in BKPyV nuclear entry. *Virology*, 474, 110, 2015.
60. Wang, X., Lv, C., Ji, X., Wang, B., Qiu, L., Yang, Z., Ivermectin treatment inhibits the replication of Porcine circovirus 2 (PCV2) *in vitro* and Ivermectin: A systematic review from antiviral effects to COVID-19 complementary regimen 601 mitigates the impact of viral infection in piglets. *Virus Res.*, 263, 80, 2019.
61. Sharun, *et al.*, Ivermectin, a new candidate therapeutic against SARS-CoV-2/COVID-19. *Ann. Clin. Microbiol. Antimicrob.*, 19, 23, 2020.

11

Toxicology, Safety, and Environmental Aspects of Ivermectin

Ayesha Akhtar, Nasreen A. Mazumdar and Afreen Inam*

Medicinal Chemistry Laboratory, Department of Chemistry, Jamia Millia Islamia, New Delhi, India

Abstract

Ivermectin discovered in 1975, and enrolled for human use in 1987, is also being useful in the global elimination programs for the eradication of neglected tropical diseases. In the recent years, several published studies have focused on expanding the role of ivermectin in a multiple number of diseases in addition to the toxicity, safety dosage, and environmental impact of this drug. This book chapter highlights the various aspects of ivermectin on animals, humans, and environment as well. Hopefully this particular chapter will be of high value to both academic and industry perspective.

***Keywords*:** Ivermectin, antiparasitic activity, neglected tropical diseases, safety

11.1 Introduction

Ivermectin (IVM), a dihydroavermectin B1 microfilaricide, is used in the treatment of strongyloidiasis, scabies, and all kinds of filariasis excluding *Dipalonema (Mansonella) perstans* infections. IVM has produced good results in the treatment of onchocerciasis (river blindness) both in controlled studies and in the WHO-sponsored program. IVM is a mixture of 22,23-dihydroavermectin B_{1a} (more than 80%) and B_{1b} (less than 20%) (Figure 11.1). It is designed to control variety of internal and exterior parasites in the animals [1].

**Corresponding author*: ainam@jmi.ac.in

Rashid Ali and Shahid ul-Islam (eds.) Chemistry and Biological Activities of Ivermectin, (247–264)

The effective dosage has been observed in the range of 50 to 200 micrograms/kg. Skin microfilariae levels are shown to be low after a single dose for up to 9 months. Mild IVM responses, caused by destruction of microfilariae, affect at least one-third of the patients. It is found that activation of neutrophil plays a critical role in the development of these responses [2]. It is well tolerated when used in combinations with albendazole to kill adult worms or with diethylcarbamazine for bancroftian filariasis [3]. IVM has a long half-life and adverse effects in some patients.

Ivermectin's route of action has been intensively studied in the recent years. It has been recognized as agonism at GABA receptors, with inhibition of the accompanying chloride channels that are particular in governing nerve cell connections. Most organisms' chloride channel activity is disrupted, resulting in parasite paralysis and death. The following are the proposed action sites:

- A postsynaptic agonist location either on the receptor or in its immediate neighborhood.
- A presynaptic location of activation of GABA release.

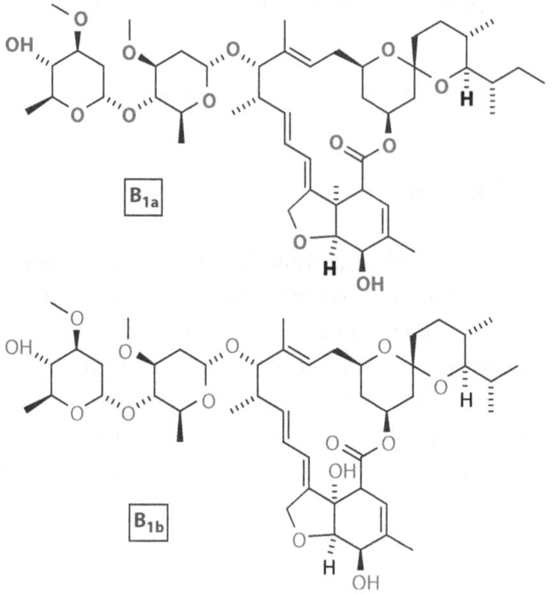

Figure 11.1 Structure of Ivermectin.

- Enhancement of GABA binding to its receptor [4].
- Binding of IVM to P glycoprotein.

Acute adverse effects of IVM include flu-like symptoms or skin irritation caused by toxins released and allergens from the killed filariae which can impact two-thirds of the patients. When these types of reactions do not occur, the medicine is thought to be useless. These appear soon after the microfilariae die, and sometimes only momentarily. The consequences are particularly severe when the number of microfilariae is considerable [5].

The late reactions caused by the IVM when used along with diethylcarbamazine for the treatment of *O. volvulus* with includes tachycardia, adenitis, pruritus, fever, hypotension, arthalgia, uticarial rash, and lymphedema. The more parasites there are, the more intense the inflammatory host response to microfilarial death. In patients with onchocerciasis, the functions of chemoattractants (eotaxin, RANTES, and MCP-3) in the recruitment of eosinophils to the site of parasite death have been examined [6]. There were some negative side effects, but none were serious. After IVM, there was endothelial production of both RANTES and eotaxin, indicating that both chemoattractants play a vital role in eosinophil recruitment in the skin during parasite destruction or degeneration [7].

11.2 Ivermectin's Antiparasitic Activity

S. Omura from the Kitasato Institute in Tokyo, Japan, discovered ivermectin (IVM) in partnership with a pharmaceutical company Merck, Sharpe and Dohme [8]. It was isolated from the fermentation broth of Streptomyces avermectins, and belongs to the avermectins class of semi-synthetic compounds. AVM and macrocyclic lactones are structurally related to macrolide antibiotics, which are ineffective against both bacteria and fungi. IVM has been shown to have antiparasitic activity in both animals and humans against a wide range of nematodes and arthropods, including intestinal nematodes and ectoparasites such as lice, scabies, mites, and botflies. These parasites include *O. volvulus* tissue microfilaria, *Loa loa*, lymphatic filariasis, and intestinal nematodes [9] (Figure 11.2 and Figure 11.3).

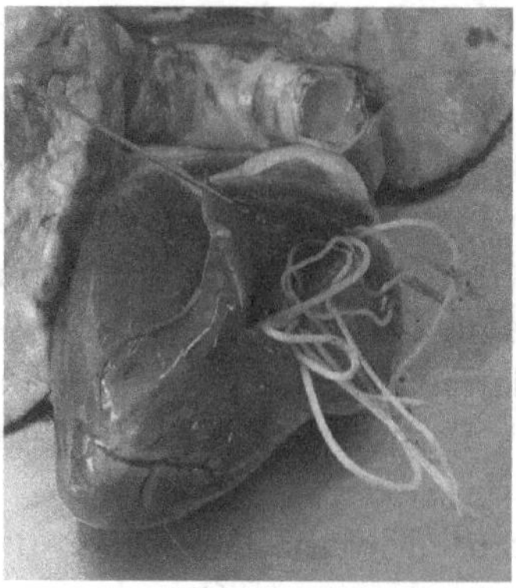

Figure 11.2 Adult worms of *D. immitis* in the heart of Dog. Picture credit: Fernando Simon *et al.* [10].

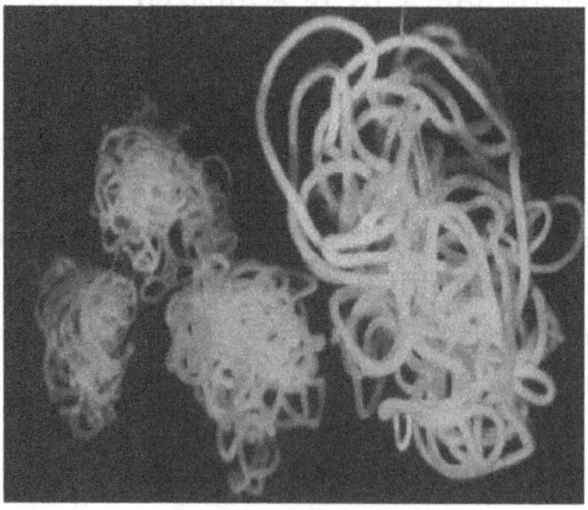

Figure 11.3 Onchocerca volvulus filarial parasite. Picture credit: Higazi, Tarig & Geary, Timothy [11].

11.3 Pharmacology

Ivermectin is transported through a glycoprotein transporter, a drug efflux transporter linked with the plasma membrane. It also promotes mortality in nematodes and arthropods by blocking GABA gated chloride channels in invertebrate muscles and nerve cells [12] (Figure 11.4). Although it has not been proved that it can cross the intact blood-brain barrier in humans, it has been related to central nervous system toxicity and quick mortality in dogs because to the blood-brain barrier's non-functional P-glycoprotein transporter [13]. It has been reported that it is a substrate for the cytochrome P450 (CYP) 3A4 enzyme. When taken orally, a single 200 mg/kg dose has a bioavailability of 50–60%, with peak levels occurring 4 to 5 hours later. The liver is in charge of metabolization, and the bulk of metabolites

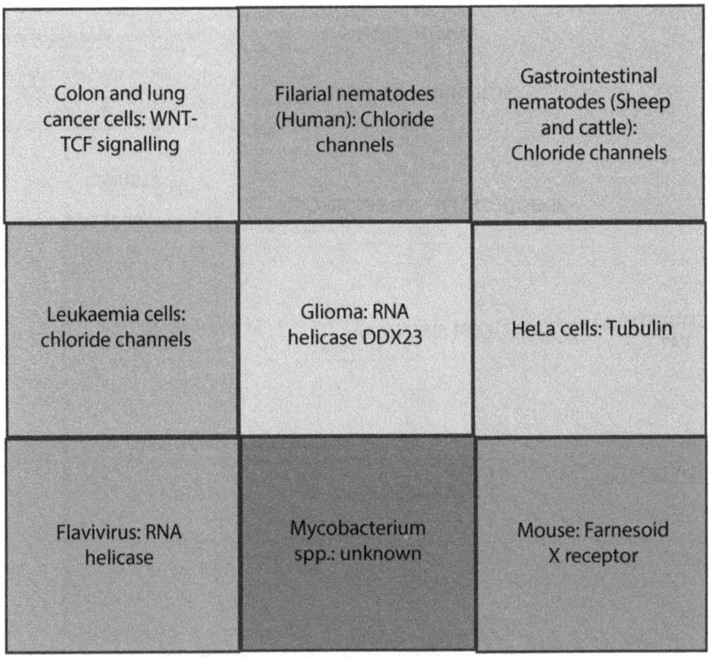

Figure 11.4 Reported predicted targets of ivermectin (IVM) [27–38].

are removed by urine and feces. The parent medication has a half-life of 12–56 hours, and its metabolites have a half-life of up to 3 days [14, 15]. Figure 11.5 depicts the general characteristics of IVM pharmacokinetics.

11.4 Adverse Effects in Humans and Animals

Every year, millions of doses of IVM are provided globally through mass medication drug administration programs in emerging and improvised countries. IVM has been proven to be safe for human consumption, albeit it is not recommended for children less than 5 years or weighing less than 15 kg, pregnant or lactating women, persons with central nervous system illness, or people who are allergic to the medicine. These restrictions are made because the medicine may be able to pass insufficiently through blood-brain barriers, thereby causing neurotoxicity [16]. When treated for scabies in children 5 years of age this drug is well tolerated without any serious consequences. When used to treat filariasis, IVM results in a Mazzotti-type reaction because it kills a lot of microfilariae [17]. IVM treatment for onchocerciasis patients with severe *Loa loa* microfilariae infection led to encephalopathy [18].

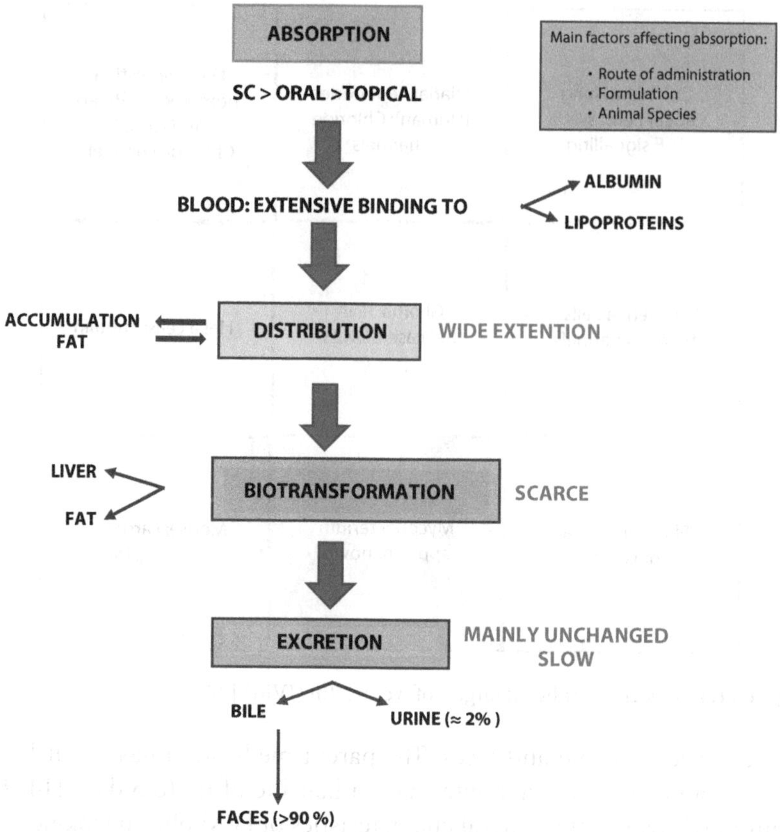

Figure 11.5 General pharmacokinetics.

IVM, unlike any other treatment, is a drug for the poor. Around 250 million people have been treated with this miraculous drug for onchocerciasis and lymphatic filariasis, two of the world's most devastating disfiguring, incapacitating, and stigma-inducing diseases. Most patients come from isolated, rural, under-resourced, and poor regions in developing countries and improvised countries, where they lack access to basic medical care and sufficient sanitation. When avermectins were discovered, they constituted a new class of compounds that could destroy a wide range of disease-causing organisms as well as infection vectors—both within and outside the body [19–21]. IVM was a revelation to humankind. It was highly efficacious against a variety of nematodes, insect, and acarine parasites at low doses. IVM is a potential drug that could be used to treat a variety of ailments and infections that primarily afflict underdeveloped countries (Table 11.1). IVM is mostly used in conjunction

Table 11.1 Reported Ivermectin dosage [39–41].

1.	Onchocerciasis	*Onchocerca volvulus*	Single dose: 150–200 µg/kg body weight as Optimal dose: 150 µg/kg Retreatment may be prerequisite every 3–12 months for 9–15 years till asymptomatic
2.	Lymphatic filariasis	*Wuchereria bancrofti*	Single dose: 150–200 µg/kg body weight (in combination with albendazole) twice annually or Single dose: 300–400 µg/kg as a single dose annually.
3.	Strongyloidiasis	*Strongyloides stercoralis*	Single dose: 150–200 µg/kg body weight (in combination with albendazole) twice annually or Single dose: 300–400 µg/kg as a single dose annually.
4.1.	Scabies	*Sarcoptes scabiei*	Single dose: 200 µg/kg (repeat dose in 7–14 days (for immunocompromised/immunocompetent patients).
4.2.	Crusted scabies (Norwegian Scabies)		Single dose: 200 µg/kg Repeat on 2, 8, 9, and 15 in combination with typical permethrin 5% cream. Severe cases may demand additional IVM treatment on the day 22 and 29.
5.	Pediculosis		Treatment generally requires > 1 dose.

(*Continued*)

Table 11.1 Reported Ivermectin dosage [39–41]. (*Continued*)

5.1.	Pediculosis humanus capis	*Pediculus capitis*	Single dose: 400 μg/kg every 7 days (2 doses).
5.2.	Pediculosis humanus corporis	*Pediculus corporis*	Single dose: 200 μg/kg every 7 days (3 doses).
5.3.	Pediculosis pubis	*Pediculus pubis*	Single dose: 250 μg/kg every 7 days (2 doses) or Single dose: 250 μg/kg every 14 days (2 doses).
6.	Demodicosis	*Demodex folliculorum and Demodex brevis*	Single dose: 200 μg/kg followed by tropical permethrin.
7.	Blepharitis	*Demodex folliculorum*	Single dose: 200 μg/kg repeat dose once in 7 days.
8.	Filariasis	*Mansonella ozzardi*	Single dose: 6 mg.
9.	Filariasis	*Mansonella streptocerca*	Single dose: 150 μg/kg.
10.	Gnathostomiasis	*Gnathostoma spinigerum*	Single dose: 200 μg/kg.
11.	Cutaneous larva migrans	*Ancylostoma braziliense*	Single dose: 200 μg/kg.
12.	Trichuriasis	*Trichuriasis trichiura*	Single dose: 200 μg/kg.
13.	Ascariasis	*Ascaris lumbricoides*	Single dose: 200 μg/kg.
14.	Enterobiasis	*Enterobius vermicularis*	Single dose: 200 μg/kg followed by a second dose for 10 days later.

with albendazole to treat lymphatic filariasis, Onchocerciasis, and Strongyloidiasis in humans. For the past three decades, IVM has been suggested as an oral treatment for mammalian parasite infections due to its outstanding safety profile and low rate of side effects. Although there have been several negative effects, they are often minor, and don't usually need stopping the medication [2, 23].

Nicolas *et al.* reported that IVM treatment during pregnancy cannot be considered safe. In 18-year, follow-up study of *O. volvulus*, about 4000 families in west Africa maternal infection was linked with f4 times augmented risk of infection in children. Children born to mothers infected with *O. volvulus* had a greater microfilarial load due to the neonate's immune system to the parasite. The risk of developing onchocerciasis - associated epilepsy is quite prominent in children weighing less than 15 kg with high microfilarial load who are eligible for IVM treatment [24].

IVM is absorbed after the oral administration in monogastric animals. It remains unchanged in excreta likely by the P-glycoprotein-dependent mechanism. A small fraction of it is metabolized in the liver through oxidation [25]. IVM is approved for the usage in the dogs and cats. Lankas and Gordon 1989 observed the acute and sub-chronic/chronic effects of IVM in few mammals such as mouse, rat, rabbit, dog, monkey, and man. The oral LD_{50} values of mouse, rat, and dog are reported to be 25 mg/kg, 50 mg/kg, and 80 mg/kg, respectively [26]. IVM in higher doses has broad range of effect on different organisms, and some them may provide the clues to its mode of action in the parasites, and may also have conceivable relevance in the treatment of humankind disease.

11.5 Ivermectin and Ectoparasites

Several ectoparasite infections have been treated with IVM which is proven to be efficacious.

11.5.1 Scabies

Oral IVM has been prescribed as a substitute for topical scabicides due to less side effects profile. IVM is more effective than first-line topical treatments including benzyl benzoate, lindane, and 5% permethrin [42]. Because IVM has no ovicidal effect, two doses of treatment are advised rather than one [43]. IVM has shown in numerous studies to be effective in treating crusted scabies in addition to ordinary scabies. Even if a single 200 mg/kg dose of the therapy might be successful, many administrations are typically needed to bring about a cure. Although IVM is effective in common and crusted scabies. yet its extensive use has been the cause of development of resistant scabies. Figure 11.6 shows symptoms of scabies on stomach hand and genital area of a child. As such there are no perfect

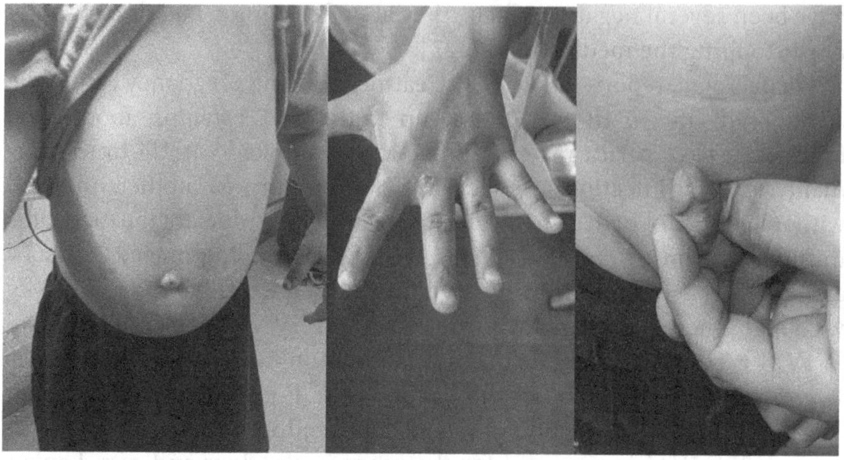

Figure 11.6 A child suffering from Scabies, Picture credit: RML Hospital, New Delhi, India.

established standards for dosing regimens. Limited information is available regarding the usage of the topical IVM (0.8%) cream to treat scabies [44].

11.5.2 Pediculosis

IVM plays a vital role in the treatment of body and crab louse infection. There are three types of lice that infest people namely *Pediculus humanus humanus*, *Pthirus pubis*, and *Pediculus humanus capitis* (body louse). According to published data, IVM play an integral part in the treatment of body and crab louse infection [45]. Foucault *et al.* noted a significant reduction in the prevalence of body lice infestation when the patients were subjected to three doses of IVM orally at seven days interval [46]. IVM's effectiveness in treating of head lice is still controversial, and further investigation is required [47].

11.5.3 Strongyloidiasis

IVM has become a popular treatment for both strongyloidiasis and disseminated infection [48–50]. There have been published data on the efficacy and safety of IVM used in veterinary science, which is not fit for humans, as a treatment for Strongyloides stercoralis hyper infection syndrome as seen in cases of intestinal ileus or severe hypoalbuminemia [51, 52]. There have been reports of a variety of dosage regimes including 200 mg/kg/day for 14 days [53], 200 mg/kg/dose every 48 h until the ileus resolves or the sputum clears of the larvae [54]. There is not much data available on the pharmacokinetics of IVM, and its tolerance, and ivermectin-related central nervous system toxicity. Furthermore, the relative efficiency of various dosing protocols for ivermectin-related clinical trials is critical [55].

11.5.4 Onchocerciasis

The recommended drug for the treatment of *O. volvulus* is IVM. Every year, approximately 50 million people are treated. IVM is a powerful microfilaricide, reducing skin microfilaria levels almost by 99% within the first few months despite having little effect on adult worm viability [56]. Genetic polymorphisms in *O. volvulus* were compared for populations for ivermectin-treated and untreated patients demonstrate selectivity of genes on *O. volvulus* [57]. Studies have shown that the selectivity occurs within the ABC transporter gene which functions as an energy-dependent efflux pump like P-glycoprotein. Recent studies signify ivermectin's impact in reduction of onchocerciasis transmission [58]. According to current data higher doses of IVM (800 mg/kg/day) are associated with an increased frequency of severe side effects [59, 60].

11.5.5 Lymphatic Filariasis

In *Wuchereria bancrofti* infection IVM act as microfilaricidal. Both IVM and albendazole has been used as a mass treatment to eradicate lymphatic filariasis in sub-Saharan African countries, where onchocerciasis is quite dominating. Several studies have been conducted to investigate the effects and mechanism of the IVM alone as well as in combination with albendazole/diethylcarbamazine on *Wuchereria bancrofti* infection [61]. *Wuchereria bancrofti* microfilaremia decreased in populations when patients treated with IVM for a long term biannually [62]. The drug requires comprehensive comparative studies for *Wuchereria bancrofti* that concisely defines the frequency of dosage, duration, and endpoints of treatment [63].

11.5.6 Loasis

IVM is prescribed as treatment to patients with increment in *Loa loa* microfilaremia (>20000/ml) and with severe post-treatment encephalopathy. The side effects of post-ivermectin are not properly understood. Risk factors of post-treatment encephalopathy include age, gender and level of exposure to ineffective larvae [64]. To understand the mechanism of IVM on *Loa loa* microfilaria, and the successive safe IVM dosage more research is needed [16].

11.6 Environmental Impact and Biodegradation of Ivermectin

The major pollutants affecting the environment are animal wastes, veterinary drugs, pesticides, and other chemicals such as detergents and disinfectants. Elevation of nutrient loads, eutrophication, and fecal contamination are the prime suspects that are affecting the environment.

IVM which is widely used to treat lice and parasitic worms in animals [65], has entered the environment through the excreta of IVM treated organisms such as cattle, horses, swine, and sheep and humans. A feces composite sample was collected 2–5 days after subcutaneous dosing. Analyses of IVM containing fractions through RIDA or collection of 1-min (1 ml) fractions of eluate for scintillation counting revealed that 40%–45% of total fecal matter contains an unaltered drug [66]. Based on the recognition of metabolites in the liver of IVM dosed steers the monosaccharides and aglycones of IVM as well as 24-hydroxy derivatives of IVM and their monosaccharides and aglycones, penetrates in the soil and are not as harmful as the parent compound [67].

Gunn and Sadd reported that the growth of earthworm affected by IVM with an EC_{50} of 4.7 mg/kg and even the production of cocoon was reduced to concentration of 4 mg/kg. Reports suggested increased unmetabolized IVM concentration in the fecal matter of cattle that were treated with IVM [68]. Unmetabolized IVM in animal feces have an effect on dung-dwelling invertebrates and insect species. The growth and development of earthworms has been affected by the presence of IVM contaminated soil which in returns the decomposition rate of organic matter. According to some published data if there is an expansion in the time of decomposition of dung from controlled cattle and IVM -treated cattle, the population of earthworms was found similar. Hence, factors like weather, climate, soil moisture, soil type, soil fauna, etc., have a more significant impact on the dung decomposition rate than the presence of IVM residue in the dung [67, 69, 70]. A detailed simulation model which focused on the mortality of insect dwelling in cattle farms under realistic sets of assumptions/approximation was provided by Sheratt *et al.* in 1998. A range of environmental factors have an impact on AVM on dung insect populations. According to some published reports the mortality of insects dwelling in a cattle farm in a farming system was unlikely to be greater than 25% [71]. Wratten and Forbes in 1996 published a data which concluded that the impact on non-target insect population as well as their ecological function is limited even under conditions of relatively high concentration of AVM use in farm cattle [72].

It has been reported that most of the drug remain unmetabolized in the feces of treated animals poses a potential threat to ivermectin usage. Initial doses of radioactive IVM 39%, 69%, and 43% had been recovered from the excreta of steer, sheep and swine when they were subjected to subcutaneous or through intraluminal injections at level of 50%, 50% and 33% above recommended dose of radioactive IVM. Approximately 351 μg unmetabolized IVM per kilogram of fecal net weight was obtained from cattle when they were subjected subcutaneously to 200 mg/kg body weight IVM through injections. The concentration IVM was approximately 5.1 μg/kg of soil when it is used as a fertilizer at rate of 15 tons/acre and plowed to depth of 15 cm [73, 74].

It has been reported that IVM contaminated dung and soil-dwelling terrestrial invertebrates and sediment-dwelling organisms inhabiting freshwater ecosystems have been adversely affected. Chemical contaminated discharge from dairy is also loading excessive nutrient in nearby water ecosystem [73].

It is used in salmon farming to control sea lice. It was found that the concentration of IVM in salmon farming was roughly 33.4 μg/kg with a half-life of approximately 100 days or a little more [75]. In 1993 Bloom and Matheson reported that freshwater fishes are less sensitive to IVM than Daphnia Magna, bluegill sunfish, and rainbow trout with an LD_{50} value of 5.3 μg/L and 3.3 μg/L [67].

IVM has more affinity for organic matter than water. The vapor pressure of IVM is approximately 1.5×109 mm Hg, indicating that it is nonvolatile and would not cause scattering in the atmosphere. IVM breaks into metabolites of more polarity and less toxicity in the environment. Biodegradation of IVM is temperature dependent with half-lives of nearly 1–2 weeks [74].

11.7 Conclusion

In this particular chapter, safety, toxicology and impact on the environment of the IVM drug have been discussed in detail. We believe that this still mature drug is yet to be explored more for various other purposes.

References

1. Njoo, F.L., Hack, C.E., Oosting, J., Stilma, J.S., Kijlstra, A., Neutrophil activation in ivermectin-treated onchocerciasis patients. *Clin. Exp. Immunol.*, 94, 330–3, 1993.

2. Eisenberger, M.A., Sinibaldi, V.J., Reyno, L.M., Sridhara, R., Jodrell, D.I., Zuhowski, E.G., Tkaczuk, K.H., Lowitt, M.H., Hemady, R.K., Jacobs, S.C., VanEcho, D., Egorin, M.J., Phase I and clinical evaluation of a pharmacologically guided regimen of suramin in patients with hormone-refractory prostate cancer. *J. Clin. Oncol.*, 13, 2174–86, 1995.
3. González-Canga, A., Fernández-Martínez, N., Sahagún-Prieto, A., Diez-Liébana, M.J., Sierra-Vega, M., García-Vieitez, J.J., A review of the pharmacological interactions of ivermectin in several animal species. *Curr. Drug Metab.*, 10, 359–68, 2009.
4. Bounias, M., Pragmatic efficacy against conceptual precaution in parasite control: The case of avermectins. *J. Environ. Biol.*, 21, 275–85, 2000.
5. Chippaux, J.P., Boussinesq, M., Gardon, J., Gardon-Wendel, N., Ernould, J.C., Severe adverse reaction risks during mass treatment with ivermectin in loiasis-endemic areas. *Parasitol. Today*, 12, 11, 448–50, 1996.
6. Cooper, P.J., Beck, L.A., Espinel, I., Deyampert, N.M., Hartnell, A., Jose, P.J., Paredes, W., Guderian, R.H., Nutman, T.B., Eotaxin and RANTES expression by the dermal endothelium is associated with eosinophil infiltration after ivermectin treatment of onchocerciasis. *Clin. Immunol.*, 95, 51–61, 2000.
7. Cross, H.F., Haarbrink, M., Egerton, G., Yazdanbakhsh, M., Taylor, M.J., Severe reactions to filarial chemotherapy and release of Wolbachia endosymbionts into blood. *Lancet*, 358, 9296, 1873–5, 2001.
8. Crump, A. and Otoguro, K., Satoshi Omura: In pursuit of nature's bounty. *Trends Parasitol.*, 21, 126–132, 2005.
9. Geary, T.G., Ivermectin 20 years on: Maturation of a wonder drug. *Trends Parasitol.*, 21, 530–532, 2005.
10. Simón, F. *et al.*, Human and animal dirofilariasis: The emergence of a zoonotic mosaic. *Clin. Microbiol. Rev.*, 25, 507–544, 2012.
11. Higazi, T. and Geary, T., Chemotherapy in the treatment, control, and elimination of human onchocerciasis. *Res. Rep. Trop. Med.*, 5, 73–99, 2014.
12. Sheriff, J.C., Kotze, A.C., Sangster, N.C., Hennessey, D.R., Effect of ivermectin on feeding by Haemonchis contortus *in vivo*. *Vet. Parasitol.*, 128, 341–346, 2005.
13. Griffin, J., Fletcher, N., Clemence, R. *et al.*, Selamectin is a potent substrate and inhibitor of human and canine P-glycoprotein. *J. Vet. Pharmacol. Ther.*, 28, 257–265, 2005.
14. Turner, S.A., Maclean, J.D., Fleckenstein, L., Greenaway, C., Parenteral administration of ivermectin in a patient with disseminated strongyloidiasis. *Am. J. Trop. Med. Hyg.*, 73, 911–914, 2005.
15. Fox, L.M., Ivermectin: Uses and impact 20 years on. *Curr. Opin. Infect. Dis.*, 19, 588–93, 2006.
16. Merck & Co. Inc., Stromectol1 (ivermectin) package insert, pp. 1–6, Merck & Co. Inc., Whitehouse Station, NJ, 2005.

17. Lawrence, G., Leafasia, J., Sheridan, J. *et al.*, Control of scabies, skin sores and haematuria in children in the Solomon Islands: Another role for ivermectin. *Bull. WHO*, 83, 34–42, 2005.
18. Gardon, J., Gardon-Wendel, N., Demanga-Ngangue *et al.*, Serious reactions after mass treatment of onchocerciasis with ivermectin in an area endemic for Loa loa infection. *Lancet*, 350, 18–22, 1997.
19. Miller, T.W. *et al.*, Avermectins, new family of potent anthelmintic agents: Isolation and chromatographic properties. *Antimicrob. Agents Chemother.*, 15, 368–371, 1979.
20. Egerton, J.R., Ostlind, D.A., Blair, L.S., Eary, C.H., Suhayda, D., Cifelli, S., Riek, R.F., Campbell, W.C., Avermectins, new family of potent anthelmintic agents: Efficacy of the B1A component. *Antimicrob. Agents Chemother.*, 15, 372–378, 1979.
21. World Health Organization, Onchocerciasis and its control report of a WHO Expert Committee on Onchocerciasis Control (WHO/TRS/852), Geneva, p. 104, 1995.
22. Zoure, H.G., Noma, M., Tekle, A.G., Amazigo, U.V., Diggle, P.J., Giorgi, E., Remme, J.H., The geographic distribution of onchocerciasis in the 20 participating countries of the African Programme for Onchocerciasis Control: (2) pre-control endemicity levels and estimated number infected. *Parasitol. Vectors*, 7, 325, 2014.
23. Remme, J.H.F., The African Programme for Onchocerciasis Control: Preparing to launch. *Trends Parasitol.*, 11, 403–406, 1995.
24. Nicolas, P., Maia, M.F., Bassat, Q. *et al.*, Safety of oral ivermectin during pregnancy: A systematic review and meta-analysis. *Lancet Glob. Health*, 8, 92–100, 2020.
25. Rendic, S.P., Metabolism and interactions of Ivermectin with human cytochrome P450 enzymes and drug transporters, possible adverse and toxic effects. *Arch. Toxicol.*, 95, 1535–1546, 2021.
26. Merola, V., Khan, S., Gwaltney-Brant, S., Ivermectin toxicosis in dogs: A retrospective study. *J. Am. Anim. Hosp. Assoc.*, 45, 106–11, 2009.
27. Taylor, H.R. and Greene, B.M., The status of ivermectin in the treatment of human onchocerciasis. *Am. J. Trop. Med. Hyg.*, 41, 460–466, 1989.
28. Campbell, W.C., An introduction to the avermectins. *N. Z. Vet. J.*, 29, 174–178, 1981.
29. Campbell, W.C. *et al.*, Ivermectin: A potent new antiparasitic agent. *Science*, 221, 823–828, 1983.
30. Sharmeen, S. *et al.*, The antiparasitic agent ivermectin induces chloride-dependent membrane hyperpolarization and cell death in leukemia cells. *Blood*, 116, 3593–3603, 2010.
31. Melotti, A. *et al.*, The river blindness drug Ivermectin and related macrocyclic lactones inhibit WNT-TCF pathway responses in human cancer. *EMBO Mol. Med.*, 6, 1263–1278, 2014.

32. Yin, J. *et al.*, DEAD-box RNA helicase DDX23 modulates glioma malignancy via elevating miR-21 biogenesis. *Brain*, 138, 2553–2570, 20152015.
33. Ashraf, S. and Prichard, R., Ivermectin exhibits potent antimitotic activity. *Vet. Parasitol.*, 226, 1–4, 2016.
34. Antebi, A., Culotti, J.G., Hedgecock, E.M., daf-12 regulates developmental age and the dauer alternative in Caenorhabditis elegans. *Dev. (Cambridge, England)*, 125, 1191–1205, 1998.
35. Ventre, E., Rozières, A., Lenief, V., Albert, F., Rossio, P., Laoubi, L., Dombrowicz, D., Staels, B., Ulmann, L., Julia, V., Vial, E., Jomard, A., Hacini-Rachinel, F., Nicolas, J.F., Vocanson, M., Topical ivermectin improves allergic skin inflammation. *Allergy*, 72, 1212–1221, 2017.
36. Mastrangelo, E., Pezzullo, M., De Burghgraeve, T., Kaptein, S., Pastorino, B., Dallmeier, K., de Lamballerie, X., Neyts, J., Hanson, A.M., Frick, D.N., Bolognesi, M., Milani, M., Ivermectin is a potent inhibitor of flavivirus replication specifically targeting NS3 helicase activity: New prospects for an old drug. *J. Antimicrob. Chemother.*, 67, 1884–1894, 2012.
37. Lim, E., Vilchèze, C., Ng, C., Jacobs, W.R., Jr, Ramón-García, S., Thompson, C.J., Anthelmintic avermectins kill Mycobacterium tuberculosis, including multidrug-resistant clinical strains. *Antimicrob. Agents Chemother.*, 57, 1040–1046, 2013.
38. Muhammed Ameen, S. and Drancourt, M., Ivermectin lacks antituberculous activity. *J. Antimicrob. Chemother.*, 68, 1936–1937, 2013.
39. González, P., González, F.A., Ueno, K., Ivermectin in human medicine, an overview of the current status of its clinical applications. *Curr. Pharm. Biotechnol.*, 13, 1103–1109, 2012.
40. Drugs for parasitic infections. *Med. Lett. Drugs Ther.*, 143, 1–31, 2013.
41. Guzzo, C.A. *et al.*, Safety, tolerability, and pharmacokinetics of escalating high doses of ivermectin in healthy adult subjects. *J. Clin. Pharmacol.*, 42, 1122–1133, 2002.
42. Heukelbach, J. and Feldmeirer, H., Scabies. *Lancet*, 367, 1767–1774, 2006.
43. Chosidow, O., Clinical practices: Scabies. *N. Engl. J. Med.*, 354, 1718–1727, 2006.
44. Karthikeyan, K., Treatment of scabies: Newer perspectives. *Postgrad Med. J.*, 81, 7–11, 2005.
45. Burkhart, C.G. and Burkhart, C.N., Oral ivermectin for Phthirus pubis. *J. Am. Acad. Dermatol.*, 51, 1037, 2004.
46. Burkhart, C.G. and Burkhart, C.N., Safety and efficacy of pediculicides for head lice. *Expert Opin. Drug Saf.*, 5, 169–179, 2006.
47. Nuesch, R., Zimmerli, L., Stockli, R. *et al.*, Imported strongyloidosis: A longitudinal analysis of 31 cases. *J. Travel Med.*, 12, 80–84, 2005.
48. World Health Organization, *WHO model list of essential medicines*, 14th edn., World Health Organization, Geneva, 2005.

49. Marty, F.M., Lowry, C.M., Rodriquez, M. *et al.*, Treatment of human disseminated strongyloidiasis with a parenteral veterinary formulation of ivermectin. *Clin. Infect. Dis.*, 41, 5–8, 2005.
50. Salluh, J.I., Feres, G.A., Velasco, E. *et al.*, Successful use of parenteral ivermectin in an immunosuppressed patient with disseminated strongyloidiasis and septic shock. *Intensive Care Med.*, 31, 1292, 2005.
51. Pacanowski, J., Santos, M.D., Roux, A. *et al.*, Subcutaneous ivermectin as a safe salvage therapy in Strongyloides stercoralis hyperinfection syndrome: A case report. *Am. J. Trop. Med. Hyg.*, 73, 122–124, 2005.
52. Gill, G.V. and Beeching, N., Comment on: Strongyloidiasis: A mistaken diagnosis and a fatal outcome in a patient with diarrhea. *Trans. R. Soc. Trop. Med. Hyg.*, 100, 83–84, 2006.
53. Rao, P.N. and Kiran, K.U., Preventing strongyloidiasis in the immunosuppressed. *Indian J. Dermatol. Venereol. Leprol.*, 71, 363–364, 2005.
54. Johnston, F.H., Morris, P.S., Speare, R. *et al.*, Strongyloidiasis: A review of the evidence for Australian practitioners. *Aust. J. Rural Health*, 13, 247–254, 2005.
55. Omura, S. and Crump, A., The life and times of ivermectin – A success story. *Nat. Rev. Microbiol.*, 12, 984–989, 2004.
56. Awadzi, K., Boakye, D.A., Edwards, G. *et al.*, An investigation of persistent microfilaridermia despite multiple treatments with ivermectin in two onchocerciasis endemic foci in Ghana. *Ann. Trop. Med. Parasitol.*, 98, 231–249, 2004.
57. Eng, J.K.L. and Prichard, R.K., A comparison of genetic polymorphism in populations of Onchocerca volvulus from untreated- and ivermectin-treated patients. *Mol. Biochem. Parasitol.*, 142, 193–202, 2005.
58. Ardelli, B.F., Guerriero, S.B., Prichard, R.K., Characterization of a half-size ATPbinding cassette transporter gene which may be a useful marker for ivermectin selection in Onchocerca volvulus. *Mol. Biochem. Parasitol.*, 145, 94–100, 2006.
59. Kamgno, J., Gardon, J., Gardon-Wendel, N. *et al.*, Adverse systemic reactions to treatment of onchocerciasis with ivermectin at normal and high doses given annually or three-monthly. *Trans. R. Soc. Trop. Med. Hyg.*, 98, 496–504, 2004.
60. Fobi, G., Gardon, J., Kamgno, J. *et al.*, A randomized, double-blind, controlled trial of the effects of ivermectin at normal and high doses, given annually or three monthly, against Onchocerca volvulus: Opthalmological results. *Trans. R. Soc. Trop. Med. Hyg.*, 99, 279–289, 2005.
61. Tisch, D.J., Michael, E., Kazura, J.W., Mass chemotherapy options to control lymphatic filariasis: A systematic review. *Lancet Infect. Dis.*, 5, 514–523, 2005.
62. Kyelem, D., Medlock, J., Sanou, S., Short communication: Impact of longterm (14 years) bi-annual ivermectin treatment on Wuchereria bancrofti microfilaremia. *Trop. Med. Int. Health*, 10, 1002–1004, 2005.

63. Gyapong, J.O., Kumaraswami, V., Biswas, G., Ottesen, E.A., Treatment strategies underpinning the global programme to eliminate lymphatic filariasis. *Expert Opin. Pharmacother.*, 6, 179–200, 2005.
64. Pion, S.D., Demanou, M., Oudin, B., Boussinesq, M., Loiasis: The individual factors associated with the presence of microfilaraemia. *Ann. Trop. Med. Parasitol.*, 99, 491–500, 2005.
65. Sands, B. and Noll, M., Toxicity of ivermectin residues in aged farmyard manure to terrestrial and freshwater invertebrates. *Insect Conserv. Divers.*, 15, 9–18, 2022.
66. Chiu, S.H.L., Buhs, R.P., Sestokas, E., Taub, R., Jacob, T.A., Determination of ivermectin residue in animal tissues by high-performance liquid chromatography-reverse isotope dilution assay. *J. Agric. Food Chem.*, 33, 99–102, 1985.
67. Halley, B.A., Jacob, T.A., Lu, A.Y.H., The environmental impact of the use of ivermectin: Environmental effects and fate. *Chemosphere*, 18, 1543–1563, 1989.
68. Gunn, A. and Sadd, J.W., The effect of ivermectin on the survival, behaviour and cocoon production of the earthworm. *Eiseniafetida. Pedobiologia*, 38, 327–333, 1994.
69. Wall, R.A. and Strong, L., Environmental consequences of treating cattle with the antiparasitic drug ivermectin. *Nature*, 327, 418–421, 1987.
70. Floate, K.D., Off-target effects of ivermectin on insects and on dung degradation in southern Alberta. *Bull. Entomol. Res.*, 88, 25–35, 1998.
71. Sherratt, T.N., Macdougall, A.D., Wratten, S., Forbes, A.B., Models to assist the evaluation of the impact of avermectins on dung insect populations. *Ecol. Modell.*, 110, 165–173, 1998.
72. Wratten, S.D. and Forbes, A.B., Environmental assessment of veterinary avermectins in temperate pastoral ecosystems. *Ann. Appl. Biol.*, 128, 329–48, 1996.
73. Tremblay, L.A., Wratten, S.D., New Zealand, *Effects of ivermectin in dairy discharges on terrestrial and aquatic invertebrates*, Dept. of Conservation, Wellington, N.Z, 2002.
74. Bloom, R.A. and Matheson III, J.C., Environmental assessment of avermectins by the US Food and Drug Administration. *Vet. Parasitol.*, 48, 281–294, 1993.
75. Davie, I.M., Gillibrand, A., Mchenery, J.G., Rae, G.H., Environmental risk of ivermectin to sediment dwelling organisms. *Aquaculture*, 163, 29–46, 1998.

Index

Also of Interest

Check out these other books by Shahid-ul-Islam published by Scrivener Publishing

Chemistry and Biological Activities of Ivermectin
Edited by Rashid Ali and Shahid-ul-Islam
Published 2023. ISBN 978-1-394-16654-1

Green Chemistry for Sustainable Water Purification
Edited by Shahid-ul-Islam, Abid Hussain Shalla and Mohammad Shahadat
Published 2023. ISBN 978-1-119-85229-2

Role of Microbes in Industrial Products and Processes
Edited by Sanjay Kumar, Narendra Kumar Shahid-ul-Islam
Published 2022. ISBN 978-1-119-90119-8

Handbook of Biomass Valorization for Industrial Applications
Edited Shahid-ul-Islam, Aabid Hussain Shalla, and Salman Ahmad Khan
Published 2021. ISBN 978-1-119-81873-1

Chemistry of Biologically Potent Natural Products and Synthetic Compounds
Edited by Shahid-ul-Islam and Javid Ahmad Banday
Published 2021. ISBN 978-1-119-64034-9

Advanced Functional Textiles and Polymers
Fabrication, Processing and Applications
Edited by Shahid-ul-Islam and B.S. Butola
Published 2020. ISBN 978-1-119-60579-9

Advances in Metallodrugs
Preparation and Applications in Medicinal Chemistry
Edited by Shahid-ul-Islam, Athar Adil Hashmi and Salman Ahmad Khan
Published 2020. ISBN 978-1-119-64042-4

Environmental Nanotechnology for Water Purification
Edited by Shahid-ul-Islam
Published 2020. ISBN 978-1-119-64045-5

Integrating Green Chemistry and Sustainable Engineering
Edited by Shahid-ul-Islam
Published 2019. ISBN 978-1-119-50982-0

Advanced Textile Engineering Materials
Edited by Shahid-ul-Islam and B.S. Butola
Published 2018. ISBN 978-1-119-48785-2

Nanomaterials in the Wet Processing of Textiles
Edited by Shahid-ul-Islam and B.S. Butola
Published 2018. ISBN 978-1-119-45980-4

Advanced Materials for Wastewater Treatment
Edited by Shahid-ul-Islam
Published 2017. ISBN 978-1-119-40776-8

Plant-Based Natural Products
Derivatives and Applications
Edited by Shahid-ul-Islam
Published 2017. ISBN 978-1-119-42383-6

Printed in the USA/Agawam, MA
September 16, 2026

910661.064